VELAN M. MUDALIAR

Composite Materials

Second Edition

Krishan K. Chawla

Composite Materials
Science and Engineering

Second Edition

With 403 Illustrations

 Springer

Krishan K. Chawla
Professor & Chairman
Dept. of Materials & Mechanical Engineering
The University of Alabama at Birmingham
Birmingham, AL 35294-4461
USA
kchawla@uab.edu

Cover illustration: Micrograph of a silicon carbide (SCS-6) fiber/Ti alloy matrix. The fiber diameter is 140 μm. (Courtesy of J.M. Baughman, NASA) See also page 175.

Library of Congress Cataloging-in-Publication Data
Chawla, Krishan Kumar
    Composite materials : science and engineering / Krishan K.
  Chawla. — 2nd ed.
       p.   cm.
    ISBN 0-387-98409-7 (hardcover : alk. paper)
    1. Composite Materials.   I. Title.   II. Series: Materials
  research and engineering (Unnumbered)
  TA418.9.C6C43   1998
  624.1e 18—dc21   97-43749

Printed on acid-free paper.

Printed in the United States of America.   (MVY)

9 8 7 6 5 4 3

ISBN 0-387-98409-7

springeronline.com

आ नो भद्राः क्रतवो यन्तु विश्वतः

Ā no bhadrāḥ kratavo yantu viśvataḥ

Let noble thoughts come to us from every side

Rigveda 1-89-i

Dedicated affectionately
to Nivi, Nikhil, and Kanika

Preface to the Second Edition

The first edition of this book came out in 1987, offering an integrated coverage of the field of composite materials. I am gratified at the reception it received at the hands of the students and faculty. The second edition follows the same format as the first one, namely, a well-balanced treatment of materials and mechanics aspects of composites, with due recognition of the importance of the processing. The second edition is a fully revised, updated, and enlarged edition of this widely used text. There are some new chapters, and others have been brought up-to-date in light of the extensive work done in the decade since publication of the first edition. Many people who used the first edition as a classroom text urged me to include some solved examples. In deference to their wishes I have done so. I am sorry that it took me such a long time to prepare the second edition. Things are happening at a very fast pace in the field of composites, and there is no question that a lot of very interesting and important work has been done in the past decade or so. Out of necessity, one must limit the amount of material to be included in a textbook. In spite of this view, it took me much more time than I anticipated. In this second edition, I have resisted the temptation to cover the whole waterfront. So the reader will find here an up-to-date treatment of the fundamental aspects. Even so, I do recognize that the material contained in this second edition is more than what can be covered in the classroom in a semester. I consider that to be a positive aspect of the book. The reader (student, researcher, practicing scientist/engineer) can profitably use this as a reference text. For the person interested in digging deeper into a particular aspect, I provide an extensive and updated list of references and suggested reading.

There remains the pleasant task of thanking people who have been very helpful and a constant source of encouragement to me over the years: M.E. Fine, S.G. Fishman, J.C. Hurt, B. Ilschner, B.A. MacDonald, A. Mortensen, J.M. Rigsbee, P. Rohatgi, S. Suresh, H. Schneider, N.S. Stoloff, and A.K. Vasudevan. Among my students and post-docs, I would like to acknowledge G. Gladysz, H. Liu, and Z.R. Xu. I am immensely grateful to my family members, Nivi, Nikhil, and Kanika. They were patient and understanding throughout. Without Kanika's help in word processing and fixing things, this work would still be unfinished. Once again I wish to record my gratitude to my parents, Manohar L. Chawla and the late Sumitra Chawla for all they have done for me!

Krishan K. Chawla
Birmingham, Alabama
February 1998

Preface to the First Edition

The subject of composite materials is truly an inter- and multidisciplinary one. People working in fields such as metallurgy and materials science and engineering, chemistry and chemical engineering, solid mechanics, and fracture mechanics have made important contributions to the field of composite materials. It would be an impossible task to cover the subject from all these viewpoints. Instead, we shall restrict ourselves in this book to the objective of obtaining an understanding of composite properties (e.g., mechanical, physical, and thermal) as controlled by their structure at micro- and macrolevels. This involves a knowledge of the properties of the individual constituents that form the composite system, the role of interface between the components, the consequences of joining together, say, a fiber and matrix material to form a unit composite ply, and the consequences of joining together these unit composites or plies to form a macrocomposite, a macroscopic engineering component as per some optimum engineering specifications. Time and again, we shall be emphasizing this main theme, that is structure–property correlations at various levels that help us to understand the behavior of composites.

In Part I, after an introduction (Chap. 1), fabrication and properties of the various types of reinforcement are described with a special emphasis on microstructure–property correlations (Chap. 2). This is followed by a chapter (Chap. 3) on the three main types of matrix materials, namely, polymers, metals, and ceramics. It is becoming increasingly evident that the role of the matrix is not just that of a binding medium for the fibers but it can contribute decisively toward the composite performance. This is followed by a general description of the interface in composites (Chap. 4). In Part II a detailed description is given of some of the important types of composites (Chap. 5), metal matrix composites (Chap. 6), ceramic composites (Chap. 7), carbon fiber composites (Chap. 8), and multifilamentary superconducting composites (Chap. 9). The last two are described separately because they are the most advanced fiber composite systems of the 1960s and 1970s. Specific characteristics and applications of these composite systems are brought out in these chapters. Finally, in Part III, the micromechanics (Chap. 10) and macromechanics (Chap. 11) of composites are described in detail, again emphasizing the theme of how structure (micro and macro) controls the properties. This is followed by a description of strength and fracture modes in composites (Chap. 12). This chapter also describes some salient points of difference, in regard to design, between conventional and fiber composite materials. This is indeed of fundamental importance in view of the fact that composite materials are not just any other new material. They represent a total departure from the way we are used to handling conventional mono-lithic materials, and, consequently, they require unconventional approaches to designing with them.

Throughout this book examples are given from practical applications of composites in various fields. There has been a tremendous increase in appli-cations of composites in sophisticated engineering items. Modern aircraft industry readily comes to mind as an ideal example. Boeing Company, for example, has made widespread use of structural components made of "advanced" composites in 757 and 767 planes. Yet another striking example is that of the Beechcraft Company's Starship 1 aircraft. This small aircraft (8–10 passengers plus crew) is primarily made of carbon and other high-performance fibers in epoxy matrix. The use of composite materials results in 19% weight reduction compared to an identical aluminum airframe. Besides this weight reduction, the use of composites made a new wing design con-figuration possible, namely, a variable-geometry forward wing that sweeps forward during takeoff and landing to give stability and sweeps back 30° in level flight to reduce drag. As a bonus, the smooth structure of composite wings helps to maintain laminar air flow. Readers will get an idea of the tremendous advances made in the composites field if they would just remind themselves that until about 1975 these materials were being produced mostly on a laboratory scale. Besides the aerospace industry, chemical, electrical, automobile, and sports industries are the other big users, in one form or another, of composite materials.

This book has grown out of lectures given over a period of more than a decade to audiences comprised of senior year undergraduate and graduate students, as well as practicing engineers from industry. The idea of this book was conceived at Instituto Militar de Engenharia, Rio de Janeiro. I am grateful to my former colleagues there, in particular, J.R.C. Guimarães, W.P. Longo, J.C.M. Suarez, and A.J.P. Haiad, for their stimulating com-panionship. The book's major gestation period was at the University of Illi-nois at Urbana-Champaign, where C.A. Wert and J.M. Rigsbee helped me to complete the manuscript. The book is now seeing the light of the day at the New Mexico Institute of Mining and Technology. I would like to thank my colleagues there, in particular, O.T. Inal, P. Lessing, M.A. Meyers, A. Miller, C.J. Popp, and G.R. Purcell, for their cooperation in many ways, tangible and intangible. An immense debt of gratitude is owed to N.J. Grant of MIT, a true gentleman and scholar, for his encouragement, corrections, and suggestions as he read the manuscript. Thanks are also due to

R. Signorelli, J. Cornie, and P.K. Rohatgi for reading portions of the manuscript and for their very constructive suggestions. I would be remiss in not mentioning the students who took my courses on composite materials at New Mexico Tech and gave very constructive feedback. A special mention should be made of C.K. Chang, C.S. Lee, and N. Pehlivanturk for their relentless queries and discussions. Thanks are also due to my wife, Nivedita Chawla, and Elizabeth Fraissinet for their diligent word processing, My son, Nikhilesh Chawla, helped in the index preparation. I would like to express my gratitude to my parents, Manohar L. and Sumitra Chawla, for their ever-constant encouragement and inspiration.

Krishan K. Chawla
Socorro, New Mexico
June 1987

About the Author

Professor Krishan K. Chawla received his B.S. degree from Banaras Hindu University and his M.S. and Ph.D. degrees from the University of Illinois at Urbana-Champaign. He has taught and/or done research work at Instituto Militar de Engenharia, Brazil; University of Illinois at Urbana-Champaign; Northwestern University; Université Laval, Canada; Ecole Polytechnique Federale de Lausanne, Switzerland; the New Mexico Institute of Mining and Technology (NMIMT) and the University of Alabama at Birmingham. Among the honors he has received are: Eshbach Distinguished Scholar at Northwestern University, U.S. Department of Energy Faculty Fellow at Oak Ridge National Laboratory, and Distinguished Researcher Award at NMIMT. In 1989–90, he served as a program director for Metals and Ceramics at the U.S. National Science Foundation (NSF). He is also a Fellow of ASM International. Among his other books are the following: *Ceramic Matrix Composites, Fibrous Materials, Mechanical Metallurgy* (coauthor), and *Mechanical Behavior of Materials* (coauthor).

Contents

PART I

CHAPTER 1

Introduction

It is a truism that technological development depends on advances in the field of materials. One does not have to be an expert to realize that the most advanced turbine or aircraft design is of no use if adequate materials to bear the service loads and conditions are not available. Whatever the field may be, the final limitation on advancement depends on materials. Composite materials in this regard represent nothing but a giant step in the ever-constant endeavor of optimization in materials.

Strictly speaking, the idea of composite materials is not a new or recent one. Nature is full of examples wherein the idea of composite materials is used. The coconut palm leaf, for example, is nothing but a cantilever using the concept of fiber reinforcement. Wood is a fibrous composite: cellulose fibers in a lignin matrix. The cellulose fibers have high tensile strength but are very flexible (i.e., low stiffness), while the lignin matrix joins the fibers and furnishes the stiffness. Bone is yet another example of a natural composite that supports the weight of various members of the body. It consists of short and soft collagen fibers embedded in a mineral matrix called apatite. A very readable description of the structure-function relationships in the plant and animal kingdoms is available in the book *Mechanical Design in Organisms* [Wainwright et al., 1976]. In addition to these naturally occurring composites, there are many other engineering materials that are composites in a very general way and that have been in use for a very long time. The carbon black in rubber, Portland cement or asphalt mixed with sand, and glass fibers in resin are common examples. Thus, we see that the idea of composite materials is not that recent. Nevertheless, one can safely mark the origin of the distinct discipline of composite materials as the beginning of the 1960s. It would not be too much off the mark to say that a concerted research and development effort in composite materials began in 1965. Since the early 1960s, there has been an increasing demand for materials that are stiffer and stronger yet lighter in fields as diverse as aerospace, energy, and civil construction. The demands made on materials for better overall performance are so great and diverse that no one material can satisfy them. This naturally led to a resurgence of the ancient concept of combining different materials in an integral–composite material to satisfy the user requirements. Such composite material systems result in a performance unattainable by the individual constituents, and they offer the great advantage of a flexible design; that is, one can, in principle, tailor-make the material as per specifications of an optimum design. This is a much more powerful statement than it might appear at first sight. It implies that, given the most efficient design of, say, an aerospace structure, an automobile, a boat, or an electric motor, we can make a composite material that meets the need. Schier and Juergens [1983] surveyed the design impact of composites on fighter aircraft. According to these authors, "composites have introduced an extraordinary fluidity to design engineering, in effect forcing the designer-analyst to create a different material for each application as he pursues savings in weight and cost." Yet another conspicuous development has been the integration of the materials science and engineering input with the manufacturing and design inputs at all levels, from conception to commissioning of an item, through the inspection during the lifetime, as well as failure analysis. More down-to-earth, however, is the fact that our society has become very energy conscious. This has led to an increasing demand for lightweight yet strong and stiff structures in all walks of life. And composite materials are increasingly providing the answers. Figure 1.1 makes a comparison, admittedly for illustrative purposes, between conventional monolithic materials, such as aluminum and steel, and composite materials [Deutsch, 1978]. This figure indicates the possibilities of improvements that one can obtain over conventional materials by the use of composite materials. As such, it describes vividly the driving force behind the large effort in the field of composite materials. Glass fiber reinforced resins have been in use since about the 1940s. Glass fiber reinforced resins are very light and strong materials, although their stiffness (modulus) is not very high, mainly because the glass fiber itself is not very stiff. The third quarter of the twentieth century saw the emergence of the so-called advanced fibers of extremely high modulus, for example, boron, carbon, silicon carbide, and alumina. These

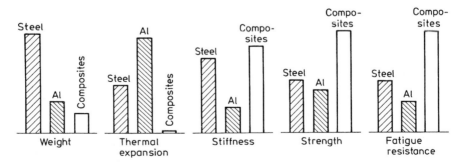

Fig. 1.1. Comparison between conventional monolithic materials and composite materials. [From Deutsch (1978), used with permission.]

fibers have been used for reinforcement of resin, metal, and ceramic matrices. Fiber reinforced composites have been more prominent than other types of composites for the simple reason that most materials are stronger and stiffer in the fibrous form than in any other form. By the same token, it must be recognized that a fibrous form results in reinforcement mainly in fiber direction. Transverse to the fiber direction, there is little or no reinforcement. Of course, one can arrange fibers in two-dimensional or even three-dimensional arrays, but this still does not gainsay the fact that one is not getting the full reinforcement effect in directions other than the fiber axis. Thus, if a less anisotropic behavior is the objective, then perhaps laminate or sandwich composites made of, say, two different materials would be more effective. A particle reinforced composite will also be reasonably isotropic. There may also be specific nonmechanical objectives for making a fibrous composite. For example, an abrasion- or corrosion-resistant surface would require the use of a laminate (sandwich) form, while in superconductors the problem of flux-pinning requires the use of extremely fine filaments embedded in a conductive matrix. In what follows, we discuss the various aspects of composites, mostly fiber composites, in greater detail, but first let us agree on an acceptable definition of a composite material. Practically everything in this world is a composite material. Thus, a common piece of metal is a composite (polycrystal) of many grains (or single crystals). Such a definition would make things quite unwieldy. Therefore, we must agree on an operational definition of *composite material* for our purposes in this text. We shall call a material that satisfies the following conditions a composite material:

1. It is manufactured (i.e., naturally occurring composites, such as wood, are excluded).
2. It consists of two or more physically and/or chemically distinct, suitably arranged or distributed phases with an interface separating them.
3. It has characteristics that are not depicted by any of the components in isolation.

References

S. Deutsch (May 1978). *23rd National SAMPE Symposium*, 34.
J.F. Schier and R.J. Juergens (Sept. 1983). *Astronautics and Aeronautics*, 44.
S.A. Wainwright, W.D. Biggs, J.D. Currey, and J.M. Gosline (1976). *Mechanical Design in Organisms*, John Wiley & Sons, New York.

CHAPTER 2

Reinforcements

2.1 Introduction

Reinforcements need not necessarily be in the form of long fibers. One can have them in the form of particles, flakes, whiskers, short fibers, continuous fibers, or sheets. It turns out that most reinforcements used in composites have a fibrous form because materials are stronger and stiffer in the fibrous form than in any other form. Specifically, in this category, we are most interested in the so-called advanced fibers, which possess very high strength and very high stiffness coupled with a very low density. The reader should realize that many naturally occurring fibers can be and are used in situations involving not very high stresses (Chawla, 1976; Chawla and Bastos, 1979). The great advantage in this case, of course, is its low cost. The vegetable kingdom is, in fact, the largest source of fibrous materials. Cellulosic fibers in the form of cotton, flax, jute, hemp, sisal, and ramie, for example, have been used in the textile industry, while wood and straw have been used in the paper industry. Other natural fibers, such as hair, wool, and silk, consist of different forms of protein. Silk fibers produced by a variety of spiders, in particular, appear to be very attractive because of their high work of fracture. Any discussion of such fibers is beyond the scope of this book. The interested reader is directed to some books that cover the vast field of fibers (Chawla, 1998; Warner, 1995). In this chapter, we confine ourselves to a variety of man-made reinforcements. Glass fiber, in its various forms, has been the most common reinforcement for polymer matrices. Kevlar aramid fiber, launched by Du Pont in the 1960s, is much stiffer and lighter than glass fiber. Gel-spun polyethylene fiber, with a stiffness comparable to Kevlar aramid fiber, was commercialized in the 1980s. Other high-performance fibers that combine high strength with high stiffness are boron, silicon carbide, carbon, and alumina. These were all developed in the second part of the twentieth century. In particular, some ceramic fibers were developed in the 1970s and 1980s by a very novel method, namely, the controlled pyrolysis of organic precursors.

Fig. 2.1. Decrease in strength (σ_f) of a carbon fiber with increase in diameter. [From de Lamotte and Perry (1970), used with permission.]

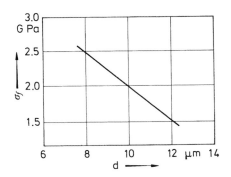

The use of fibers as high-performance engineering materials is based on three important characteristics (Dresher, 1969):

1. A small diameter with respect to its grain size or other microstructural unit. This allows a higher fraction of the theoretical strength to be attained than is possible in a bulk form. This is a direct result of the so-called size effect; that is, the smaller the size, the lower the probability of having imperfections in the material. Figure 2.1 shows that the strength of a carbon fiber decreases as its diameter increases (de Lamotte and Perry, 1970). Although this figure shows a linear drop in strength with increasing fiber diameter, a nonlinear relationship is not uncommon. Figure 2.1 should be taken only as a general trend indicator.
2. A high aspect ratio (length/diameter, l/d), which allows a very large fraction of the applied load to be transferred via the matrix to the stiff and strong fiber (see Chap. 10).
3. A very high degree of flexibility, which is really a characteristic of a material that has a high modulus and a small diameter. This flexibility permits use of a variety of techniques for making composites with these fibers.

Next we consider the concept of flexibility, and then we describe the general fiber spinning processes.

Flexibility

The flexibility of a given material is a function of its elastic modulus E and the moment of inertia of its cross section I. The elastic modulus of a material is quite independent of its form or size. It is generally a constant for a given chemical composition and density. Thus, for a given composition and density, the flexibility of a material is determined by shape, size of the cross section, and its radius of curvature, which is a function of its strength. We can use the inverse of the product of bending moment (M) and the radius of curvature (R) as a measure of flexibility. From the simple bending beam

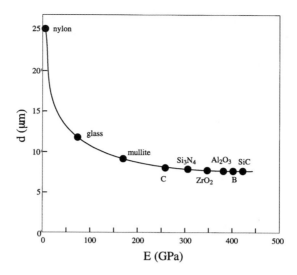

Fig. 2.2. Fiber diameter of materials with flexibility equal to that of a 25-μm-diameter nylon fiber.

theory, we have

$$\frac{M}{I} = \frac{E}{R}$$

$$MR = EI = \frac{E\pi d^4}{64}$$

$$\frac{1}{MR} = \frac{64}{E\pi d^4} \tag{2.1}$$

where d is the equivalent diameter. Equation (2.1) indicates that flexibility $1/MR$ is a very sensitive function of diameter d. Figure 2.2 shows the diameter of various materials in fibrous form with flexibility ($1/MR$) equal to that of a 25-μm-diameter nylon fiber (a typical flexible fiber) as a function of the elastic modulus. Note that given a sufficiently small diameter, it is possible for a metal or ceramic to have the same degree of flexibility as that of a 25-μm-diameter nylon; it is another matter that obtaining such a small diameter can be prohibitively expensive.

Fiber Spinning Processes

Fiber spinning is the process of extruding a liquid through small holes in a spinneret to form solid filaments. In nature, silkworms and spiders produce continuous filaments by this process. There exists a variety of different fiber spinning techniques. We give a brief description of these here:

Wet spinning: A solution is extruded into a coagulating bath. The jets of liquid freeze or harden in the coagulating bath as result of chemical or physical changes.

Dry spinning: A solution consisting of a fiber-forming material and a solvent is extruded through a spinneret. A stream of hot air impinges on the jets of solution emerging from the spinneret, evaporates the solvent, and leaves the solid filaments.

Melt spinning: The fiber-forming material is heated above its melting point and the molten material is extruded through a spinneret. The liquid jets harden into solid filaments in air on emerging from the spinneret holes.

Dry-jet wet-spinning: This is a special process devised for spinning of aramid fibers. In this process, an appropriate polymer is extruded through spinneret holes, passes through an air gap before entering a coagulation bath, and then goes on a spool for winding. We describe this process in detail in Sec. 2.5.2.

Stretching and Orientation

The process of extrusion through a spinneret results in some chain orientation in the filament. Generally, the molecules in the surface region undergo more orientation than the ones in the interior because the edges of the spinneret hole affect the near-surface molecules more. This is known as the *skin effect,* and it can affect many other properties of the fiber, such as the adhesion with a polymeric matrix or the ability to be dyed. Generally, the as-spun fiber is subjected to some stretching, causing further chain orientation along the fiber axis and consequently better tensile properties, such as stiffness and strength, along the fiber axis. The amount of stretch is generally given in terms of a draw ratio, which is the ratio of the initial diameter to the final diameter. For example, nylon fibers are typically subjected to a draw ratio of 5 after spinning. A high draw ratio results in a high elastic modulus. Increased alignment of chains means a higher degree of crystallinity in a fiber. This also affects the ability of a fiber to absorb moisture. The higher the degree of crystallinity, the lower the moisture absorption. In general, the higher degree of crystallinity translates into a higher resistance to penetration by foreign molecules, i.e., a greater chemical stability. The stretching treatment serves to orient the molecular structure along the fiber axis. It does not, generally, result in complete elimination of molecular branching; that is, one gets molecular orientation but not extension. Such stretching treatments do result in somewhat more efficient packing than in the unstretched polymer, but there is a limit to the amount of stretch that can be given to a polymer because the phenomenon of necking can intervene and cause rupture of the fiber.

Table 2.1. Approximate chemical compositions of some glass fibers (wt. %)

Composition	E Glass	C Glass	S Glass
SiO_2	55.2	65.0	65.0
Al_2O_3	8.0	4.0	25.0
CaO	18.7	14.0	—
MgO	4.6	3.0	10.0
Na_2O	0.3	8.5	0.3
K_2O	0.2	—	—
B_2O_3	7.3	5.0	—

2.2   Glass Fibers

Glass fiber is a generic name like carbon fiber or steel. A variety of different chemical compositions is commercially available. Common glass fibers are silica based (~ 50–60% SiO_2) and contain a host of other oxides—of calcium, boron, sodium, aluminum, and iron, for example. Table 2.1 gives the compositions of some commonly used glass fibers. The designation E stands for electrical because E glass is a good electrical insulator in addition to having good strength and a reasonable Young's modulus; C stands for corrosion because C glass has a better resistance to chemical corrosion; S stands for the high silica content that makes S glass withstand higher temperatures than other glasses. It should be pointed out that most of the continuous glass fiber produced is of the E glass type but, notwithstanding the designation E, electrical uses of E glass fiber are only a small fraction of the total market.

2.2.1   Fabrication

Figure 2.3 shows schematically the conventional fabrication procedure for glass fibers (specifically, the E glass fibers that constitute the workhorse of the resin reinforcement industry) (Loewenstein, 1983; Parkyn, 1970; Lowrie, 1967). The raw materials are melted in a hopper and the molten glass is fed into the electrically heated platinum bushings or crucibles; each bushing contains about 200 holes at its base. The molten glass flows by gravity through these holes, forming fine continuous filaments; these are gathered together into a strand and a *size* is applied before it is a wound on a drum. The final fiber diameter is a function of the bushing orifice diameter; viscosity, which is a function of composition and temperature; and the head of glass in the hopper. In many old industrial plants the glass fibers are not produced directly from fresh molten glass. Instead, molten glass is first turned into marbles, which after inspection are melted in the bushings.

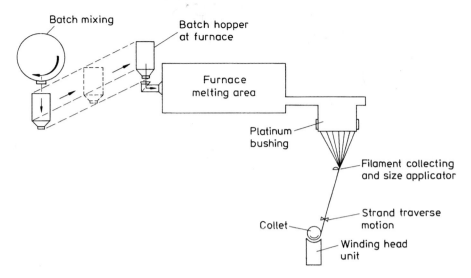

Fig. 2.3. Schematic of glass fiber manufacture.

Modern plants do produce glass fibers by direct drawing. Figure 2.4 shows some forms in which glass fiber is commercially available.

The conventional methods of making glass or ceramic fibers involve drawing from high-temperature melts of appropriate compositions. This route has many practical difficulties such as the high temperatures required, the immiscibility of components in the liquid state, and the easy crystallization during cooling. Several techniques have been developed for preparing glass and ceramic fibers (Chawla, 1998). An important technique is called the sol-gel technique (Brinker and Scherer, 1990; Jones, 1989). We shall come back to this sol-gel technique at various places in this book. Here we will just provide a brief description. A *sol* is a colloidal suspension in which the individual particles are so small (generally in the nm range) that they show no sedimentation. A *gel*, on the other hand, is a suspension in which the liquid medium has become viscous enough to behave more or less like a solid. The sol-gel process of making a fiber involves a conversion of fibrous gels, drawn from a solution at a low temperature, into glass or ceramic fibers at several hundred degrees Celsius. The maximum heating temperature in this process is much lower than that in conventional glass fiber manufacture. The sol-gel method using metal alkoxides consists of preparing an appropriate homogeneous solution, changing the solution to a sol, gelling the sol, and converting the gel to glass by heating. The sol-gel technique is a very powerful technique for making glass and ceramic fibers. The 3M Company produces a series of alumina and silica-alumina fibers, called the Nextel fibers, from metal alkoxide solutions (see Sect. 2.6). Figure 2.5 shows an example of drawn silica fibers (cut from a continuous fiber spool) obtained by the sol-gel technique (Sakka, 1985).

Fig. 2.4. Glass fiber is available in a variety of forms: **a** chopped strand, **b** continuous yarn, **c** roving, **d** fabric. (Courtesy of Morrison Molded Fiber Glass Company.)

Glass filaments are easily damaged by the introduction of surface defects. To minimize this and to make handling of these fibers easy, a sizing treatment is given. The size, or coating, protects and binds the filaments into a strand.

2.2.2 Structure

Inorganic, silica-based glasses are analogous to organic glassy polymers in that they are amorphous, i.e., devoid of any long-range order that is characteristic of a crystalline material. Pure, crystalline silica melts at 1800 °C. However, by adding some metal oxides, we can break the Si–O bonds and

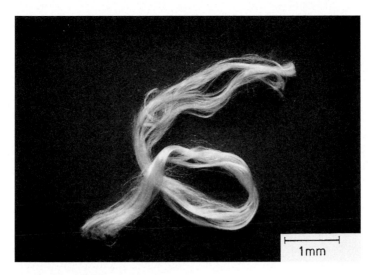

Fig. 2.5. Continuous glass fibers (cut from a spool) obtained by the sol-gel technique. [From Sakka (1985), used with permission.]

obtain a series of amorphous glasses with rather low glass transition temperatures. Figure 2.6a shows a two-dimensional network of silica glass. Each polyhedron consists of oxygen atoms bonded covalently to silicon. What happens to this structure when Na_2O is added to the glass is shown in Figure 2.6b. Sodium ions are linked ionically with oxygen but they do not join the network directly. Too much Na_2O will impair the tendency for glassy structure formation. The addition of other metal oxide types (Table 2.1) serves to alter the network structure and the bonding and, consequently, the properties. Note the isotropic, three-dimensional network structure of glass (Fig. 2.6); this leads to the more or less isotropic properties of glass fibers. That is, for the glass fiber, Young's modulus and thermal expansion coefficients are

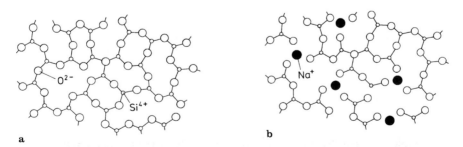

Fig. 2.6. Amorphous structure of glass: **a** a two-dimensional representation of silica glass network and **b** a modified network that results when Na_2O is added to **a**. Note that Na^+ is ionically linked with O^{2-} but does not join the network directly.

Table 2.2. Typical properties of E glass fibers

Density $(g\ cm^{-3})$	Tensile strength (MPa)	Young's modulus (GPa)	Coefficient of thermal expansion (K^{-1})
2.55	1750	70	4.7×10^{-6}

the same along the fiber axis and perpendicular to it. This is unlike many other fibers, such as aramid and carbon, which are highly anisotropic.

2.2.3 Properties and Applications

Typical mechanical properties of E glass fibers are summarized in Table 2.2. Note that the density is quite low and the strength is quite high; Young's modulus, however, is not very high. Thus, while the strength-to-weight ratio of glass fibers is quite high, the modulus-to-weight ratio is only moderate. It is this latter characteristic that led the aerospace industry to other so-called advanced fibers (e.g., boron, carbon, Al_2O_3, and SiC). Glass fibers continue to be used for reinforcement of polyester, epoxy, and phenolic resins. It is quite cheap, and it is available in a variety of forms (Fig. 2.4). Continuous strand is a group of 204 individual fibers; roving is a group of parallel strands; chopped fibers consists of strand or roving chopped to lengths between 5 and 50 mm. Glass fibers are also available in the form of woven fabrics or nonwoven mats.

Moisture decreases glass fiber strength. Glass fibers are also susceptible to what is called static fatigue; that is, when subjected to a constant load for an extended time period, glass fibers can undergo subcritical crack growth. This leads to failure over time at loads that might be safe when considering instantaneous loading.

Glass fiber reinforced resins are used widely in the building and construction industry. Commonly, these are called glass-reinforced plastics, or GRP. They are used in the form of a cladding for other structural materials or as an integral part of a structural or non-load-bearing wall panel; window frames, tanks, bathroom units, pipes, and ducts are common examples. Boat hulls, since the mid-1960s, have primarily been made of GRP. Use of GRP in the chemical industry (e.g., as storage tanks, pipelines, and process vessels) is fairly routine. The rail and road transportation industry and the aerospace industry are other big users of GRP.

2.3 Boron Fibers

Boron is an inherently brittle material. It is commercially made by chemical vapor deposition of boron on a substrate, that is, boron fiber as produced is itself a composite fiber.

In view of the fact that rather high temperatures are required for this deposition process, the choice of substrate material that goes to form the core of the finished boron fiber is limited. Generally, a fine tungsten wire is used for this purpose. A carbon substrate can also be used. The first boron fibers were obtained by Weintraub (1911) by means of reduction of a boron halide with hydrogen on a hot wire substrate.

The real impulse in boron fiber fabrication, however, came in 1959, when Talley (Talley, 1959; Talley et al., 1960) used the process of halide reduction to obtain amorphous boron fibers of high strength. Since then, interest in the use of strong but light boron fibers as a possible structural component in aerospace and other structures has been continuous, although it must be admitted that this interest has periodically waxed and waned in the face of rather stiff competition from other so-called advanced fibers, in particular, carbon fibers.

2.3.1 Fabrication

Boron fibers are obtained by chemical vapor deposition (CVD) on a substrate. There are two processes:

1. *Thermal decomposition of a boron hydride.* This method involves low temperatures, and, thus, carbon-coated glass fibers can be used as a substrate. The boron fibers produced by this method, however, are weak because of a lack of adherence between the boron and the core. These fibers are much less dense owing to the trapped gases.
2. *Reduction of boron halide.* Hydrogen gas is used to reduce boron trihalide:

$$2\,BX_3 + 3\,H_2 \rightarrow 2\,B + 6\,HX \qquad (2.2)$$

where X denotes a halogen: Cl, Br, or I.

In this process of halide reduction, the temperatures involved are very high, and, thus, one needs a refractory material, for example, a high-melting-point metal such as tungsten, as a substrate. It turns out that such metals are also very heavy. This process, however, has won over the thermal reduction process despite the disadvantage of a rather high-density substrate (the density of tungsten is 19.3 g cm^{-3}) mainly because this process gives boron fibers of a very high and uniform quality. Figure 2.7 shows a schematic of boron filament production by the CVD technique, and Figure 2.8 shows a commercial boron filament production facility; each vertical reactor shown in this picture produces continuous boron monofilament.

In the process of BCl$_3$ reduction, a very fine tungsten wire (10–12 μm diameter) is pulled into a reaction chamber at one end through a mercury seal and out at the other end through another mercury seal. The mercury seals act as electrical contacts for resistance heating of the substrate wire when gases (BCl$_3$ + H$_2$) pass through the reaction chamber, where they react on the incandescent wire substrate. The reactor can be a one- or multistage,

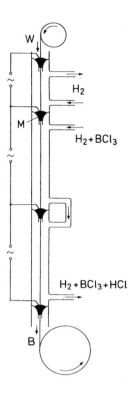

Fig. 2.7. Schematic of boron (B) fiber production by halide decomposition on a tungsten (W) substrate. [From van Maaren et al. (1975), used with permission.]

vertical or horizontal, reactor. BCl_3 is an expensive chemical, and only about 10% of it is converted into boron in this reaction. Thus, an efficient recovery of the unused BCl_3 can result in a considerable lowering of the boron filament cost.

There is a critical temperature for obtaining a boron fiber with optimum properties and structure (van Maaren et al., 1975). The desirable amorphous form of boron occurs below this critical temperature while above this temperature crystalline forms of boron also occur that are undesirable from a mechanical properties viewpoint, as we shall see in Sect. 2.3.2. With the substrate wire stationary in the reactor, this critical temperature is about 1000 °C. In a system where the wire is moving, this critical temperature is higher, and it increases with the speed of the wire. One generally has a diagram of the type shown in Figure 2.9, which shows the various combinations of wire temperature and wire drawing speed to produce a certain diameter of boron fiber. Fibers formed in the region above the dashed line are relatively weak because they contain undesirable forms of boron as a result of recrystallization. The explanation for this relationship between critical temperature and wire speed is that boron is deposited in an amorphous state and the more rapidly the wire is drawn out from the reactor, the higher the allowed temperature is. Of course, higher wire drawing speed also results in an increase in production rate and lower costs.

Fig. 2.8. A boron filament production facility. (Courtesy of AVCO Specialty Materials Co.)

Boron deposition on a carbon monofilament (~35-µm diameter) substrate involves precoating the carbon substrate with a layer of pyrolytic graphite. This coating accommodates the growth strains that result during boron deposition (Krukonis, 1977). The reactor assembly is slightly different from that for boron on tungsten substrate, because pyrolitic graphite is applied online.

2.3.2 Structure and Morphology

The structure and morphology of boron fibers depend on the conditions of deposition: temperature, composition of gases, gas dynamics, and so on. While theoretically the mechanical properties are limited only by the strength of the atomic bond, in practice, there are always structural defects and morphological irregularities present that lower the mechanical properties. Temperature gradients and trace concentrations of impurity elements inevitably cause process irregularities. Even greater irregularities are caused by fluctuations in electric power, instability in gas flow, and any other operator-induced variables.

Structure

Depending on the conditions of deposition, the elemental boron can exist in various crystalline polymorphs. The form produced by crystallization from

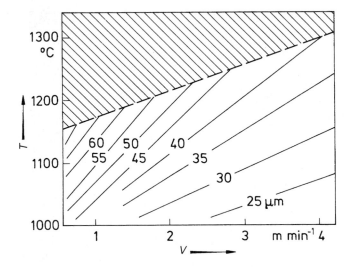

Fig. 2.9. Temperature (T) versus wire speed (V) for a series of boron filament diameters. Filaments formed in the gray region (above the dashed line) contain crystalline regions and are undesirable. [From van Maaren et al. (1975), used with permission.]

the melt or chemical vapor deposition above 1300 °C is β-rhombohedral. At temperatures lower than this, if crystalline boron is produced, the most commonly observed structure is α-rhombohedral.

Boron fibers produced by the CVD method described earlier have a microcrystalline structure that is generally called *amorphous*. This designation is based on the characteristic X-ray diffraction pattern produced by the filament in the Debye-Scherrer method, that is, large and diffuse halos with d spacings of 0.44, 0.25, 0.17, 1.4, 1.1, and 0.091 nm, typical of amorphous material (Vega-Boggio and Vingsbo, 1978). Electron diffraction studies, however, lead one to conclude that this "amorphous" boron is really a nanocrystalline phase with a grain diameter of the order of 2 nm (Krukonis, 1977).

Based on X-ray and electron diffraction studies, one can conclude that amorphous boron is really nanocrystalline β-rhombohedral. In practice, the presence of microcrystalline phases (crystals or groups of crystals observable in the electron microscope) constitutes an imperfection in the fiber that should be avoided. Larger and more serious imperfections generally result from surpassing the critical temperature of deposition (see Sect. 2.3.1) or the presence of impurities in the gases.

When boron fiber is made by deposition on a tungsten substrate, as is generally the case, then depending on the temperature conditions during deposition, the core may consist of, in addition to tungsten, a series of compounds, such as W_2B, WB, W_2B_5, and WB_4 (Galasso et al., 1967). A boron fiber cross section (100 μm diameter) is shown in Figure 2.10a, while Figure

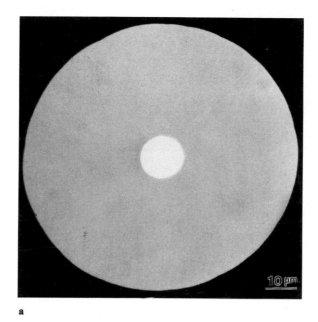

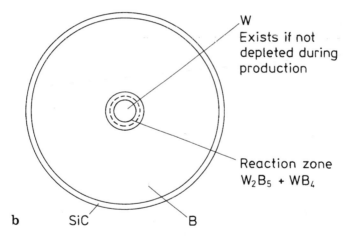

Fig. 2.10. a Cross section of a 100-μm-diameter boron fiber. **b** Schematic of the cross section of a boron fiber with SiC barrier layer.

2.10b shows schematically the various subparts of the cross section. The various tungsten boride phases are formed by diffusion of boron into tungsten. Generally, the fiber core consists only of WB_4 and W_2B_5. On prolonged heating, the core may be completely converted into WB_4. As boron diffuses into the tungsten substrate to form borides, the core expands from its original 12.5 μm (original tungsten wire diameter) to 17.5 μm. The SiC coating

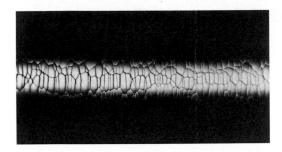

Fig. 2.11. Characteristic corn-cob structure of boron fiber. [From van Maaren et al. (1975), used with permission.]

shown in Figure 2.10b is a barrier coating used to prevent any adverse reaction between B and the matrix, such as Al, at high temperatures. The SiC barrier layer is vapor deposited onto boron using a mixture of hydrogen and methyldichlorosilane.

Morphology

The boron fiber surface shows a "corn-cob" structure consisting of nodules separated by boundaries (Fig. 2.11). The nodule size varies during the course of fabrication. In a very general way, the nodules start as individual nuclei on the substrate and then grow outward in a conical form until a filament diameter of 80–90 μm is reached, above which the nodules seem to decrease in size. Occasionally, new cones may nucleate in the material, but they always originate at an interface with a foreign particle or inclusion.

2.3.3 Residual Stresses

Boron fibers have inherent residual stresses that have their origin in the process of chemical vapor deposition. Growth stresses in the nodules of boron, stresses induced by the diffusion of boron into the tungsten core, and stresses generated by the difference in the coefficient of expansion of deposited boron and tungsten boride core, all contribute to the residual stresses and thus can have a considerable influence on the fiber mechanical properties. The residual stress pattern across the transverse section of a boron fiber is shown in Figure 2.12 (Vega-Boggio and Vingsbo, 1978). The compressive stresses on the fiber surface are due to the quenching action involved in pulling the fiber out from the chamber (Vega-Boggio and Vingsbo, 1978). Morphologically, the most conspicuous aspect of these internal stresses is the frequently observed radial crack in the transverse section of these fibers. The crack runs from within the core to just inside the external surface. Some workers, however, doubt the preexistence of this radial crack (Krukonis, 1977). They think that the crack appears during the process of boron fiber fracture.

Fig. 2.12. Schematic of residual stress pattern across the transverse section of a boron fiber. [From Vega-Boggio and Vingsbo (1978), used with permission.]

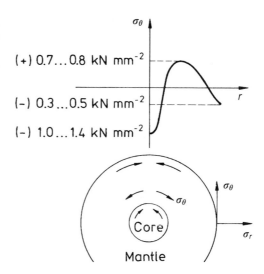

$(+)\ 0.7\ldots 0.8\ \text{kN mm}^{-2}$

$(-)\ 0.3\ldots 0.5\ \text{kN mm}^{-2}$

$(-)\ 1.0\ldots 1.4\ \text{kN mm}^{-2}$

2.3.4 Fracture Characteristics

It is well known that brittle materials show a distribution of strengths rather than a single value. Imperfections in these materials lead to stress concentrations much higher than the applied stress levels. Because the brittle material is not capable of deforming plastically in response to these stress concentrations, fracture ensues at one or more such sites. Boron fiber is indeed a very brittle material, and cracks originate at preexisting defects located at either the boron-core interface or the surface. Figure 2.13 shows the charac-

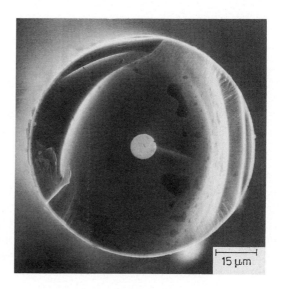

Fig. 2.13. Fracture surface of a boron fiber showing a characteristically brittle fracture and a radial crack

teristic brittle fracture of a boron fiber and the radical crack. The surface defects are due to the nodular surface that results from the growth of boron cones. In particular, when a nodule coarsens due to an exaggerated growth around a contaminating particle, a crack can result from this large nodule and weaken the fiber.

2.3.5   Properties and Applications of Boron Fibers

Many researchers have investigated the mechanical properties of boron fibers (Krukonis, 1977; Vega-Boggio and Vingsbo, 1978; Galasso et al., 1967; Galasso and Paton, 1966; DeBolt, 1982; Wawner, 1967; DiCarlo, 1985). Due to the composite nature of the boron fiber, complex internal stresses and defects such as voids and structural discontinuities result from the presence of a core and the deposition process. Thus, one would not expect boron fiber to show the intrinsic strength of boron. The average tensile strength of boron fiber is 3–4 GPa, while its Young's modulus is between 380 and 400 GPa.

An idea of the intrinsic strength of boron can be obtained in a flexure test (Wawner, 1967). In flexure, assuming the core and interface to be near the neutral axis, critical tensile stresses would not develop at the core or interface. Flexure tests on boron fibers lightly etched to remove any surface defects gave a strength of 14 GPa. Without etching, the strength was half this value. There has been some effort at NASA Lewis Research Center to improve the tensile strength and toughness (or fracture energy) of boron fibers by making them larger in diameter (DiCarlo, 1985). We shall discuss the requirements for metal matrix composites (MMC) in terms of the fiber diameter and other parameters in Chapter 6. Here, we restrict ourselves to boron fibers. Table 2.3 provides a summary of the characteristics of boron fiber. Commercially produced 142-µm-diameter boron fiber shows tensile strengths less than 3.8 GPa. The tensile strength and fracture energy values of the as-received and some limited-production-run larger-diameter fibers showed improvement after chemical polishing, as shown in Table 2.3. Fibers showing strengths greater than 4 GPa had their fracture controlled by a tungsten boride core, while fibers with strengths of 4 GPa were controlled by fiber surface flaws. The high-temperature treatment, listed in Table 2.3, improved the fiber properties by putting a permanent axial compressive strain in the sheath.

Boron has a density of 2.34 g cm^{-3} (about 15% less than that of aluminum). Boron fiber with a tungsten core has a density of 2.6 g cm^{-3} for a fiber with 100-µm diameter. Its melting point is 2040 °C, and it has a thermal expansion coefficient of 8.3×10^{-6} °C^{-1} up to 315 °C.

Boron fiber composites are in use in a number of U.S. military aircraft, notably the F-14 and F-15, and in the U.S. space shuttle. They are also used for stiffening golf shafts, tennis rackets, and bicycle frames. One big obstacle

Table 2.3. Strength properties of improved large-diameter boron fibers

Diameter (μm)	Treatment	Strength		Relative fracture energy
		Average[a] (GPa)	COV[b] (%)	
142	As-produced	3.8	10	1.0
406	As-produced	2.1	14	0.3
382	Chemical polish	4.6	4	1.4
382	Heat treatment plus polish	5.7	4	2.2

[a] Gauge length = 25 mm.
[b] Coefficient of variation = standard deviation/average value.
Source: Reprinted with permission from *Journal of Metals*, **37**, No. 6, 1985, a publication of The Metallurgical Society, Warrendale, PA.

to the widespread use of boron fiber is its high cost compared to other fibers. A major portion of this high price is the cost of the tungsten substrate.

2.4   Carbon Fibers

Carbon is a very light element with a density equal to 2.268 g cm^{-3}. Carbon can exist in a variety of crystalline forms. Our interest here is in the so-called graphitic structure wherein the carbon atoms are arranged in the form of hexagonal layers. The other well-known form of carbon is the covalent diamond structure wherein the carbon atoms are arranged in a three-dimensional configuration with little structural flexibility. Another form of carbon is Buckminster Fullerene (or Bucky ball), with a molecular composition of C_{60} or C_{70}. Carbon in the graphitic form is highly anisotropic, with a theoretical Young's modulus in the layer plane being equal to about 1000 GPa, while that along the c-axis is equal to about 35 GPa. The graphite–structure (Fig. 2.14a) has a very dense packing in the layer planes. The lattice structure is shown more clearly with only lattice planes in Figure 2.14b. As we know, the bond strength determines the modulus of a material. Thus, the high-strength bond between carbon atoms in the layer plane results in an extremely high modulus while the weak van der Waals–type bond between the neighboring layers results in a lower modulus in that direction. Consequently, in a carbon fiber one would like to have a very high degree of preferred orientation of hexagonal planes along the fiber axis.

Carbon fibers of extremely high modulus can be made by carbonization of organic precursor fibers followed by graphitization at high temperatures. The organic precursor fiber, that is, the raw material for carbon fiber, is generally a special textile polymeric fiber that can be carbonized without melting. The precursor fiber, like any polymeric fiber, consists of long-chain

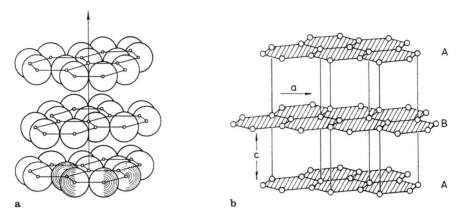

Fig. 2.14. a The densely packed graphitic layer structure. **b** The hexagonal lattice structure of graphite.

molecules (0.1–1 μm when fully stretched) arranged in a random manner. Such polymeric fibers generally have poor mechanical properties and typically show rather large deformations at low stresses mainly because the polymeric chains are not ordered. A commonly used precursor fiber is polyacrylonitrile (PAN). Other precursor fibers include rayon and the ones obtained from pitches, polyvinyl alcohol, polyimides, and phenolics.

Carbon fiber is a generic term representing a family of fibers (Chawla, 1981). As pointed out earlier, unlike the rigid diamond structure, graphitic carbon has a lamellar structure. Thus, depending on the size of the lamellar packets, their stacking heights, and the resulting crystalline orientations, one can obtain a range of properties. Most of the carbon fiber fabrication processes involve the following essential steps:

1. A *fiberization* procedure to make a precursor fiber. This generally involves wet-, dry-, or melt-spinning followed by some drawing or stretching.
2. A *stabilization* treatment that prevents the fiber from melting in the subsequent high-temperature treatments.
3. A thermal treatment called *carbonization* that removes most noncarbon elements.
4. An optional thermal treatment called *graphitization* that improves the properties of carbon fiber obtained in step 3.

It should be clear to the reader by now that in order to make a high-modulus fiber, one must improve the orientation of graphitic crystals or lamellas. This is achieved by various kinds of thermal and stretching treatments involving rather rigorous controls. If a constant stress were applied for a long time, for example, it would result in excessive fiber elongation and the accompanying reduction in area may lead to fiber fracture.

2.4.1 Preparation

Shindo (1961) in Japan was the first to prepare high-modulus carbon fiber starting from PAN. He obtained a Young's modulus of about 170 GPa. In 1963, British researchers at Rolls Royce discovered that a high elastic modulus of carbon fiber was obtained by stretching. They obtained, starting from PAN, a carbon fiber with an elastic modulus of about 600 GPa. Since then, developments in the technology of carbon fibers have occurred in rapid strides. The minute details of the conversion processes from precursor fiber to a high-modulus carbon fiber continue to be proprietary secrets. All the methods, however, exploit the phenomenon of thermal decomposition of an organic fiber under well-controlled conditions of rate and time of heating, environment, and so on. Also, in all processes the precursor is stretched at some stage of pyrolysis to obtain the high degree of alignment of graphitic basal planes.

Ex-PAN Carbon Fibers

Carbon fibers made from PAN are called *ex-PAN carbon fibers*. The polyacrylonitrile fibers are stabilized in air (a few hours at 250 °C) to prevent melting during the subsequent higher-temperature treatment. The fibers are prevented from contracting during this oxidation treatment. The black fibers obtained after this treatment are heated slowly in an inert atmosphere to 1000–1500 °C. Slow heating allows the high degree of order in the fiber to be maintained. The rate of temperature increase should be low so as not to destroy the molecular order in the fibers. The final heat treatment consists of holding the fibers for very short duration at temperatures up to 3000 °C. This improves the fiber texture and thus increases the elastic modulus of the fiber. Figure 2.15 shows, schematically, this PAN-based carbon fiber production process (Baker, 1983). Typically, the carbon fiber yield is about 50%.

    Figure 2.16a shows the flexible PAN molecular structure. Note the all-carbon backbone. This structure is essentially that of polyethylene with a nitrile (CN) group on every alternate carbon atom. The structural changes occurring during the conversion of PAN to carbon fiber are as follows. The initial stretching treatment of PAN improves the axial alignment of the polymer molecules. During this oxidation treatment, the fibers are maintained under tension to keep the alignment of PAN while it transforms into rigid ladder polymer (Fig. 2.16b). In the absence of tensile stress in this step, a relaxation will occur, and the ladder polymer structure will become disoriented with respect to the fiber axis. After the stabilizing treatment, the resulting ladder-type structure (also called *oriented cyclic structure*) has a high glass transition temperature so that there is no need to stretch the fiber during the next stage, which is carbonization. There are still considerable quantities of nitrogen and hydrogen present, which are eliminated as gaseous waste products during carbonization, that is, heating to 1000–1500 °C (Fig. 2.15). The carbon atoms remaining after this treatment are mainly in the

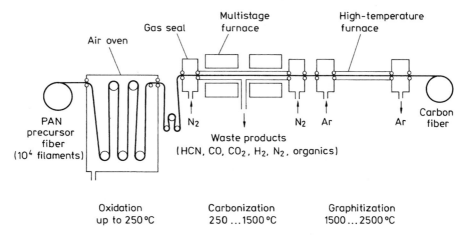

Fig. 2.15. Schematic of PAN-based carbon fiber production. [Reprinted with permission from Baker (1983).]

form of a network of extended hexagonal ribbons, which has been called *turbostratic* graphite structure in the literature. Although these strips tend to align parallel to the fiber axis, the degree of order of one ribbon with respect to another is relatively low. This can be improved further by heat treatment at still higher temperatures (up to 3000 °C). This is the graphitization treatment (Fig. 2.15). The mechanical properties of the resultant carbon fiber may vary over a large range depending mainly on the temperature of the

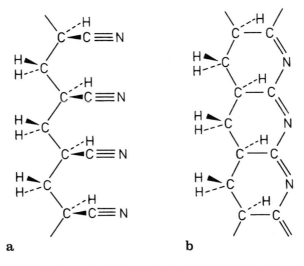

Fig. 2.16. a Flexible polyacrylonitrile molecule. **b** Rigid ladder (or oriented cyclic) molecule.

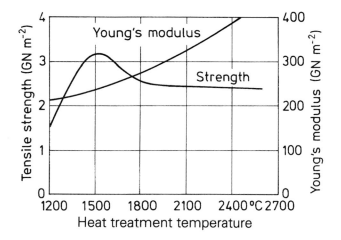

Fig. 2.17. Strength and elastic modulus of carbon fiber as a function of final heat treatment temperature. [After Watt (1970), used with permission.]

final heat treatment (Fig. 2.17) (Watt, 1970). Hot stretching above 2000 °C results in plastic deformation of fibers, leading to an improvement in properties.

Ex-Cellulose Carbon Fibers

Cellulose is a natural polymer that is frequently found in a fibrous form. In fact, cotton fiber, which is cellulosic, was one of the first to be carbonized. Thomas Edison did that to obtain carbon filament for an incandescent lamp. Cotton has the desirable property of decomposing before melting. It is inappropriate, however, for high-modulus carbon fiber manufacture because it has a rather low degree of orientation along the fiber axis, although it is highly crystalline. It is also not available as a tow of continuous filaments and is quite expensive. These difficulties have been overcome in the case of rayon fiber, which is made from wood pulp, a cheap source. The cellulose is extracted from wood pulp, and continuous filament tows are produced by wet spinning.

Rayon is a thermosetting polymer. The process used for the conversion of rayon into carbon fiber involves the same stages: fiberization, stabilization in a reactive atmosphere (air or oxygen, <400 °C), carbonization (<1500 °C), and graphitization (>2500 °C). Various reactions occur during the first stage, causing extensive decomposition and evolution of H_2O, CO, CO_2, and tar. The stabilization is carried out in a reactive atmosphere to inhibit tar formation and improve yield (Bacon, 1973). Chain fragmentation or depolymerization occurs in this stage. Because of this depolymerization, stabilizing under tension, as in the case of PAN precursor, does not work (Bacon, 1973). The carbonization treatment involves heating to about 1000 °C in

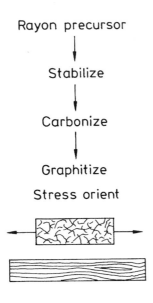

Rayon precursor

↓

Stabilize

↓

Carbonize

↓

Graphitize

Stress orient

Fig. 2.18. Schematic of rayon-based carbon fiber production. [After Diefendorf and Tokarsky (1975), used with permission.]

nitrogen. Graphitization is carried out at 2800 °C under stress. This orienting stress at high temperature results in plastic deformation via multiple-slip system operation and diffusion. Figure 2.18 shows the process schematically. The carbon fiber yield from rayon is between 15 and 30% by weight, compared to a yield of about 50% in the case of PAN precursors.

Ex-Pitch Carbon Fibers

There are various sources of pitch, but the three commonly used sources are polyvinyl chloride (PVC), petroleum asphalt, and coal tar. Pitch-based carbon fibers are attractive because of the cheap raw material, high yield of carbon fiber, and a highly oriented carbon that is obtained from mesophase pitch precursor fiber.

The same sequence of oxidation, carbonization, and graphitization is required for making carbon fibers out of pitch precursors. Orientation in this case is obtained by spinning. An isotropic but aromatic pitch is subjected to melt spinning at very high strain rates and quenched to give a highly oriented fiber. This thermoplastic fiber is then oxidized to form a cross-linked structure that makes the fiber nonmelting. This is followed by carbonization and graphitization.

Commercial pitches are mixtures of various organic compounds with an average molecular weight between 400 and 600. Prolonged heating above 350 °C results in the formation of a highly oriented, optically anisotropic liquid crystalline phase (mesophase). When observed under polarized light, anisotropic mesophase dispersed in an isotropic pitch appears as microspheres floating in pitch. The liquid crystalline mesophase pitch can be melt spun into a precursor for carbon fiber. The melt spinning process involves shear and elongation in the fiber axis direction, and thus a high degree of preferred orientation is achieved. This orientation can be further developed during conversion to carbon fiber. The pitch molecules (aromatics of low molecular weight) are stripped of hydrogen, and the aromatic molecules coalesce to form larger bidimensional molecules. Very high values of Young's modulus can be obtained. It should be appreciated that one must have the pitch in a state amenable to spinning in order to produce the precursor fiber, which is made infusible to allow carbonization to occur without melting. Thus, the pitches obtained from petroleum asphalt and coal tar need pretreatments. This pretreatment can be avoided in the case of PVC by means of a carefully controlled thermal degradation of PVC. The molecular weight controls the viscosity of the melt polymer and the melting range. Thus, it also controls the temperature and the spinning speed. Pitches are polydispersoid systems, and thus they have a large range of molecular weights, which can be adjusted by solvent extraction or distillation. Figure 2.19 shows the process of pitch-based carbon fiber manufacture starting from an isotropic pitch and a mesophase pitch (Diefendorf and Tokarsky, 1975).

2.4.2 Structural Changes Occurring During Processing

The thermal treatments for all precursor fibers serve to remove noncarbon elements in the form of gases. For this, the precursor fibers are stabilized to ensure that they decompose before melting. Generally, they become black

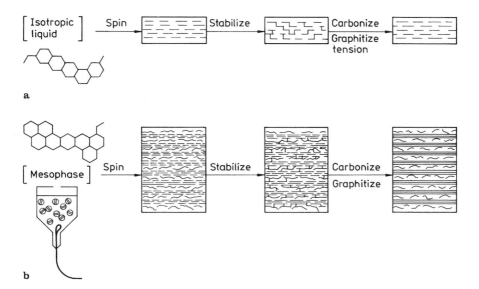

Fig. 2.19. Schematic of pitch-based carbon fiber production: **a** isotropic pitch process, **b** mesophase pitch process. [With permission from Diefendorf and Tokarsky (1975).]

Fig. 2.20. Scanning electron micrograph of PAN-based carbon fiber (fiber diameter is 8 µm).

after this treatment. Carbon fibers obtained after carbonization contain many "grown-in" defects because the thermal energy supplied at these low temperatures is not enough to break the already-formed carbon-carbon bonds. That is why these carbon fibers are very stable up to 2500–3000 °C when they change to graphite. The decomposition of the precursor fiber invariably results in a weight loss and a decrease in fiber diameter. The weight loss can be considerable—from 40 to 90%, depending on the precursor and treatment (Ezekiel and Spain, 1967). The external morphology of the fiber, however, is generally maintained. Thus, precursor fibers with transverse sections in the form of a kidney bean, dog bone, or circle maintain their form after conversion to carbon fiber. Figure 2.20 shows a scanning electron micrograph of a PAN-based carbon fiber. Note the surface markings that appear during the fiber drawing process.

At the microscopic level, carbon fibers possess a rather heterogeneous microstructure. Not surprisingly, many workers (Diefendorf and Tokarsky, 1975; Watt and Johnson, 1969; Johnson and Tyson, 1969; Perret and Ruland, 1970; Bennett and Johnson, 1978, 1979; Inal et al., 1980) have attempted to characterize the structure of carbon fibers, and one can find a number of models in the literature. There is a better understanding of the structure of PAN-based carbon fibers. Essentially, a carbon fiber consists of many graphitic lamellar ribbons oriented roughly parallel to the fiber axis with a complex interlinking of layer planes both longitudinally and laterally. Based on high-resolution lattice fringe images of longitudinal and transverse sections in TEM, a schematic two-dimensional representation is given in Figure 2.21 (Bennett and Johnson, 1979), while a three-dimensional model is shown in Figure 2.22 (Bennett and Johnson, 1978). The structure is typically defined in terms of crystallite dimensions, L_a and L_c in directions a and c, respectively, as shown in Figure 2.23. The degree of alignment and the parameters L_a and L_c vary with the graphitization temperature. Both L_a and L_c increase with increasing heat treatment temperature.

2.4.3 Properties and Applications

The density of the carbon fiber varies with the precursor and the thermal treatment given. It is generally in the range of 1.6–2.0 g cm^{-3}. Note that the

Fig. 2.21. Two-dimensional representation of PAN-based carbon fiber. [After Bennett and Johnson (1979), used with permission.]

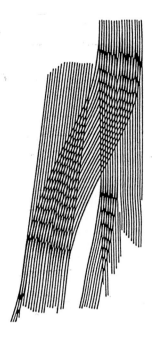

density of the carbon fiber is more than that of the precursor fiber; the density of the precursor is generally between 1.14 and 1.19 g cm^{-3} (Bennett et al., 1983). There are always flaws of various kinds present, which may arise from impurities in the precursors or may simply be the misoriented layer planes. A mechanism of tensile failure of carbon fiber based on the presence of misoriented crystallites is shown in Figure 2.23 (Bennett et al., 1983). Figure 2.23a shows a misoriented crystallite linking two crystallites parallel to the fiber axis. Under the action of applied stress, basal plane rupture occurs in the misoriented crystallite in the L_c direction, followed by crack development along L_a and L_c (Fig. 2.23b). Continued stressing causes complete failure of the misoriented crystallite (Fig. 2.23c). If the crack size is greater than the critical size in the L_a and L_c directions, catastrophic failure results.

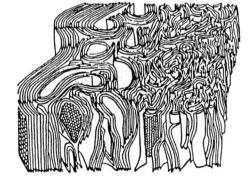

Fig. 2.22. Three-dimensional representation of PAN-based carbon fiber. [From Bennett and Johnson (1978), used with permission.]

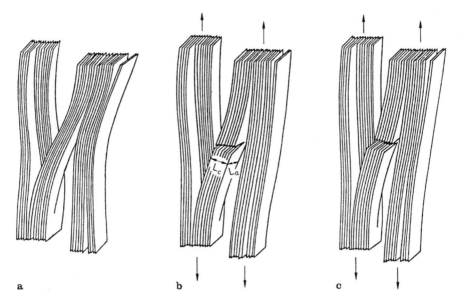

Fig. 2.23. Model for tensile failure of carbon fiber: **a** a misoriented crystallite linking two crystallites parallel to the fiber axis, **b** basal plane rupture under the action of applied stress, **c** complete failure of the misoriented crystallite. [From Bennett et al. (1983), used with permission.]

As mentioned earlier, the degree of order, and consequently the modulus in the fiber axis direction, increases with increasing graphitization temperature. Fourdeux et al. (1971) measured the preferred orientation of various carbon fibers and plotted it against an orientation parameter q (Fig. 2.24). The parameter q has a value of -1 for perfect orientation and zero for the isotropic case. In Figure 2.24 we have plotted the absolute value of q. Note also that the modulus has been corrected for porosity. The theoretical curve fits the experimental data very well.

Even among the PAN carbon fibers, we can have a series of carbon fibers: for example, high tensile strength but medium Young's modulus (HT) fiber (200–300 GPa); high Young's modulus (HM) fiber (400 GPa); extra- or superhigh tensile strength (SHT); and superhigh modulus type (SHM) carbon fibers. The mesophase pitch-based carbon fibers show rather high modulus but low strength levels (2 GPa). Not unexpectedly, the HT-type carbon fibers show a much higher strain-to-failure value than the HM type. The mesophase pitch-based carbon fibers are used for reinforcement, while the isotropic pitch-based carbon fibers (very low modulus) are more frequently used as insulation and fillers. Table 2.4 compares the properties of some commonly obtainable carbon fibers and graphite monocrystal (Singer, 1979). For high-temperature applications involving carbon fibers, it is important to take into account the variation of inherent oxidation resistance of

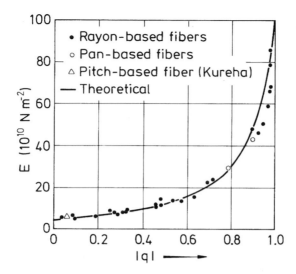

Fig. 2.24. Variation of longitudinal elastic modulus for various carbon fibers with the degree of preferred orientation. The value of the orientation parameter, q, is 1 for perfect orientation and zero for the isotropic case. [From Fourdeux et al. (1971), used with permission.]

carbon fibers with modulus. Figure 2.25 shows that the oxidation resistance of carbon fiber increases with the modulus value (Riggs, 1985). The modulus, as we know, increases with the final heat treatment temperature during processing.

We note from Table 2.4 that the carbon fibers produced from various precursor materials are fairly good electrical conductors. Although this led to some work toward a potential use of carbon fibers as current carriers for electrical power transmission (Murday et al., 1984), it has also caused extreme concern in many quarters. The reason for this concern is that if the extremely fine carbon fibers accidentally become airborne (during manufacture or service) they can settle on electrical equipment and cause short circuiting. An interesting characteristic of ex-mesophase pitch carbon fiber is the extremely high thermal conductivity it can have. Ex-pitch carbon fibers with a suitably oriented microstructure can have thermal conductivity as high as 1100 W/mK. The figure for an ex-PAN carbon fiber is generally less than 50 W/mK.

Anisotropic as the carbon fibers are, they have two principal coefficients of thermal expansion, namely, transverse or perpendicular to the fiber axis, α_t, and parallel to the fiber axis α_l. Typical values are

$$\alpha_t \simeq 5.5 \text{ to } 8.4 \times 10^{-6} \text{ K}^{-1}$$

$$\alpha_l \simeq -0.5 \text{ to } -1.3 \times 10^{-6} \text{ K}^{-1}$$

Compressive properties of ex-mesophase carbon fibers are about half their

Table 2.4. Comparison of properties of different carbon fibers

Precursor	Density (g cm⁻³)	Young's modulus (GPa)	Electrical resistivity (10⁻⁴ Ω cm)
Rayon[a]	1.66	390	10
Polyacrylonitrile[b] (PAN)	1.74	230	18
Pitch (Kureha)			
LT[c]	1.6	41	100
HT[d]	1.6	41	50
Mesophase pitch[e]			
LT	2.1	340	9
HT	2.2	690	1.8
Single-crystal[f] graphite	2.25	1000	0.40

[a] Union Carbide, Thornel 50.
[b] Union Carbide, Thornel 300.
[c] LT, low-temperature heat-treated.
[d] HT, high-temperature heat-treated.
[e] Union Carbide type P fibers.
[f] Modulus and resistivity are in-plane values.
Source: Adapted with permission from Singer (1979).

tensile properties! Still, they are an order of magnitude better than aramid-type fibers (see Sect. 2.5).

Carbon fibers are used in a variety of applications in the aerospace and sporting goods industries. Cargo bay doors and booster rocket casings in the U.S. space shuttle are made of carbon fiber–reinforced epoxy composites. Modern commercial aircraft also use carbon fiber–reinforced composites. With the ever-decreasing price of carbon fibers, applications of carbon fibers in other areas have also increased, for example, various machinery items such as turbine, compressor, and windmill blades and flywheels; in the field of medicine the applications include both equipment and implant materials (e.g., ligament replacement in knees and hip joint replacement). We discuss these in more detail in Chapter 8.

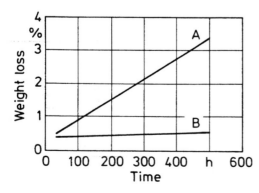

Fig. 2.25. Oxidation resistance, measured as weight loss in air at 350 °C, of carbon fibers having different moduli: (A) Low modulus Celion 3000 (240 GPa) and (B) High modulus Celion G-50 (345 GPa). [After J.P. Riggs, *Encyclopedia of Polymer Science and Engineering*, 2e, Vol. 2, 1985, John Wiley and Sons, New York, reprinted with permission.]

2.5   Organic Fibers

In general, polymeric chains assume a random coil configuration, i.e., they have the so-called cooked-spaghetti structure (see Chap. 3). In this random coil structure, the macromolecular chains are neither aligned in one direction nor stretched out. Thus, they have predominantly weak van der Waals interactions rather than strong covalent interactions, resulting in a low strength and stiffness. Because the covalent carbon-carbon bond is very strong, one would expect that linear chain polymers, such as polyethylene, would be potentially very strong and stiff. Conventional polymers show a Young's modulus, E, of about 10 GPa or less. Highly drawn polymers with a Young's modulus of about 70 GPa can be obtained easily. However, if one wants strong and stiff organic fibers, one must obtain oriented molecular chains with full extension. Thus, in order to obtain high-stiffness and -strength polymers, we must extend these polymer chains and pack them in a parallel array. The orientation of these polymer chains with respect to the fiber axis and the manner in which they fit together (i.e., order or crystallinity) are controlled by their chemical nature and the processing route. There are two ways of achieving molecular orientation, one without high molecular extension (Fig. 2.26a) and the other with high molecular extension (Fig. 2.26b). It is the kind of chain structure shown in Figure 2.26b, i.e., molecular chain orientation coupled with molecular chain extension, that is needed for high stiffness and strength. To get a Young's modulus value greater than 70 GPa, one needs rather high draw ratios, i.e., a very high degree of elongation must be carried out under such conditions that macroscopic elongation results in a corresponding elongation at a molecular level. It turns out that the Young's modulus, E, of a polymeric fiber increases linearly with the deformation ratio (draw ratio in tensile drawing or die drawing and extrusion ratio in hydrostatic extrusion). The drawing behavior of a polymer is a sensitive function of (i) its molecular weight and molecular weight distribution and (ii) deformation conditions (temperature and strain rate). Too low a drawing temperature produces voids, while too high a drawing temperature results in flow drawing, i.e., the macroscopic elongation of the material does not result in a molecular alignment, and consequently, no stiffness enhancement results. An oriented and extended macromolecular chain structure, however, is not

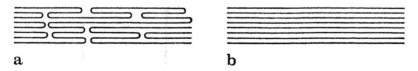

a   b

Fig. 2.26. Two types of molecular orientation: **a** oriented without high molecular extension and **b** oriented with high molecular extension. [From Barham and Keller (1985), used with permission.]

easy to achieve in practice. Nevertheless, considerable progress in this area has been made during the last quarter of the twentieth century. Organic fibers, such as aramid and polyethylene, possessing high strength and modulus are the fruits of this realization. Two very different approaches have been taken to make high-modulus organic fibers. These are:

1. Processing the conventional flexible-chain polymers in such a way that the internal structure takes a highly oriented and extended-chain arrangement. Structural modification of "conventional" polymers such as high-modulus polyethylene was developed by choosing appropriate molecular weight distributions, followed by drawing at suitable temperatures to convert the original folded-chain structure into an oriented, extended chain structure.
2. The second, radically different, approach involves synthesis, followed by extrusion of a new class of polymers, called liquid crystal polymers. These have a rigid rod molecular chain structure. The liquid crystalline state, as we shall see, has played a very significant role in providing highly ordered, extended chain fibers.

These two approaches have resulted in two commercialized high-strength and high-stiffness fibers, polyethylene and aramid. Next, we will describe the processing, structure, and properties of these two fibers.

2.5.1   Oriented Polyethylene Fibers

The ultrahigh-molecular-weight polyethylene fiber is a highly crystalline fiber with very high stiffness and strength. This results from some innovative processing and control of the structure of polyethylene.

Processing of Polyethylene Fibers

Drawing of melt crystallized polyethylene (molecular mass between 10^4 and 10^5) to very high draw ratios can result in moduli of up to 70 GPa. Tensile drawing, die drawing, or hydrostatic extrusion can be used to obtain the high permanent or plastic strains required to obtain a high modulus. It turns out that modulus is dependent on the draw ratio but independent of how the draw ratio is obtained (Capaccio et al., 1979). In all these drawing processes, the polymer chains become merely oriented without undergoing molecular extension, and we obtain the kind of structure shown in Figure 2.26a. Later developments led to solution and gel spinning of very high molecular weight polyethylene ($>10^6$) with moduli as high as 200 GPa. The gel spinning method of making polyethylene fibers has become technologically and commercially most successful. Pennings and coworkers (Kalb and Pennings, 1980; Smook and Pennings, 1984) made high-modulus polyethylene fiber by solution spinning. Their work was followed by Smith and Lemstra (1976, 1980), who made polyethylene fiber by gel spinning. The gel spinning process of making polyethylene was industrialized in the 1980s. *Gels* are swollen networks in which crystalline regions form the network junctions. An appropriate polymer solution is converted into gel, which is drawn to give the fiber. At least three commercial firms produce oriented polyethylene fiber using similar techniques. DSM (Dutch State Mines) produces a fiber called *Dyneema*; AlliedSignal, a U.S. company, produces *Spectra* fiber under license from DSM; and Mitsui, a Japanese company, produces a polyethylene fiber with the trade name *Tekmilon*. Next, we will describe the gel spinning process of making the high-stiffness polyethylene fiber.

Gel Spinning of Polyethylene Fiber

Polyethylene (PE) is a particularly simple, linear macromolecule, with the following chemical formula

$$[-CH_2-CH_2-CH_2-CH_2-CH_2-CH_2-CH_2-CH_2-]_n$$

Thus, compared to other polymers, it is easier to obtain an extended and oriented chain structure in polyethylene. High-density polyethylene (HDPE) is preferred to other types of polyethylene because HDPE has fewer branch points along its backbone and a high degree of crystallinity. These characteristics of linearity and crystallinity are important from the point of getting a high degree of orientational order and obtaining an extended chain structure in the final fiber.

Figure 2.27 provides a flow diagram of the gel spinning process for making the high-modulus polyethylene fiber. The three companies mentioned earlier use different solvents, such as decalin, paraffin oil, and paraffin wax, to make a dilute (5–10%) solution of polymer in solvent at about 150 °C. A dilute solution is important in that it allows for a lesser chain entanglement, which makes it easier for the final fiber to be highly oriented. A polyethylene gel is produced when the solution coming out of the spinneret is quenched by air. The as-spun gelled fiber enters a cooling bath. At this stage, the fiber is thought to have a structure consisting of folded chain lamellae with solvent between them and a swollen network of entanglements. These entanglements allow the as-spun fiber to be drawn to very high draw ratios, which can be as high as 200. The maximum draw ratio is related to the average distance between the entanglements, i.e., the solution concen-

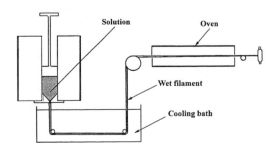

Fig. 2.27. Gel spinning process used to make high-modulus polyethylene fiber.

Table 2.5. Properties of polyethylene fibers*

Property	Spectra 900	Spectra 1000
Density (g cm^{-3})	0.97	0.97
Diameter (μm)	38	27
Tensile Strength (GPa)	2.7	3.0
Tensile modulus (GPa)	119	175
Tensile strain to fracture (%)	3.5	2.7

* Manufacturer's data; indicative values.

tration. The gelled fibers are drawn at 120 °C. One problem with this gel route is the rather low spinning rates of 1.5 m min^{-1}. At higher rates, the properties obtained are not very good (Kalb and Pennings 1980; Smook and Pennings 1984).

Structure and Properties of Polyethylene Fiber

The unit cell of a single crystal (orthorhombic) of polyethylene has the dimensions of 0.741, 0.494, and 0.255 nm. There are four carbon and eight hydrogen atoms per unit cell. One can compute the theoretical density of polyethylene, assuming a 100% single-crystal polyethylene. If one does that, the theoretical density of polyethylene comes out to be 0.9979 g cm^{-3}; of course, in practice, one can only tend toward this theoretical value. As it turns out, the highly crystalline, UHMWPE fiber has a density of 0.97 g cm^{-3}, which is very near the theoretical value. Thus, polyethylene fiber is very light; in fact, it is lighter than water and thus floats on water. A summary of some commercially available polyethylene fibers is provided in Table 2.5.

Its strength and modulus are slightly lower than those of aramid fibers but on a per-unit-weight basis, i.e., specific property values are about 30% to 40% higher than those of aramid. As is true of most organic fibers, both polyethylene and aramid fibers must be limited to low-temperature (lower than 150 °C) applications.

Another effect of the high degree of chain alignment in these fibers is manifested when they are put in a polymeric matrix to form a fiber reinforced composite. High-modulus polyethylene fibers, such as Spectra or Dyneema, are hard to bond with any polymeric matrix. Some kind of surface treatment must be given to the polyethylene fiber to bond with resins such as epoxy and PMMA. By far, the most successful surface treatment involves a cold gas (such as air, ammonia, or argon) plasma (Kaplan et al., 1988). A plasma consists of gas molecules in an excited state, i.e., highly reactive, dissociated molecules. When the polyethylene, or any other fiber, is treated with a plasma, surface modification occurs by removal of any surface contaminants and highly oriented surface layers, addition of polar and functional groups on the surface, and introduction of surface roughness;

all these factors contribute to an enhanced fiber/matrix interfacial strength
(Biro et al.,1992; Brown et al., 1992; Hild and Schwartz, 1992a, 1992b;
Kaplan et al., 1988; Li et al., 1992). Exposure to the plasma for just a few
minutes is enough.

Polyethylene fiber is crystalline 90–95% and has a density of 0.97 g cm^{-3}.
There is a linear relationship between density and crystallinity for polyethylene. A 100% crystalline polyethylene will have a theoretical density,
based on an orthorhombic unit cell, of about 1 g cm^{-3}. A totally amorphous
polyethylene (0% crystallinity) will have a density of about 0.85 g cm^{-3}.
Raman spectroscopy has been used to study the deformation behavior of
polyethylene fiber.

2.5.2 Aramid Fibers

Aramid fiber is a generic term for a class of synthetic organic fibers called
aromatic polyamide fibers. The U.S. Federal Trade Commission gives a
good definition of an aramid fiber as "a manufactured fiber in which the
fiber-forming substance is a long-chain synthetic polyamide in which at least
85% of the amide linkages are attached directly to two aromatic rings."
Commercial names of aramid fibers include Kevlar and Nomex (Du Pont),
Teijinconex and Technora (Teijin), and Twaron (Akzo). *Nylon* is a generic
name for any long-chain polyamide. Aramid fibers such as Nomex or Kevlar,
however, are ring compounds based on the structure of benzene as opposed
to the linear compounds used to make nylon. The basic difference between
Kevlar and Nomex is that Kevlar has *para*-oriented aromatic rings, i.e., it
has a symmetrical molecule, with bonds from each aromatic ring being
parallel, while Nomex is *meta*-oriented, with bonds at 120-degree angles to
each other. The basic chemical structure of aramid fibers consists of oriented
para-substituted aromatic units, which makes them rigid rodlike polymers.
The rigid rodlike structure results in a high glass transition temperature and
poor solubility, which makes fabrication of these polymers, by conventional
drawing techniques, difficult. Instead, they are melt spun from liquid crystalline polymer solutions, as described in the next section.

Processing of Aramid Fibers

Processing of aramid fibers involves solution-polycondensation of diamines
and diacid halides at low temperatures. Hodd and Turley (1978), Morgan
(1979), and Magat (1980) have given simplified accounts of the theory
involved in the fabrication of aramid fibers. The most important point is that
the starting spinnable solutions that give high-strength and high-modulus
fibers have liquid crystalline order. Figure 2.28 shows schematically various
states of a polymer in solution. Figure 2.28a shows two-dimensional, linear,
flexible chain polymers in solution. These are called *random coils*, as the
figure suggests. If the polymer chains can be made of rigid units, that is,

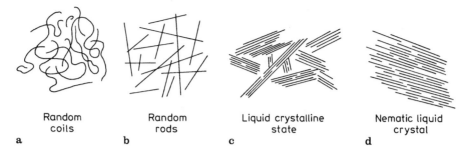

Random coils	Random rods	Liquid crystalline state	Nematic liquid crystal
a	b	c	d

Fig. 2.28. Various states of polymer in solution: **a** two-dimensional, linear, flexible chains (random coils), **b** random array of rods, **c** partially ordered liquid crystalline state, and **d** nematic liquid crystal (randomly distributed parallel rods).

rodlike, we can represent them as a random array of rods (Figure 2.28b). Any associated solvent may contribute to the rigidity and to the volume occupied by each polymer molecule. With increasing concentration of rod-like molecules, one can dissolve more polymer by forming regions of partial order, that is, regions in which the chains form a parallel array. This partially ordered state is called a *liquid crystalline state* (Figure 2.28c). When the rodlike chains become approximately arranged parallel to their long axes, but their centers remain unorganized or randomly distributed, we have what is called a *nematic* liquid crystal (Figure 2.28d). It is this kind of order that is found in the extended-chain polyamides.

Liquid crystal solutions, because of the presence of the ordered domains, are optically anisotropic, that is, *birefringent*. Figure 2.29 shows the anisotropic Kevlar aramid and sulfuric acid solution at rest between crossed polarizers. The parallel arrays of polymer chains in the liquid crystalline state become even more ordered when these solutions are subjected to shear as, for example, in extruding through a spinneret hole. It is this inherent property of liquid crystal solutions that is exploited in the manufacture of aramid fibers. The characteristic fibrillar structure of aramid fibers is due

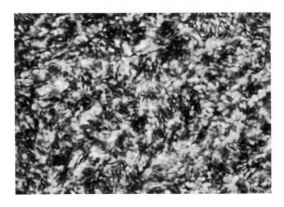

Fig. 2.29. Anisotropic Kevlar aramid and sulfuric acid solution at rest between crossed polarizers. (Courtesy of Du Pont Co.)

to the alignment of polymer crystallites along the fiber axes. Para-oriented aromatic polyamides form liquid crystal solutions under certain conditions of concentration, temperature, solvent, and molecular weight. Figure 2.30a shows a phase diagram of the system poly-*p*-benzamide in tetramethylurea-LiCl solutions (Magat, 1980). Only under certain conditions do we get the desirable anisotropic state. There also occurs an anomalous relationship between viscosity and polymer concentration in liquid crystal solutions. Initially, an increase in viscosity occurs as the concentration of polymer in solution increases, as it would in any ordinary polymer solution. At a critical point where it starts assuming an anisotropic liquid crystalline shape, a sharp drop in the viscosity occurs; see Figure 2.30b. The drop in viscosity of liquid crystal polymers at a critical concentration was predicted by Flory (1956). The drop in viscosity occurs due to the formation of a lyotropic nematic structure. The liquid crystalline regions act like dispersed particles and do not contribute to solution viscosity. With increasing polymer concentration, the amount of liquid crystalline phase increases up to a point after which the viscosity tends to rise again. There are other requirements for forming a liquid crystalline solution from aromatic polyamides. The molecular weight must be greater than some minimum value and the solubility must exceed the critical concentration required for liquid crystallinity. Thus, starting from liquid crystalline spinning solutions containing highly ordered arrays of extended polymer chains, we can spin fibers directly into an extremely oriented, chain-extended form. These as-spun fibers are quite strong and, because the chains are highly extended and oriented, the use of conventional drawing techniques becomes optional.

Para-oriented rigid diamines and dibasic acids give polyamides that yield, under appropriate conditions of solvent, concentration, and polymer molecular weight, the desired nematic liquid crystal structure. One would like to have, for any solution spinning process, a high molecular weight in order to have improved mechanical properties, a low viscosity to ease processing conditions, and a high polymer concentration for high yield. For para-aramid, poly-*p*-phenylene terephthalamide (PPTA), tradename Kevlar, the nematic liquid crystalline state is obtained in 100% sulfuric acid at a polymer concentration of about 20%. The polymer solution, often referred to as the *dope*, has concentrated sulfuric acid as the solvent for PPTA.

For aramid fibers, the dry jet–wet spinning method is used. The process is illustrated in Figure 2.31. Solution-polycondensation of diamines and diacid halides at low temperatures (near $0\,°C$) gives the aramid forming polyamides. Low temperatures are used to inhibit any by-product generation and promote linear polyamide formation. The resulting polymer is pulverized, washed, and dried; mixed with concentrated H_2SO_4; and extruded through a spinneret at about $100\,°C$. The jets from the orifices pass through about 1 cm of air layer before entering a cold water ($0–4\,°C$) bath. The fiber solidifies in the air gap, and the acid is removed in the coagulation bath. The spinneret capillary and air gap cause rotation and alignment of the domains, resulting in highly crystalline and oriented as-spun fibers. The air gap also allows the

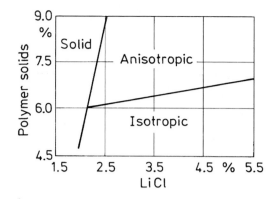

a

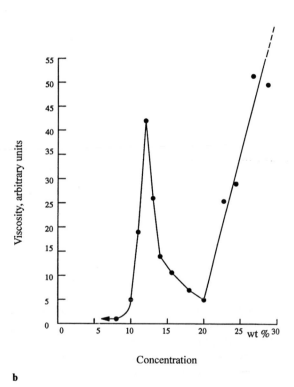

Concentration

b

Fig. 2.30. a. Phase diagram of poly-*p*-benzamide in tetramethylurea–LiCl solutions. Note that the anisotropic state is obtained under certain conditions. [With permission from Magat (1980).] **b** Viscosity versus polymer concentration in solution. A sharp drop in viscosity occurs where the solution starts becoming anisotropic liquid crystal. [After Kikuchi (1982).]

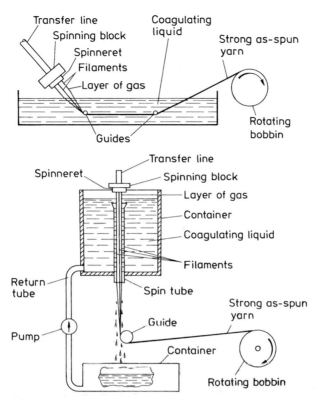

Fig. 2.31. The dry jet–wet spinning process of producing aramid fibers. [After Chiao and Chiao (1982), used with permission.]

dope to be at a higher temperature than is possible without the air gap. The higher temperature allows a more concentrated spinning solution to be used, and higher spinning rates are possible. Spinning rates of several hundred meters per minute are not unusual. Figure 2.32 compares the dry jet–wet spinning method used with nematic liquid crystals and the spinning of a conventional polymer. The oriented chain structure, together with molecular extension, is achieved with dry jet–wet spinning. The conventional wet or dry spinning gives precursors that need further processing for a marked improvement in properties (Jaffe and Jones, 1985). The as-spun fibers are washed in water, wound on a bobbin, and dried. Fiber properties are modified by the use of appropriate solvent additives, by changing the spinning conditions, and by means of some post-spinning heat treatments, if necessary.

Teijin aramid fiber, known as Technora (formerly HM-50), is made slightly differently from the liquid crystal route just described. Three monomers—terephthalic acid, *p*-phenylenediamine (PDA), and 3,4-diamino

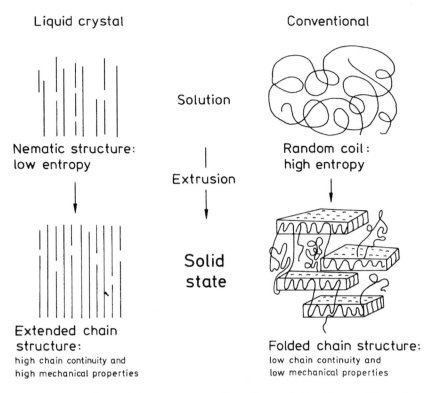

Fig. 2.32. Comparison of dry jet–wet spinning of nematic liquid crystalline solution and conventional spinning of a polymer. [Reprinted from Jaffe and Jones (1985), p. 349, courtesy of Marcel Dekker, Inc.]

dephenyl ether—are used. The ether monomer provides more flexibility to the backbone chain, which results in a fiber that has slightly better compressive properties than PPTA aramid fiber made via the liquid crystal route. An amide solvent with a small amount of salt (calcium chloride or lithium chloride) is used as a solvent (Ozawa et al., 1978). The polymerization is done at 0–80 °C in 1 to 5 hours and with a polymer concentration of 6–12%. The reaction mixture is spun from a spinneret into a coagulating bath containing 35–50% $CaCl_2$. Draw ratios between 6 and 10 are used.

Structure of Aramid Fibers

Kevlar aramid fiber is the most studied of all aramid fibers. Thus, our description of structure will be mostly from Kevlar. The chemical formula of Kevlar aramid is given in Figure 2.33. Chemically, the Kevlar- or Twaron-type aramid fiber is poly (p-phenyleneterephthalamide), which is a poly-condensation product of terephthaloyl chloride and p-phenylene diamine. The aromatic rings impart the rigid rodlike characteristics of aramid. These chains are highly oriented and extended along the fiber axis, with the resul-

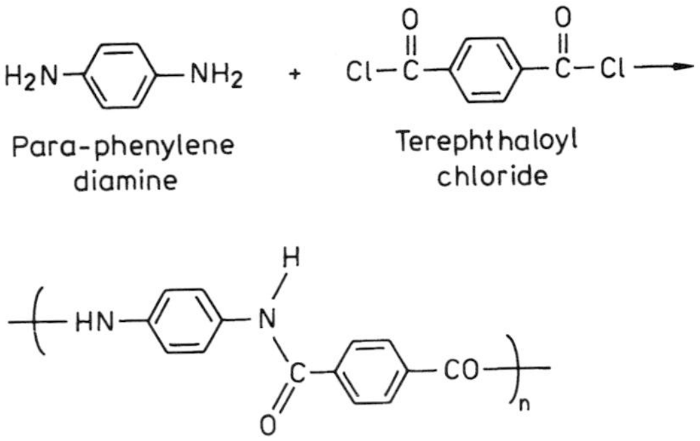

Fig. 2.33. Chemical structure of aramid fiber.

tant high modulus. Aramid fiber has a highly crystalline structure, and the para orientation of the aromatic rings in the polymer chains results in a high packing efficiency. Strong covalent bonding in the fiber direction and weak hydrogen bonding in the transverse direction (see Figure 2.34) result in highly anisotropic properties.

The structure of Kevlar aramid fiber has been investigated by electron microscopy and diffraction. A schematic representation of the supramolecular structure of Kevlar 49 is shown in Figure 2.35 (Dobb et al., 1980). It shows radially arranged, axially pleated crystalline supramolecular sheets. The molecules form a planar array with interchain hydrogen bonding. The stacking sheets form a crystalline array, but between the sheets the bonding is rather weak. Each pleat is about 500 nm long, and the pleats are separated by transitional bands. The adjacent components of a pleat make an angle of

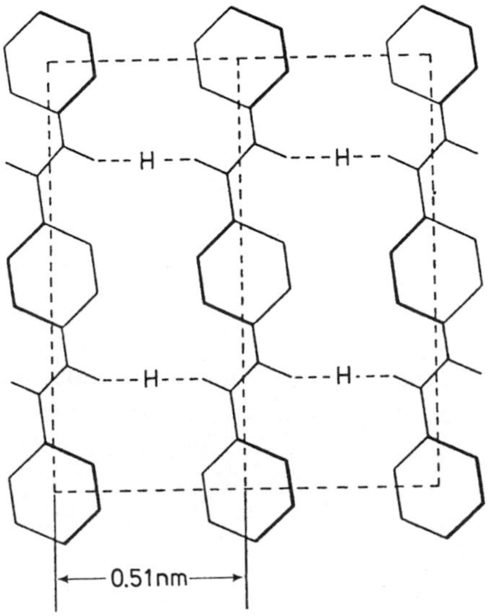

Fig. 2.34. Strong covalent bonding in the fiber direction and weak hydrogen bonding (indicated by *H*) in the transverse direction.

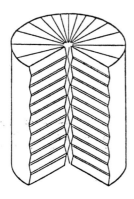

Fig. 2.35. Schematic representation of the supramolecular structure of aramid fiber, Kevlar 49. The structure consists of radially arranged, axially pleated crystalline sheets. [From Dobb et al. (1980), used with permission.]

170°. Such a structure is consistent with the experimentally observed rather low-longitudinal shear modulus and poor properties in compression and transverse to the Kevlar fiber axis. A correlation between good compressive characteristics and a high glass transition temperature (or melting point) has been suggested (Northolt, 1981). Thus, since the glass transition temperature of organic fibers is lower than that of inorganic fibers, the former would be expected to show poorer properties in compression. For aramid- and polyethylene-type high stiffness fibers, compression results in the formation of kink bands leading to an eventual ductile failure. Yielding is observed at about 0.5% strain; this is thought to correspond to a molecular rotation of the amide carbon-nitrogen bond shown in Figure 2.33 from "the normal extended transconfiguration to a kinked cis configuration" (Tanner et al., 1986). This causes a 45° bend in the chain. This bend propagates across the unit cell, the microfibrils, and a kink band results in the fiber. This anisotropic behavior of Kevlar fiber is revealed in the SEM micrograph in Figure 2.36. Figure 2.36 shows buckling or kink marks on the compressive side of a knotted Kevlar aramid fiber. Note the absence of such markings on

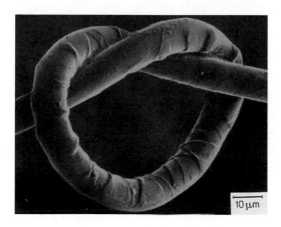

Fig. 2.36. Knotted Kevlar aramid fiber showing buckling marks on the compressive side. The tensile side is smooth. (Courtesy of Fabric Research Corp.)

10 μm

Table 2.6. Properties of Kevlar aramid fiber yarns*

Property	K 29	K 49	K 68	K 119	K 129	K 149
Density (g cm^{-3})	1.44	1.45	1.44	1.44	1.45	1.47
Diameter (μm)	12	12	12	12	12	12
Tensile strength (GPa)	2.8	2.8	2.8	3.0	3.4	2.4
Tensile strain to fracture (%)	3.5–4.0	2.8	3.0	4.4	3.3	1.5–1.9
Tensile modulus (GPa)	65	125	101	55	100	147
Moisture regain (%) at 25 °C, 65% RH	6	4.3	4.3	—	—	1.5
Coefficient of expansion (10^{-6} K^{-1})	−4.0	−4.9	—	—	—	—

*All data from Du Pont brochures. Indicative values only. 25-cm yarn length was used in tests (ASTM D-885). *K* stands for Kevlar, a trademark of Du Pont.

the tensile side. Such markings on the aramid fiber surface have also been reported by, among others, DeTeresa et al. (1984) when the aramid fiber is subjected to uniform compression or torsion.

Properties and Applications of Aramid Fibers

Some of the important properties of Kevlar aramid fibers are summarized in Table 2.6. As can be seen from this table, the Kevlar aramid fiber is very light and has very high stiffness and strength in tension. The two well-known varieties are Kevlar 49 and Kevlar 29. Kevlar 29 has about half the modulus but double the strain to failure of Kevlar 49. It is this high strain to failure of Kevlar 29 that makes it useful for making vests that are used for protection against small arms. It should be emphasized that aramid fiber, like most other high-performance organic fibers, has rather poor characteristics in compression, its compressive strength being only about 1/8 of its tensile strength. This follows from the anisotropic nature of the fiber as previously discussed. In tensile loading, the load is carried by the strong covalent bonds while in compressive loading, weak hydrogen bonding and van der Waals bonds come into play, which lead to rather easy local yielding, buckling, and kinking of the fiber. Thus, aramid-type high-performance fibers are not recommended for applications involving compressive forces.

Kevlar aramid fiber has good vibration-damping characteristics. Dynamic (commonly sinusoidal) perturbations are used to study the damping behavior of a material. The material is subjected to an oscillatory strain. We can characterize the damping behavior in terms of a quantity called the called logarithmic decrement, Δ, which is defined as the natural logarithm of the ratio of amplitudes of successive vibrations, i.e.,

$$\Delta = \ln \frac{\theta_n}{\theta_{n+1}}$$

where θ_n and θ_{n+1} are the two successive amplitudes. The logarithmic decre-

Table 2.7. Properties of Technora fiber*

Density (g cm^{-3})	Diameter (μm)	Tensile Strength (GPa)	Tensile Modulus (GPa)	Tensile strain to fracture (%)
1.39	12	3.1	71	4.4

* Manufacturer's data; indicative values only.

ment is proportional to the ratio of maximum energy dissipated per cycle/ maximum energy stored in the cycle. Composites of Kevlar aramid fiber/ epoxy matrix show about 5 times the loss decrement of glass fiber/epoxy.

Like other polymers, aramid fibers are sensitive to ultraviolet (UV) light. When exposed to ultraviolet light, aramid fibers discolor from yellow to brown and lose mechanical properties. Radiation of a particular wavelength can cause degradation because of absorption by the polymer and breakage of chemical bonds. Near-UV and part of the visible spectrum should be avoided for outdoor application involving use of unprotected aramid fibers. A small amount of such light emanates from incandescent and fluorescent lamps or sunlight filtered by window glass. Du Pont Co. recommends that Kevlar aramid yarn should not be stored within one foot (0.3 m) of fluorescent lamps or near windows.

The Technora fiber of Teijin shows properties that are a compromise between conventional fibers and rigidrod fibers; Table 2.7 provides a summary of these. In terms of its stress-strain behavior, it can be said that Technora fiber lies between Kevlar 49 and Kevlar 29.

Kevlar aramid fibers provide an impressive array of properties and applications. The fibers are available in three types; each type is meant for specific applications (Magat, 1980):

1. *Kevlar*. This is meant mainly for use as rubber reinforcement for tires (belts or radial tires for cars and carcasses of radial tires for trucks) and, in general, for mechanical rubber goods.
2. *Kevlar 29*. This is used for ropes, cables, coated fabrics for inflatables, architectural fabrics, and ballistic protection fabrics. Vests made of Kevlar 29 have been used by law-enforcement agencies in many countries.
3. *Kevlar 49*. This is meant for reinforcement of epoxy, polyester, and other resins for use in the aerospace, marine, automotive, and sports industries.

2.6   Ceramic Fibers

Continuous ceramic fibers present an attractive package of properties. They combine rather high strength and elastic modulus with high-temperature capability and a general freedom from environmental attack. These characteristics make them attractive as reinforcements in high-temperature structural materials.

There are three ceramic fiber fabrication methods: chemical vapor deposition, polymer pyrolysis, and sol-gel techniques. The latter two involve rather novel techniques of obtaining ceramics from organometallic polymers. The sol-gel technique was mentioned in Section 2.2 regarding the manufacture of glass fibers. The sol-gel technique is also used to produce a variety of oxide fibers commercially. Another breakthrough in the ceramic fiber area is the concept of pyrolyzing, under controlled conditions, polymers containing silicon and carbon or nitrogen to produce high-temperature ceramic fibers. This idea is nothing but an extension of the polymer pyrolysis route to produce a variety of carbon fibers wherein a suitable carbon-based polymer (e.g., PAN and pitch) is subjected to controlled heating to produce carbon fibers (see Sect. 2.4). The pyrolysis route of producing ceramic fibers has been used with polymers containing silicon, carbon, nitrogen, and boron, with the end products being SiC, Si_3N_4, B_4C, and BN in fiber form, foam, or coating. We will describe some important ceramic fibers.

Oxide Fibers

Many alumina-type oxide fibers are available commercially. Alumina has many allotropical forms: γ, δ, η, and α. Of these, α-alumina is the most stable form. 3M Co. produces a number of alumina-based continuous fibers by sol-gel processing. The sol-gel process of making fibers involves the following steps, which are common to all fibers:

1. Formulate sol.
2. Concentrate to form a viscous gel.
3. Spin the precursor fiber.
4. Calcine to obtain the oxide fiber.

Boric salts of aluminum decompose into transition aluminum oxide spinels such as η-Al_2O_3 above 400 °C. These transition cubic spinels convert to hexagonal α-Al_2O_3 on heating to between 1000° and 1200 °C. The problem is that the nucleation rate of pure α-Al_2O_3 is too low and results in rather large grains. Also, during the transformation to α phase, a large shrinkage results in a rather large porosity. The α-Al_2O_3 fiber, 3M Co.'s tradename Nextel 610, is obtained by seeding the high-temperature α-alumina with a very fine hydrous colloidal iron oxide. The fine iron oxide improves the nucleation rate of α-Al_2O_3, resulting in a high-density, ultrafine, homogeneous α-Al_2O_3 fiber (Wilson, 1990). α-Fe_2O_3 is isostructural with α-Al_2O_3 (5.5% lattice mismatch). 3M's hydrous colloidal iron oxide sol appears to be an efficient nucleating agent. Without the seeding with iron oxide, the η-alumina-to-α-alumina transformation occurred at about 1100 °C. With 1% Fe_2O_3, the transformation temperature was decreased to 1010 °C, while with 4% Fe_2O_3, the transformation temperature came down to 977 °C. Con-

Table 2.8. Composition and properties of Nextel series fibers

Fiber Type	Composition (wt%)	Diameter (μm)	Density (g cm^{-3})	Tensile Strength (MPa)	Young's Modulus (GPa)
Nextel 312	Al_2O_3-62, SiO_2-24, B_2O_3-14	10–12	2.7	1700	152
Nextel 440	Al_2O_3-70, SiO_2-28, B_2O_3-2	10–12	3.05	2000	186
Nextel 550	Al_2O_3-73, SiO_2-27	10–12	3.03	2000	193
Nextel 610	Al_2O_3-99+, SiO_2-.2–.3, Fe_2O_3-.4–.7	10–12	3.75	1900	370
Nextel 720	Al_2O_3-85, SiO_2-15	10–12	3.4	2130	260

comitantly, the grain size was refined. In addition to Fe_2O_3, about 0.5 wt. % SiO_2 is added to reduce the final grain size, although SiO_2 inhibits the transformation to the α phase. The SiO_2 addition also reduces grain growth during soaking at 1400 °C.

3M Co. produces a series of ceramic fibers, called the Nextel fibers, that are mainly $Al_2O_3 + SiO_2$ and some B_2O_3. The compositions and properties of some Nextel fibers are given in Table 2.8. The sol-gel manufacturing process used by 3M Co. has metal alkoxides as the starting materials. The reader will recall that metal alkoxides are M(OR)n-type compounds, where *M* is the metal, *n* is the metal valence, and *R* is an organic compound. Selection of an appropriate organic group is very important. It should provide sufficient stability and volatility to the alkoxide so that M—OR bonds are broken and MO—R is obtained to give the desired oxide ceramics. Hydrolysis of metal alkoxides results in sols that are spun and gelled. The gelled fiber is then densified at relatively low temperatures. The high surface free energy available in the pores of the gelled fiber allows for relatively low-temperature densification. The sol-gel process provides close control over solution composition and rheology of fiber diameter. The disadvantage is that rather large dimensional changes must be accommodated and fiber integrity conserved. Figure 2.37 shows an optical micrograph of Nextel 312 fibers. Sowman (1988) has provided details of the process used by 3M Co. for making the Nextel oxide fibers. Aluminum acetate [$Al(OH)_2$ ($OOCCH_3$) · $1/3\,H_3BO_3$], e.g., "Niaproof," from Niacet Corp., is the starting material. Aluminum acetate with an Al_2O_3/B_2O_3 ratio of 3 to 1 becomes spinnable after water removal from an aqueous solution. In the fabrication of 3M continuous fibers, a 37.5% solution of basic aluminum acetate in water is concentrated in a rotating flask partially immersed in a water bath at 32–36 °C. After concentration to an equivalent Al_2O_3 content of 28.5%, a viscous solution with viscosity, η, between 100 and 150 Pa s is obtained. This is extruded through a spinneret under a pressure of 800–1000 kPa. Shiny, colorless fibers

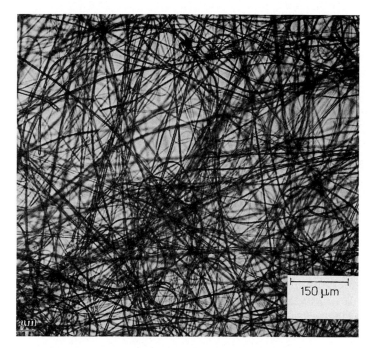

Fig. 2.37. Optical micrograph of Nextel 312 ($Al_2O_3 + B_2O_3 + SiO_2$) fiber.

are obtained on firing to 1000 °C. The microstructure shows cube- and lath-shaped crystals. The boria addition lowers the temperature required for mullite formation and retards the transformation of alumina to α-Al_2O_3. One needs boria in an amount equivalent to or greater than a 9 Al_2O_3 : 2 B_2O_3 ratio in Al_2O_3-B_2O_3-SiO_2 compositions to prevent the formation of crystalline alumina.

Many other alumina- or alumina-silica-type fibers are available. Most of these are made by the sol-gel process. Sumitomo Chemical Co. produces a fiber that is a mixture of alumina and silica. The flow diagram of this process is shown in Figure 2.38. Starting from an organoaluminum (polyaluminoxanes or a mixture of polyaluminoxanes and one or more kinds of Si-containing compounds), a precursor fiber is obtained by dry spinning. This precursor fiber is calcined to produce the final fiber. The fiber structure consists of fine crystallites of spinel. SiO_2 serves to stabilize the spinel structure and prevents it from transforming to α-Al_2O_3 (Saitow, 1992). Yet another variety of alumina fiber available commercially is a δ-Al_2O_3, short staple fiber produced by ICI (tradename *Saffil*). This fiber has about 4% SiO_2 and a very fine diameter (3 μm). The aqueous phase contains an oxide sol and an organic polymer. The sol is extruded as filaments into a coagulating (or precipitating) bath in which the extruded shape gels. The gelled fiber is then dried and calcined to produce the final oxide fiber. Aluminum oxychloride

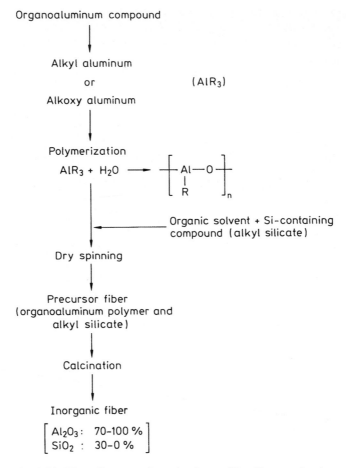

Fig. 2.38. Flow diagram of an alumina + silica fiber production.

[$Al_2(OH)_5Cl$] is mixed with a medium-molecular-weight polymer such as 2 wt.% polyvinyl alcohol. This solution is slowly evaporated in a rotary evaporator until a viscosity of about 80 Pa s (800 P) is attained. This solution is extruded through a spinneret, then the fibers are wound on a drum and fired to about 800 °C. The organic material is burned away, and a fine-grain alumina fiber with 5–10% porosity and a diameter of 3–5 μm is obtained. The fibers as produced at this stage are suitable for filtering purposes because of their high porosity. By heating them to 1400–1500 °C, which causes 3–4% of linear shrinkage, one obtains a refractory alumina fiber suitable for reinforcement purposes.

A technique called *edge-defined film-fed growth* (EFG) has been used to make continuous, monocrystalline sapphire (Al_2O_3) fiber (LaBelle and Mlavsky, 1967; LaBelle, 1971; Pollack, 1972; Hurley and Pollack, 1972). LaBelle and Mlavsky (1967) grew sapphire (Al_2O_3) single-crystal fibers using a modified Czochralski puller and radio frequency heating. The process is called the edge-defined film-fed growth method because the external edge of the die defines the shape of the fiber and the liquid is fed in the form of a film. Fiber growth rates as high as 200 mm/min have been attained. The die material must be stable at the melting point of alumina; a molybdenum die is commonly used. A capillary supplies a constant liquid level at the crystal interface. A sapphire seed crystal is used. Molten alumina wets both molybdenum and alumina. The crystal grows from a molten film between the growing crystal and the die. The crystal shape is defined by the external shape of the die rather than the internal shape. A fiber produced by this method, called Saphikon, has a hexagonal structure with its c-axis parallel to the fiber axis. The diameter is rather large, between 125 and 250 μm.

A *laser-heated floating zone method* has been devised to make a variety of ceramic fibers. Gasson and Cockayne (1970) used laser heating for crystal growth of Al_2O_3, Y_2O_3, $MgAl_2O_4$, and Na_2O_3. Haggerty (1972) used a four-beam laser-heated float zone method to grow single-crystal fibers of Al_2O_3, Y_2O_3, TiC, and TiB_2. A CO_2 laser is focused on the molten zone. A source rod is brought into the focused laser beam. A seed crystal, dipped into the molten zone, is used to control the orientation. Crystal growth starts by moving the source and seed rods simultaneously. Mass conservation dictates that the diameter is reduced as the square root of the feed rate/pull rate ratio. It is easy to see that, in this process, the fiber purity is determined by the purity of the starting material.

Another technique of making oxide fibers, called the *inviscid melt technique*, uses melt drawing of fiber (Wallenberger et al., 1992). In principle, any material that can be made molten can be drawn into a fibrous shape. Organic polymeric fibers, such as nylon, aramid, and a variety of glasses, are routinely converted into a fibrous form by passing a molten material with an appropriate viscosity through an orifice. The *inviscid* (meaning low viscosity) melt technique uses this principle. Essentially, the technique involves the extrusion of a low-viscosity molten jet through an orifice into a chemically reactive environment. The low-viscosity jet is unstable with respect to surface tension because of a phenomenon called *Rayleigh waves*. These waves are surface waves that form on the surface of the low-viscosity jet stream. Rayleigh waves grow exponentially in amplitude and tend to break up the jet into droplets. The key here is to stabilize the molten jet against breakup by the Rayleigh waves. In the case of glasses and organic polymers, the melts have high viscosity (more than 10^5 poise), and the high viscosity delays the Rayleigh breakup until the molten jet freezes. In the case of a low-viscosity melt, one can avoid the breakup of the molten jet by chemically stabilizing it. For example, in a process used at the Du Pont Co. to make fibers of alumina + calcia, a low-viscosity (10 poise) jet was chemically stabilized with propane before the Rayleigh waves could break the stream into droplets. The inviscid jet must be stabilized in about 10^{-3} s or it will break up into droplets. Yet another point to note in regard to this process is that small-diameter fibers are difficult to make. For example, the smallest-diameter alumina-calcia fiber produced by Du Pont is 105 μm. These fibers, according to Du Pont, have tensile strengths close to 1 GPa.

2.7   Nonoxide Fibers

Nonoxides such as silicon carbide and silicon nitride have very attractive properties. Silicon carbide, in particular, is commercially available in a fibrous form. In this section, we describe some of these nonoxide fibers.

2.7.1   Silicon Carbide Fibers

Silicon carbide fiber must be regarded as a major development in the field of ceramic reinforcements during the last quarter of the twentieth century. In particular, a process developed by the late Professor Yajima in Japan, involving controlled pyrolysis of a polycarbosilane precursor to yield a flexible fiber, must be considered to be the harbinger of the manufacture of ceramic fibers from polymeric precursors. In this section we describe the processing, microstructure, and properties of these and other silicon carbide fibers in some detail.

We can easily classify the fabrication methods for SiC as conventional or nonconventional. The former category would include chemical vapor deposition while the latter would include controlled pyrolsis of polymeric precursors. There is yet another important type of SiC available for reinforcement purposes: SiC whiskers.

CVD Silicon Carbide Fibers

Silicon carbide fiber can be made by chemical vapor deposition on a substrate heated to approximately 1300 °C. The substrate can be tungsten or carbon. The reactive gaseous mixture contains hydrogen and alkyl silanes. Typically, a gaseous mixture consisting of 70% hydrogen and 30% silanes is introduced at the reactor top (Fig. 2.39), where the tungsten substrate (~ 13 μm diameter) also enters the reactor. Mercury seals are used at both ends as contact electrodes for the filament. The substrate is heated by combined direct current (250 mA) and very high frequency (VHF ~ 60 MHz) to obtain an optimum temperature profile. To obtain a 100-μm SiC monofilament, it generally takes about 20 seconds in the reactor. The filament is wound on a spool at the bottom of the reactor. The exhaust gases (95% of the original mixture + HCl) are passed around a condenser to recover the unused silanes. Efficient reclamation of the unused silanes is very important for a cost-effective production process. This CVD process of making SiC fiber is very similar to that of B fiber manufacture. The nodules on the surface of SiC are smaller than those seen on B fibers. Such CVD processes result in composite

Fig. 2.39. CVD process for SiC monofilament fabrication. [From J.V. Milewski et al. (1974), reproduced with permission.]

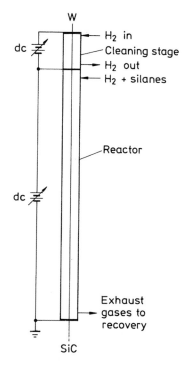

monofilaments that have built-in residual stresses. The process is, of course, very expensive. Methyltrichlorosilane is an ideal raw material, as it contains one silicon and one carbon atom, i.e., one would expect a stoichiometric SiC to be deposited. The chemical reaction is:

$$CH_3SiCl_3(g) \xrightarrow{H_2} SiC(s) + 3\,HCl(g)$$

An optimum amount of hydrogen is required. If the hydrogen is less than sufficient, chlorosilanes will not be reduced to Si and free carbon will be present in the mixture. If too much hydrogen is present, the excess Si will form in the end product. Generally, solid (free) carbon and solid or liquid silicon are mixed with SiC. The final monofilament (100–150 μm) consists of a sheath of mainly β-SiC with some α-SiC on the tungsten core. The {111} planes in SiC deposit are parallel to the fiber axis. The cross section of SiC monofilament resembles closely that of a boron fiber. Properties of a CVD SiC monofilament are given in Table 2.9.

Textron Specialty Materials Co. developed a series of surface-modified silicon carbide fibers, called SCS fibers. These special fibers have a complex through the thickness gradient structure. SCS-6, for example, is a thick fiber (diameter = 142 μm) that is produced by chemical vapor deposition of silicon- and carbon-containing compounds onto a pyrolytic graphite–coated carbon fiber core. The pyrolytic graphite coating is applied to a carbon

Table 2.9. Properties of CVD SiC monofilament

Composition	Diameter (μm)	Density (g cm^{-3})	Tensile strength (MPa)	Young's modulus (GPa)
β-SiC	140	3.3	3500	430

monofilament to give a substrate of 37 μm. This is then coated with SiC by CVD to give a final monofilament of 142 μm diameter. The surface modification of the SCS-6 fibers consists of the following. The bulk of the 1-μm-thick surface coating consists of C-doped Si. Figure 2.40 shows schematically the Textron SCS-6 silicon carbide fiber and its characteristic surface compositional gradient. Zone I at and near the surface is a carbon-rich zone. In zone II, Si content decreases. This is followed by zone III in which the Si content increases back to the stoichiometric SiC composition. Thus, the SCS-6 silicon carbide fiber has a carbon rich surface and back to stoichiometric SiC a few μm from the surface.

Another CVD-type silicon carbide fiber is called *sigma fiber*. Sigma fiber filament is a continuous silicon carbide monofilament by CVD on a tungsten substrate.

Nonoxide Fibers via Polymers

As pointed out earlier, the SiC fiber obtained via CVD is very thick and not very flexible. Work on alternate routes of obtaining fine, continuous, and flexible fiber had been ongoing, when in the mid-seventies the late Professor

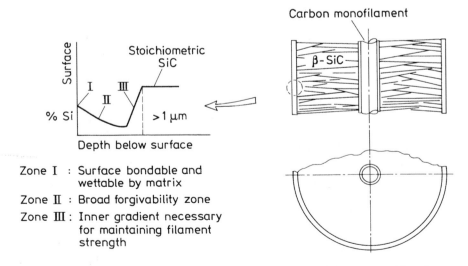

Fig. 2.40. Schematic of SCS-6 silicon carbide fiber and its surface compositional gradient. (Courtesy of Textron Specialty Materials Co.)

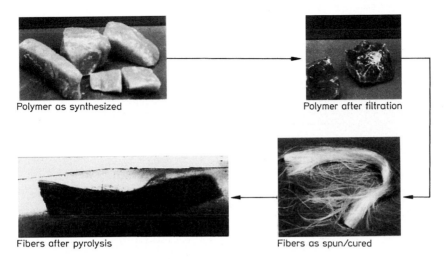

Fig. 2.41. Schematic of ceramic fiber production starting from silicon-based polymers. [Adapted from Wax (1985), used with permission.]

Yajima and his colleagues (1976; 1980) in Japan developed a process of making such a fiber by controlled pyrolysis of a polymeric precursor. This method of using silicon-based polymers to produce a family of ceramic fibers with good mechanical properties, good thermal stability, and oxidation resistance has enormous potential. The various steps involved in this polymer route, shown in Figure 2.41 (Wax, 1985), are:

1. Polymer characterization (yield, molecular weight, purity, and so on).
2. Melt spin polymer into a precursor fiber.
3. Cure the precursor fiber to crosslink the molecular chains, making it infusible during the subsequent pyrolysis.

Specifically, the Yajima process of making SiC involves the following steps and is shown schematically in Figure 2.42. Polycarbosilane a high-molecular-weight polymer containing Si and C is synthesized. The starting material is commercially available dimethylchlorosilane. Solid polydimethylsilane is obtained by dechlorination of dimethylchlorosilane by reacting it with sodium. Polycarbosilane is obtained by thermal decomposition and polymerization of polydimethylsilane. This is carried out under high pressure in an autoclave at 470 °C in an argon atmosphere for 8 to 14 hours. A vacuum distillation treatment at up to 280 °C follows. The average molecular weight of the resulting polymer is about 1500. This is melt spun from a 500-hole nozzle at about 350 °C under N_2 gas to obtain the so-called preceramic continuous, precursor fiber. The precursor fiber is quite weak (tensile strength ~ 10 MPa). This is converted to inorganic SiC by curing in air, heating to about 1000 °C in N_2 gas, followed by heating to 1300 °C in N_2 under stretch. This is basically the Nippon Carbon Co. manufacture process for Nicalon fibers (Simon and Bunsell, 1984). Clearly, small variations exist

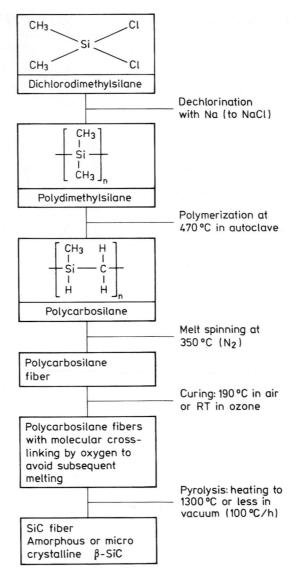

Fig. 2.42. Schematic of SiC (Nicalon) production. [Adapted from Andersson and Warren (1984), used with permission.]

between this and the various laboratory processes. During this pyrolysis, the first stage of conversion occurs at approximately 550 °C when cross-linking of polymer chains occurs. Above this temperature, the side chains containing hydrogen and methyl groups decompose. Fiber density and mechanical properties improve sharply. The conversion to SiC is done above about 850 °C.

The properties of Nicalon start degrading above about 600 °C because of the thermodynamic instability of composition and microstructure. A ceramic-grade Nicalon, called NLP-201, having low oxygen content is also available. Yet another version of multifilament silicon carbide fiber is Tyranno, produced by Ube Industries in Japan. This is made by pyrolysis of poly (titano carbosilane), and it contains between 1.5 and 4 wt.% titanium.

Fine-diameter, polymer-derived silicon carbide fibers generally have high oxygen content. This results from the curing of the precursor fibers in an oxidizing atmosphere to introduce cross-linking. Cross-linking is required to make the precursor fiber infusible during the subsequent pyrolysis step. One way around this is to use electron beam curing. Other techniques involve dry spinning of high-molecular-weight carbosilane polymers (Sacks et al., 1995). In this case, the as-spun fiber does not require a curing step because of the high-molecular-weight polycarbosilane polymer used, i.e., it does not melt during pyrolysis without requiring curing. Sacks et al. (1995) used dopant additions to produce low-oxygen, near-stoichiometric, small-diameter (10–15 μm) SiC fibers. They reported an average tensile strength for these fibers of about 2.8 GPa. Their fiber has a C-rich and near stoichiometric SiC and low oxygen content (Sacks et al., 1995; Toreki et al., 1994). They start with polydimethylsilane (PDMS), which has an Si–Si backbone. This is subjected to pressure pyrolysis to obtain polycarbosilane (PCS), which has an Si–C backbone. The key point in their process is to have a molecular weight of PCS between 5000 and 20,000. This is a high molecular weight compared to that used in other processes. The spinning dope is obtained by adding suitable spinning aids and a solvent. This dope is dry spun to produce what are called the *green fibers*, which are heated under controlled conditions to produce SiC fiber.

Laine and coworkers (1993, 1995) and Zhang et al. (1995) have used a polyemethylsilane (PMS) ($[CH_3SiH]_x$) as the precursor polymer for making a fine-diameter, silicon carbide fiber. Spinning aids were used to stabilize the polymer solution and the precursor fiber was extruded from a 140-μm orifice extruder into an argon atmosphere. The precursor fiber was pyrolyzed at 1800 °C in Ar. Boron, added as a sintering aid, helped to obtain a dense product.

Another silicon carbide multifilament fiber, made via a polymeric precursor by Dow Corning Corp., USA, is called *Sylramic*. According to the manufacturer, this textile-grade silicon carbide fiber has a nanocrystalline, stoichiometric silicon carbide (crystallite size of 0.5 μm). Its density is 3.0 g cm^{-3}, and it has a tensile strength and modulus of 3.15 GPa and 405 GPa, respectively.

Structure and Properties

The structure of Nicalon fiber has been studied by many researchers. Figure 2.43 shows a high-resolution transmission electron micrograph of Nicalon-

Fig. 2.43. High-resolution transmission electron micrograph of Nicolon fiber showing its amorphous structure. (Courtesy of K. Okamura.)

type SiC produced in laboratory, indicating the amorphous nature of the SiC produced by the Yajima process. The commercial variety of Nicalon has an amorphous structure while another, noncommercial variety, showed a microcrystalline structure (SiC grain radius of 1.7 nm) (Laffon et al., 1989). The microstructural analysis shows that both fibers contain SiO_2 and free carbon in addition to SiC. The density of the fiber is about 2.6 g cm^{-3}, which is low compared to that of pure β–SiC, which is not surprising because the composition is a mixture of SiC, SiO_2, and C.

The properties of Nicalon fiber are summarized in Table 2.10. A comparison of Nicalon SiC fiber with CVD SiC fiber will show that the CVD fiber is superior in properties, mainly because it is mostly β-SiC while the Nicalon fiber is a mixture of SiC, SiO_2, and free carbon. Figure 2.44 shows a comparison of the creep strain in Nicalon and CVD SiC fiber. Notice the superior performance of CVD fiber.

2.8 Whiskers

Whiskers are monocrystalline, short fibers with extremely high strength. This high strength, approaching the theoretical strength, comes about because of the absence of crystalline imperfections such as dislocations. Being monocrystalline, there are no grain boundaries either. Typically, whiskers have a diameter of a few μm and a length of a few mm. Thus, their aspect ratio (length/diameter) can vary from 50 to 10,000. Whiskers do not have uniform dimensions or properties. This is perhaps their greatest disadvantage, i.e., the

Table 2.10. Typical properties of Nicalon fiber

Density	Diameter	Modulus	Strength at 20°C		Strength at 1400°C (Oxygen)	Creep strain at 1300°C, 0.6 GPa, 20 h
			As-Produced	After 1400°C (Argon)		
2.6 g cm^{-3}	10–20 μm (500 per yarn)	180 GPa	2GPa	<1 GPa	<0.5 GPa	4.5%

Source: Reprinted with permission from *Journal of Metals,* **37**, No. 6, 1985, a publication of The Metallurgical Society, Warrendale, PA.

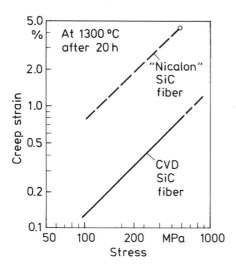

Fig. 2.44. Comparison of creep strain in CVD SiC and Nicalon fibers. (Reprinted with permission from *Journal of Metals*, **37**, No. 6, 1985, a publication of the Metallurgical Society, Warrendale, PA.)

spread in properties is extremely large. Handling and alignment of whiskers in a matrix to produce a composite are other problems.

Whiskers are normally obtained by vapor phase growth. Early in the 1970s, a new process was developed, beginning with rice hulls, to produce SiC particles and whiskers (Lee and Cutler, 1975; Milewski et al., 1974). The SiC particles produced by this process are of a finer size. Rice hulls are a waste by-product of rice milling. For each 100 kg of rice milled, about 20 kg of rice hull is produced. Rice hulls contain cellulose, silica, and other organic and inorganic materials. Silica from soil is dissolved and transported in the plant as monosilicic acid. This is deposited in the cellulosic structure by liquid evaporation. It turns out that most of the silica ends up in hull. It is the intimate mixture of silica within the cellulose that gives the near-ideal amounts of silica and carbon for silicon carbide production. Raw rice hulls are heated in the absence of oxygen at about 700 °C to drive out the volatile compounds. This is called *coking*. Coked rice hulls, containing about equal amounts of SiO_2 and free C, are heated in inert or reducing atmosphere (flowing N_2 or NH_3 gas) at a temperature between 1500° and 1600 °C for about 1 hour to form silicon carbide as per the following reaction:

$$3\,C + SiO_2 \longrightarrow SiC + 2\,CO$$

Figure 2.45 shows the schematic of the process. When the reaction is over, the residue is heated to 800 °C to remove any free C. Generally, both particles and whiskers are produced, together with some excess free carbon. A wet process is used to separate the particles and the whiskers. Typically, the average aspect ratio of the as-produced whiskers is 75.

Exceptionally strong and stiff silicon carbide whiskers were grown at Los Alamos National Laboratory (LANL) using the so-called VLS process

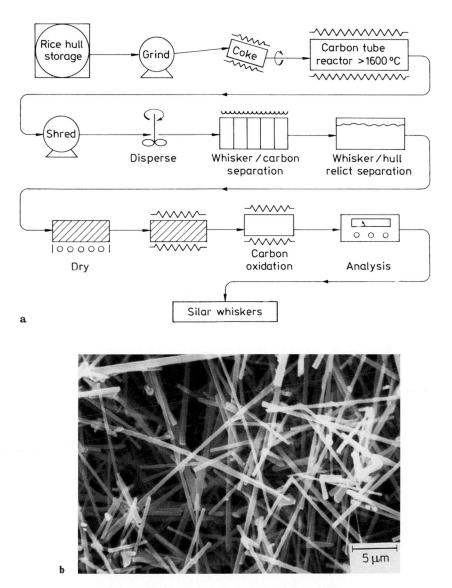

Fig. 2.45. a Schematic of SiC whisker production process starting from rice hulls. **b** Scanning electron micrograph of SiC whiskers obtained from rice hulls. [Courtesy of Advanced Composite Materials Corporation (formerly Arco).]

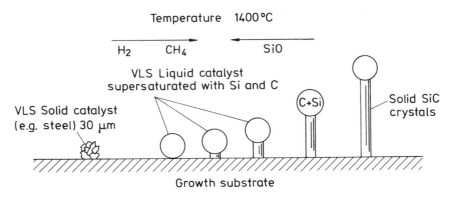

Fig. 2.46. The VLS process for SiC whisker growth. [After Milewski et al. (1985), used with permission.]

(Milewski et al., 1985; Petrovic et al., 1985). The average tensile strength and modulus were 8.4 GPa and 581 GPa, respectively. The acronym *VLS* stands for *vapor* feed gases, *liquid* catalyst, and *solid* crystalline whiskers. In this process, the catalyst forms a liquid solution interface with the growing crystalline phase, while elements are fed from the vapor phase through the liquid-vapor interface. Whisker growth takes place by precipitation from the supersaturated liquid at the solid-liquid interface. The catalyst must take in solution the atomic species of the whisker to be grown. For SiC whiskers, transition metals and iron alloys meet this requirement. Figure 2.46 illustrates the LANL process for SiC whisker growth. Steel particles (~ 30 µm) are used as the catalyst. At 1400 °C, the solid steel catalyst particle melts and forms a liquid catalyst ball. From the vapor feed of SiC, H_2, and CH_4, the liquid catalyst extracts C and Si atoms and forms a supersaturated solution. The gaseous silicon monoxide is generated as per the following reaction:

$$SiO_2(g) + C(s) \longrightarrow SiO(g) + CO(g)$$

The supersaturated solution of C and Si in the liquid catalyst precipitates out solid SiC whisker on the substrate. As the precipitation continues, the whisker grows (Figure 2.46). The researchers at LANL observed and identified a range of whisker morphologies. The tensile strength values ranged from 1.7 to 23.7 GPa in 40 tests. Whisker lengths were about 10 mm; the equivalent circular diameter averaged 5.9 µm.

Other Nonoxide Fibers

In addition to silicon carbide–based ceramic fibers, there are other promising ceramic fibers, e.g., silicon nitride, boron carbide, and boron nitride.

Silicon nitride (Si_3N_4) fibers can be prepared by reactive chemical vapor deposition (CVD) using volatile silicon compounds. The reactants are generally SiCl$_4$ and NH$_3$. Si$_3$N$_4$ is deposited on a carbon or tungsten substrate. Again, as in other CVD processes, the resultant fiber has good properties, but the diameter is very large and it is expensive. In the polymer route, organosilzane polymers with methyl groups on Si and N have been used as silicon nitride precursors. Such carbon-containing, silicon-nitrogen precursors on pyrolysis give silicon carbide as well as silicon nitride, i.e., the resulting fiber is not an SiC-free silicon nitride fiber. Wills et al. (1983) have discussed the mechanisms involved in the conversion of various organometallic compounds into ceramics.

Silicon Carbide in a Particulate Form

SiC in particulate form has been available quite cheaply and abundantly for abrasive, refractories, and chemical uses. In this conventional process, silica in the form of sand and carbon in the form of coke are made to react at 2400 °C in an electric furnace. The SiC produced in the form of large granules is subsequently comminuted to the desired size.

2.9   Effect of High-Temperature Exposure on the Strength of Ceramic Fibers

Carbon fiber is excellent at high temperatures in an inert atmosphere. In air, at temperatures above 400–450 °C, it starts oxidizing. SiC and Si$_x$N$_y$ show reasonable oxidation resistance for controlled composition fibers. SiC starts oxidizing above 1300–1400 °C in air. The high-temperature strength of the SiC-type fibers is limited by oxidation and internal void formation, while in the case of oxide fibers and intergranular glassy phase leads to softening. Mah et al. (1984) studied the degradation of Nicalon fiber after heat treatment in different environments. The strength degradation of this fiber at temperatures greater than 1200 °C was because of CO evaporation from the fiber as well as β-SiC grain growth. Lipowitz et al. (1990) observed that ceramic fibers made via pyrolysis of polymeric precursors, especially with compositions Si-C-O and Si-N-C-O, have lower densities than the theoretical values. The theoretical density, ρ_t, can be calculated using the relationship

$$\rho_t = \rho_i V_i$$

where ρ is the density, V is the volume fraction, the subscript i indicates the ith phase, and summation over all the phases present is implied. Lipowitz et al. (1990) used X-ray scattering techniques to show that porosity present in such fibers was due to globular pores of nm size and that the pore fractions ranged from 5 to 25%. According to these authors, nanochannels form during the early states of pyrolysis, when rather large volumes of gases are given out. In the later stages of pyrolysis, smaller volumes of gases are given out. In the later stages of pyrolysis, i.e., during densification, these nano-channels suffer a viscous collapse and nanopores are formed. The volume fraction of nanopores decreases with increasing pyrolysis temperature. The reader should note that a higher density and a lower void fraction will lead to a higher elastic modulus of these ceramic fibers.

2.10   Comparison of Fibers

A comparison of some important characteristics of creep performance rein-forcements discussed individually in Sections 2.2 through 2.7 is made in Table 2.11, and a plot of strength versus modulus is shown in Figure 2.47. We compare and contrast some salient points of these fibers.

First of all, we note that all these high-performance fibers have very low density values. Given the general low density of these fibers, the best of these fibers group together in the top-right-hand corner of Figure 2.46. The reader will also recognize that the elements comprising these fibers pertain to the first two rows of the periodic table. Also to be noted is the fact that, irre-spective of whether in compound or elemental form, they are mostly cova-lently bonded, which is the strongest bond. Generally, such light, strong, and stiff materials are very desirable in most applications, but particularly in aerospace, land transportation, the energy-related industry, housing and civil construction, and so on.

Fiber flexibility is associated with the Young's modulus and the diameter (see Sec. 2.1). In the general area of high-modulus (i.e., high-E) fibers, the diameter becomes the dominant parameter controlling the flexibility. For a given E, the smaller the diameter the more flexible it is. Fiber flexibility is a very desirable characteristic if one wants to bend, wind, and weave a fiber in order to make a complex-shaped final product.

Some of these fibers have quite anisotropic characteristics. Consider the situation in regard to thermal properties; in particular, the thermal expan-sion coefficient of carbon is quite different in the radial and longitudinal directions. This would also be true of any single-crystal fiber or whisker, e.g., alumina single-crystal fiber, which has a hexagonal structure. In this respect, ceramic (SiC and Al_2O_3), C, and B fibers are quite good. Ceramic matrix composites can go to very high temperatures, indeed. An important problem that comes up at these very high temperatures ($>1500\,°C$) is that of fiber and matrix oxidation. Carbon fiber, for example, does not have good oxidation resistance at these high temperatures. SiC- or Si_3N_4-type ceramic fibers are the only suitable candidates for reinforcement at very high temperatures (>1200–$1300\,°C$) and in air. It would appear that oxide fibers would be the likely candidates, because of their inherent stability in air, for applications at temperatures higher than $1300\,°C$ in air.

Table 2.11. Properties of reinforcement fibers

Characteristic	PAN-Based Carbon		Kevlar 49	E Glass	SiC		Al$_2$O$_3$	Boron (W)
	HM	HS			CVD	Nicalon		
Diameter (μm)	7–10	7.6–8.6	12	8–14	100–200	10–20	20	100–200
Density (g cm^{-3})	1.95	1.75	1.45	2.55	3.3	2.6	3.95	2.6
Young's modulus (GPa)								
Parallel to fiber axis	390	250	125	70	430	180	379	385
Perpendicular to fiber axis	12	20	—	70	—	—	—	—
Tensile strength (GPa)	2.2	2.7	2.8–3.5	1.5–2.5	3.5	2	1.4	3.8
Strain to fracture (%)	0.5	1.0	2.2–2.8	1.8–3.2	—	—	—	—
Coefficient of thermal expansion (10^{-6} K^{-1})								
Parallel to fiber axis	−0.5–0.1	0.1––0.5	−2––5	4.7	5.7	—	7.5	8.3
Perpendicular to fiber axis	7–12	7–12	59	4.7	—	—	—	—

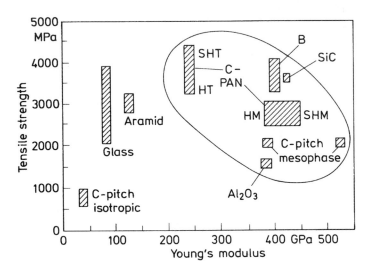

Fig. 2.47. Comparison of different fibers.

Another important characteristic of these high-performance fibers is their rather low values of strain to fracture, generally less than 2–3%. This means that in a CMC, the reinforcement and the matrix may not be much different in terms of strain to fracture. Also, the modulus ratio of the reinforcement and the matrix may be 2–3 or as low as 1. This is a very different situation from that encountered in PMCs and MMCs.

References

C.-H. Andersson and R. Warren (1984). *Composites*, **15**, 16.

R. Bacon (1973). In *Chemistry and Physics of Carbon*, vol. 9, Marcel Dekker, New York, p. 1.

A.A. Baker (1983). *Metals Forum*, **6**, 81.

P.J. Barham and A. Keller (1985). *J. Mater. Sd.*, **20**, 2281.

S.C. Bennett and D.J. Johnson (1978). In *Fifth International Carbon and Graphite Conference*, Society of the Chemical Industry, London, p. 377.

S.C. Bennett and D.J. Johnson (1979). *Carbon*, **17**, 25.

S.C. Bennett, D.J. Johnson, and W. Johnson (1983). *J. Mater. Sc.*, **18**, 3337.

D.A. Biro, G. Pleizier and Y. Deslandes (1992). *J. Mater. Sci. Lett.*, **11**, 698.

C.J. Brinker and G. Scherer (1990). *The Sol-Gel Science*, Academic Press, New York.

J.R. Brown, P.J.C. Chappell and Z. Mathys (1992) *J. Mater. Sci.*, **27**, 3167.

G. Capaccio, A.G. Gibson, and I.M. Ward (1979). In *Ultra-High Modulus Polymers*, Applied Science Publishers, London, p. 1.

K.K. Chawla (1976). In *Proceedings of the International Conference on the Mechanical Behavior of Materials II*, ASM, Metals Park, Ohio, p. 1920.

K.K. Chawla (1981). *Mater. Sci. Eng.*, **48**, 137.

K.K. Chawla (1998). *Fibrous Materials*, Cambridge University Press, Cambridge.

K.K. Chawla and A.C. Bastos (1979). In *Proceedings of the International Conference on the Mechanical Behavior of Materials III*, Pergamon Press, Oxford, p. 191.

C.C. Chiao and T.T. Chiao (1982). In *Handbook of Composites*, Van Nostrand Reinhold, New York, p. 272.

E. de Lamotte and A.J. Perry (1970). *Fibre Sci. Tech.*, **3**, 157.

H.E. DeBolt (1982). In *Handbook of Composites*, Van Nostrand Reinhold, New York, p. 171.

S.J. DeTeresa, S.R. Allen, R.J. Farris, and R.S. Porter (1984). *J. Mater. Sd.*, **19**, 57.

J.A. DiCarlo (June 1985). *J. Met.* **37**, 44.

R.J. Diefendorf and E. Tokarsky (1975). *Polym. Eng. Sci.*, **15**, 150.

M.G. Dobb, D.J. Johnson, and B.P. Saville (1980). *Philos. Trans. R. Soc. London*, **A294**, 483.

W.H. Dresher (April, 1969). *J. Metals*, **21**, 17.

H.N. Ezekiel and R.G. Spain (1967). *J. Polym. Sci. C*, **19**, 271.

P.J. Flory (1956). *Proc. Roy. Soc. (London)*, **234A**, 73.

A. Fourdeux, R. Perret, and W. Ruland (1971). In *Carbon Fibres: Their Composites and Applications*, The Plastics Institute, London, p. 57.

F. Galasso and A. Paton (1966). *Trans. Met. Soc. AIME*, **236**, 1751.

F. Galasso, D. Knebl, and W. Tice (1967). *J. Appl. Phys.*, **38**, 414.

D.G. Gasson and B. Cockayne (1970) *J. of Mater. Sci.*, **5**, 100.

J.S. Haggerty (1972). *NASA-CR-120948*, NASA Lewis Res. Center, Cleveland, OH.

D.N. Hild and P. Schwartz (1992a) *J. Adhes. Sci. Technol.*, **6**, 879.

D.N. Hild and P. Schwartz (1992b) *J. Adhes. Sci. Technol.*, **6**, 897.

K.A. Hodd and D.C. Turley (1978). *Chem. Br.* **14**, 545.

G.F. Hurley and J.T.A. Pollack (1972). *Met. Trans.*, **7**, 397.

O.T. Inal, N. Leca, and L. Keller (1980). *Phys. Status Solidi*, **62**, 681.

M. Jaffe and R.S. Jones (1985). In *Handbook of Fiber Science & Technology*, vol. 111, *High Technology Fibers*, Part A, Marcel Dekker, New York, p. 349.

J. Johnson and C.N. Tyson (1969). *Br. J. Appl. Phys.*, **2**, 787.

R.W. Jones (1989). *Fundamental Principles of Sol-Gel Technology*, The Institute of Metals, London.

B. Kalb and A.J. Pennings (1980). *J. Mater. Sci.*, **15**, 2584.

S.L. Kaplan, P.W. Rose, H.X. Nguyen and H.W. Chang (1988). *SAMPE Quarterly*, **19**, 55.

T. Kikuchi (1982). *Surface*, **20**, 270.

V. Krukonis (1977). In *Boron and Refractory Borides*, Springer-Verlag, Berlin, p. 517.

S.L. Kwolek, P.W. Morgan, J.R. Schaefgen, and L.W. Gultich (1977). *Macromolecules*, **10**, 1390.

H.E. LaBelle (1971). *Mater. Res. Bull.*, **6**, 581.

H.E. LaBelle and A.I. Mlavsky (1970). *Mater. Res. Bull.*, **6**, 571.

C. Laffon, A.M. Flank, P. Lagarde, M. Laridjani, R. Hagege, P. Olry, J. Cotteret, J. Dixmier, J.L. Niquel, H. Hommel and A.P. Legrand (1989). *J. Mater Sci.*, **24**, 1503.

R.M. Laine and F. Babonneau (1993). *Chem. Mater.*, **5**, 260.

R.M. Laine, Z-F. Zhang, K.W. Chew, M. Kannisto and C. Scotto (1995). In *Ceramic Processing Science and Technology*, Am. Ceram. Soc., Westerville, OH, p. 179.

J.-G. Lee and I. B. Cutler (1975). *Am. Ceram. Soc. Bull.*, **54**, 195.

Z.F. Li, A.N. Netravali and W. Sachse (1992). *J. Mater. Sci.*, **27**, 4625.

J. Lipowitz, J.A. Rabe, and L.K. Frevel (1990). *J. Mater. Sci.*, **25**, 2118.

K.L. Loewenstein (1983). *The Manufacturing Technology of Continuous Glass Fibers*, 2nd ed., Elsevier, New York.

R.E. Lowrie (1967). In *Modern Composite Materials*, Addison-Wesley, Reading, MA, p.270.

E.E. Magat (1980). *Philos. Trans. R. Soc. London*, **A296**, 463.

T. Mah, N.L. Hecht, D.E. McCullum, J.R. Hoenigman, H.M. Kim, A.P. Katz, and H.A. Lipsitt (1984). *J. Mater. Sci.*, **19**, 1191.

J.V. Milcwski, J.L. Sandstrom, and W.S. Brown (1974). In *Silicon Carbide 1973*, University of South Carolina Press, Columbia, p. 634.

J.V. Milewski, F.D. Gac, J.J. Petrovic, S.R. Skaggs (1985). *J. Mater. Sci.*, **20**, 1160.

P.W. Morgan (1979). *Plast. Rubber: Mater. Appl.*, **4**, 1.

J.S. Murday, D.D. Dominguez, L.A. Moran, W.D. Lee, and R. Eaton (1984). *Synth. Met.* **9**, 397.

M.G. Northolt (1981). *J. Mater. Sci.*, 16, 2025.

S. Ozawa, Y. Nakagawa, K. Matsuda, T. Nishihara and H. Yunoki (1978). US patent 4,075,172

B. Parkyn (Ed.) (1970). *Glass Reinforced Plastics*, Butterworth, London.

R. Perret and W. Ruland (1970). *J. Appl. Crystallogr.*, **3**, 525.

J.J. Petrovic. J.V. Milewski. D.L. Rohr, and F.D. Gac (1985). *J. Mater. Sd.*, **20**, 1167.

J.T.A. Pollack (1972) *J. Mater. Sci.*, **7**, 787.

J.P. Riggs (1985). In *Encyclopedia of Polymer Science & Engineering*, 2nd ed., vol. 2, John Wiley & Sons, New York, p. 640.

M.D. Sacks, G.W. Scheiffele, M. Saleem, G.A. Staab, A.A. Morrone and T.J. Williams (1995). *Ceramic Matrix Composites: Advanced High-Temperature Structural Materials*, MRS, Pittsburgh, PA, p. 3.

S. Sakka (1985). *Am. Ceram. Soc. Bull.*, **64**, 1463.

G. Simon and A.R. Bunsell (1984). *J. Mater. Sci.*, **19**, 3649.

L.S. Singer (1979). In *Ultra-High Modulus Polymers*, Applied Science Publishers, Essex, England, p.251.

A. Shindo (1961). *Rep. Osaka Ind. Res. Inst.* No. 317.

P. Smith and P.J. Lemstra (1976). *Colloid Polymer Sci.*, **15**, 258.

W.D. Smith (1977). In *Boron and Refractory Borides*, Springer-Verlag, Berlin, p. 541.

J. Smook and A.J. Pennings (1984). *J. Mater. Sci.*, **19**, 31.

H.G. Sowman (1988). In *Sol-Gel Technology*, L.J. Klein (ed), Noyes Pub., Park Ridge, NJ, p. 162.

C.P. Talley (1959). *J. Appl. Phys.*, **30**, 1114.

C.P. Talley, L. Line, and O. Overman (1960). In *Boron: Synthesis, Structure, and Properties*, Plenum Press, New York, p. 94.

D. Tanner, A.K. Dhingra, and J.J. Pigliacampi (March, 1986). *J. Met.*, **38**, 21.

W. Toreki, C.D. Batich, M.D. Sacks, M. Saleem, G.J. Choi, and A.A. Morrone (1994). *Compos. Sci. and Technol.*, **51**, 145.

A.C. van Maaren, O. Schob, and W. Westerveld (1975). *Philips Tech. Rev.*, **35**, 125.

J. Vega-Boggio and O. Vingsbo (1978). In *1978 International Conference on Composite Materials, ICCM/2*, TMS-AIME, New York, p. 909.

F.T. Wallenberger, N.E. Weston, K. Motzfeldt and D.G. Swartzfager (1992). *J. Amer. Ceram. Soc.*, **75**, 629.

S.B. Warner (1995). *Fiber Science*, Prentice Hall, Englewood Cliffs, NJ.

R. Warren and C.-H. Andersson (1984). *Composites*, **15**, 101.

W. Watt (1970). *Proc. R. Soc.*, **A319**, 5.

W. Watt and W. Johnson (1969). *Appl. Polym. Symp.*, **9**, 215.

F.W. Wawner (1967). In *Modern Composite Materials*, Addison-Wesley, Reading, MA, p. 244.

S.G. Wax (1985). *Am. Ceram. Soc. Bull*, **64**, 1096.

E. Weintraub (1911). *J. Ind. Eng. Chem.*, **3**, 299.

R.R. Wills, R.A. Mankle, and S.P. Mukherjee (1983). *Am. Ceram. Soc. Bull.* **62**, 904.

D.M. Wilson (1990). In *Proc. 14th Conf. on Metal Matrix, Carbon, and Ceramic Matrix Composites, Cocoa Beach, FL, Jan. 17–19, 1990*, NASA Conference Publication 3097, Part 1, p. 105.

K.J. Wynne and R.W. Rice (1984). *Ann. Rev. Mater. Sci.*, **15**, 297.

S. Yajima (1980). *Philos. Trans. R. Soc. London*, **A294**, 419.

S. Yajima, K. Okamura, J. Hayashi, and M. Omori (1976). *J. Am. Ceram. Soc.*, **59**, 324.

Z.-F. Zhang, S. Scotto and R.M. Laine (1994) in *Ceram. Eng. Sci. Proc.*, **15**, 152.

Suggested Reading

A.R. Bunsell (Ed.) (1988). *Fibre Reinforcements for Composite Materials*, Elsevier, Amsterdam.

K.K. Chawla (1998). *Fibrous Materials*, Cambridge University Press, Cambridge.

J.B. Donnet and R.C. Bansal (1984). *Carbon Fibers*, second edition, Marcel Dekker, New York.

E. Fitzer (1985). *Carbon Fibres and Their Composites*, Springer-Verlag, Berlin.

L. H Peebles (1995). *Carbon Fibers*, CRC Press, Boca Raton, FL.

S.B. Warner (1995). *Fiber Science*, Prentice Hall, Englewood Cliffs, NJ.

W. Watt and B. V. Perov (Eds.) (1985). *Strong Fibres*, vol. 1. in the Handbook of Composites series, North-Holland, Amsterdam.

Yang, H.H. (1993). *Kevlar Aramid Fiber*, John Wiley, Chichester, UK.

CHAPTER 3

Matrix Materials

A brief description of the various matrix materials, polymers, metals, and ceramics, is given in this chapter. We emphasize the characteristics that are relevant to composites. The reader should consult the references listed under Suggested Reading for greater details regarding any particular aspect of these materials.

3.1 Polymers

Polymers are structurally much more complex than metals or ceramics. They are cheap and can be easily processed. On the other hand, polymers have lower strength and modulus and lower temperature use limits. Prolonged exposure to ultraviolet light and some solvents can cause the degradation of polymer properties. Because of predominantly covalent bonding, polymers are generally poor conductors of heat and electricity. Polymers, however, are generally more resistant to chemicals than are metals. Structurally, polymers are giant chainlike molecules (hence the name *macromolecules*) with covalently bonded carbon atoms forming the backbone of the chain. The process of forming large molecules from small ones is called *polymerization*; that is, polymerization is the process of joining many monomers, the basic building blocks, together to form polymers. There are two important classes of polymerization:

1. *Condensation polymerization*: In this process a stepwise reaction of molecules occurs and in each step a molecule of a simple compound, generally water, forms as a by-product.
2. *Addition polymerization*: In this process monomers join to form a polymer without producing any by-product. Addition polymerization is generally carried out in the presence of catalysts. The linear addition of ethylene molecules (CH_2) results in polyethylene (a chain of ethylene molecules),

with the final mass of polymer being the sum of monomer masses:

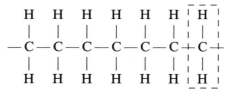

Based on their behavior, there are two major classes of polymers, produced by either condensation or addition polymerization, i.e., thermosetting and thermoplastic polymers. Thermosets undergo a curing reaction that involves cross-linking of polymeric chains. They harden on curing, hence the term *thermoset*. The curing reaction can be initiated by appropriate chemical agents or by application of heat and pressure, or by exposing the monomer to an electron beam. Thermoplastics are polymers that flow under the application of heat and pressure, i.e., they soften or become plastic on heating. Cooling to room temperature hardens thermoplastics. Their different behavior, however, stems from their molecular structure and shape, molecular size or mass, and the amount and type of bond (covalent or van der Waals). We first describe the basic molecular structure in terms of the configurations of chain molecules. Figure 3.1 shows the different chain configuration types.

1. *Linear polymers*. As the name suggests, this type of polymer consists of a long chain of atoms with attached side groups. Examples include polyethylene, polyvinyl chloride, and polymethyl methacrylate. Figure 3.1a shows the configuration of linear polymers; note the coiling and bending of chains.

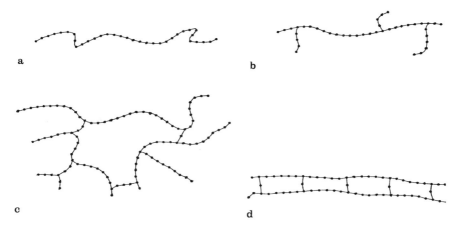

a

b

c

d

Fig. 3.1. Different molecular chain configurations: **a** linear, **b** branched, **c** cross-linked, **d** ladder.

2. *Branched polymers*. Polymer branching can occur with linear, crosslinked, or any other type of polymer; see Figure 3.lb.
3. *Cross-linked polymers*. In this case, molecules of one chain are bonded with those of another; see Figure 3.lc. Crosslinking of molecular chains results in a three-dimensional network. Crosslinking makes sliding of molecules past one another difficult, thus such polymers are strong and rigid.
4. *Ladder polymers*. If we have two linear polymers linked in a regular manner (Fig. 3.ld) we get a ladder polymer. Not unexpectedly, ladder polymers are more rigid than linear polymers.

3.1.1 Glass Transition Temperature

Pure crystalline materials have well-defined melting temperatures. The melting point is the temperature at which crystalline order is completely destroyed on heating. Polymers, however, show a range of temperatures over which crystallinity vanishes. Figure 3.2 shows specific volume (volume/unit mass) versus temperature curves for amorphous and semicrystalline polymers. When a polymer liquid is cooled, it contracts. The contraction occurs because of a decrease in the thermal vibration of molecules and a reduction in the free space; that is, the molecules occupy the space less loosely. In the case of amorphous polymers, this contraction continues below T_m, the melting point of crystalline polymer, to T_g the glass transition temperature where the supercooled liquid polymer becomes extremely rigid owing to extremely high viscosity. Unlike the melting point, T_m, where there occurs a transformation from the liquid phase to a crystalline phase, the structure of a glassy or amorphous material below T_g is essentially that of a liquid, albeit a very viscous one. Such a phenomenon is commonly observed in silica-based inorganic glasses. In the case of amorphous polymers, we are dealing with a glassy structure made of organic molecules. The glass transition temperature T_g, although it does not represent a thermodynamic phase trans-

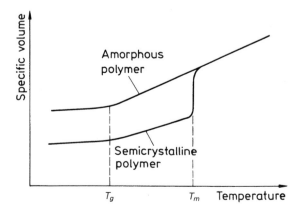

Fig. 3.2. Specific volume versus temperature for an amorphous and a semicrystalline polymer.

formation, is in many ways akin to the melting point for the crystalline solids. Many physical properties (e.g., viscosity, heat capacity, elastic modulus, and expansion coefficient) change abruptly at T_g. Polystyrene, for example, has a T_g of about 100 °C and is therefore rigid at room temperature. Rubber, on the other hand, has a T_g of about -75 °C and therefore is flexible at room temperature. T_g is a function of the chemical structure of the polymer. For example, if a polymer has a rigid backbone structure and/or bulky branch groups, then T_g will be quite high.

Although both amorphous polymers and inorganic silica-based glasses have a glass transition temperature T_g, generally, the T_g of inorganic glasses is several hundred degrees Celsius higher than that of polymers. The reason for this is the different types of bonding and the amount of crosslinking in the polymers and glasses. Inorganic glasses have mixed covalent and ionic bonding and are highly crosslinked. This gives them a higher thermal stability than polymers, which have covalent and van der Waals bonding and a lesser amount of crosslinking than found in inorganic glasses.

3.1.2 Thermoplastics and Thermosets

Polymers that soften or melt on heating are called thermoplastic polymers and are suitable for liquid flow forming. Examples include low- and high-density polyethylene, polystyrene, and polymethyl methacrylate (PMMA). When the structure is amorphous, there is no apparent order among the molecules and the chains are arranged randomly; see Figure 3.3a. Small, platelike single crystalline regions called *lamellae* or *crystallites* can be obtained by precipitation of the polymer from a dilute solution. In the lamellae, long molecular chains are folded in a regular manner; see Figure 3.3b. Many crystallites group together and form spherulites, much like grains in metals.

When the molecules in a polymer are crosslinked in the form of a network, they do not soften on heating. We call such cross-linked polymers *thermosetting polymers*. Thermosetting polymers decompose on heating.

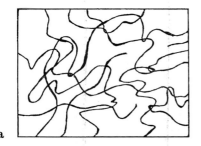
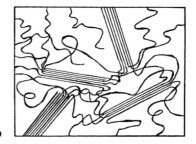

a   b

Fig. 3.3. Possible arrangements of polymer molecules: **a** amorphous, **b** semicrystalline.

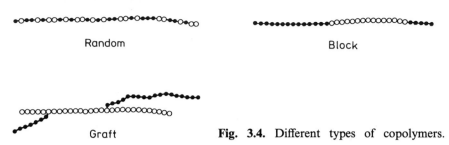

Fig. 3.4. Different types of copolymers.

Crosslinking makes sliding of molecules past one another difficult, making the polymer strong and rigid. A typical example is that of rubber crosslinked with sulfur, that is, vulcanized rubber. Vulcanized rubber has 10 times the strength of natural rubber. Common examples of thermosetting polymers include epoxy, phenolic, polyester, polyurethane, and silicone.

3.1.3   Copolymers

There is another type of classification of polymers based on the type of repeating unit. When we have one type of repeating unit forming the polymer chain, we call it a *homopolymer*. *Copolymers*, on the other hand, are polymer chains with two different monomers. If the two different monomers are distributed randomly along the chain, we have a *regular*, or *random*, *copolymer*. If, however, a long sequence of one monomer is followed by a long sequence of another monomer, we have a *block copolymer*. If we have a chain of one type of monomer and branches of another type, we have a *graft copolymer*. Figure 3.4 shows schematically the different types of copolymers.

3.1.4   Molecular Weight

Molecular weight (MW) is a very important parameter for characterization of polymers. Generally, strength increases but strain-to-failure decreases with increasing molecular weight. Of course, concomitant with increasing molecular weight, the processing of polymers becomes more difficult. The degree of polymerization (DP) indicates the number of basic units (mers) in a polymer. These two parameters are related as follows:

$$MW = DP \times (MW)_u$$

where $(MW)_u$ is the molecular weight of the repeating unit. In general, polymers do not have exactly identical molecular chains but may consist of a mixture of different species, each of which has a different molecular weight or DP. Thus, the molecular weight of the polymer is characterized by a distribution function. Clearly, the narrower this distribution function is, the more homogeneous the polymer is. That is why one speaks of an average molecular weight or degree of polymerization.

It is instructive to compare the molecular weights of some common polymeric materials vis à vis monomeric materials. A molecule of water, H_2O, has a molecular weight of 18. Benzene, a low-molecular-weight organic solvent, has a molecular weight of 78. Compared to these, natural rubber has a molecular weight of about 10^6. Polyethylene, a common synthetic polymer, can have molecular weights greater than 10^5. The molecular size of these high-molecular-weight solids is also very large. The molecular diameter of water, for example, is 40 nm, while that of polyethylene can be as large as 6400 nm (Mandelkern, 1983).

3.1.5 Degree of Crystallinity

Polymers can be amorphous or partially crystalline; see Figure 3.3. A 100% crystalline polymer is difficult to obtain. In practice, depending on the polymer type, molecular weight, and crystallization temperature, the amount of crystallinity in a polymer can vary from 30 to 90%. The inability to attain a fully crystalline structure is mainly due to the long chain structure of polymers. Some twisted and entangled segments of chains that get trapped between crystalline regions never undergo the conformational reorganization necessary to achieve a fully crystalline state. Molecular architecture also has an important bearing on the polymer crystallization behavior. Linear molecules with small or no side groups crystallize easily. Branched chain molecules with bulky side groups do not crystallize as easily. For example, linear high-density polyethylene can be crystallized to 90%, while branched polyethylene can be crystallized to only about 65%. Generally, the stiffness and strength of a polymer increase with the degree of crystallinity. It should be mentioned that deformation processes such as slip and twinning, as well as phase transformations that take place in monomeric crystalline solids, can also occur in polymeric crystals.

3.1.6 Stress-Strain Behavior

Characteristic stress-strain curves of an amorphous polymer and of an elastomer (a rubbery polymer) are shown in Figure 3.5a and b, respectively. Note that the elastomer does not show a Hookean behavior; its behavior is characterized as nonlinear elastic. The characteristically large elastic range shown by elastomers results from an easy reorganization of the tangled chains under the action of an applied stress.

Yet another point in which polymers differ from metals and ceramics is the extreme temperature dependence of their elastic moduli. Figure 3.6 shows schematically the variation of the elastic modulus of an amorphous polymer with temperature. In the temperature range below T_g, the polymer is hard and a typical value of elastic modulus would be about 5 GPa. Above T_g, the modulus value drops significantly and the polymer shows a rubbery behavior. Above T_f (the temperature at which the polymer becomes fluid),

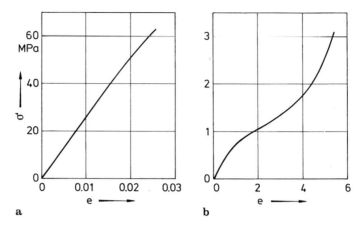

Fig. 3.5. a Hookean elastic behavior of a glassy polymer. **b** Nonlinear elastic behavior of an elastomer.

the modulus drops abruptly. It is in this region of temperatures above T_f that polymers are subjected to various processing operations.

3.1.7   Thermal Expansion

Polymers generally have higher thermal expansivities than metals and ceramics. Furthermore, their thermal expansion coefficients are not truly constants; that is, the polymers expand markedly in a nonlinear way with temperature. Epoxy resins have coefficient of linear expansion values between 50×10^{-6} and 100×10^{-6} K^{-1} while polyesters show values between 100×10^{-6} and 200×10^{-6} K^{-1}. Small compositional changes can have a marked influence on the polymer expansion characteristics.

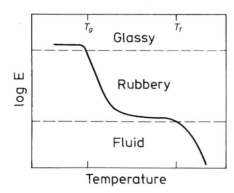

Fig. 3.6. Variation of elastic modulus of an amorphous polymer with temperature (schematic).

3.1.8 Fire Resistance

Fire performance of any polymer depends on a number of variables, such as surface spreading of flame, fuel penetration, and oxygen index. The oxygen index is the minimum amount of oxygen that will support combustion.

The degree of flammability of a polymer is a function of the following parameters:

- Matrix type and amount (dominant!).
- Quantity of fire-retardant additives.
- Type and amount of reinforcement (if any).

The order of increasing fire resistance of common polymer matrix materials is: polyester, vinyl ester, epoxy, phenolic. Phenolic has very low smoke emission and gives out no toxic by-products. In the case of thermosets, such as polyester, it is common practice to add fire-retardant additives. Addition of high glass fiber also helps.

3.1.9 Common Polymeric Matrix Materials

Thermosets and thermoplastic polymers are common matrix materials. A brief description of each is provided here.

Common Thermoset Matrix Materials

Among the common polymer matrices used with continuous fibers are polyester and epoxy resins. We provide a summary of thermosetting resins commonly used as matrix materials.

Epoxy

This is one of the major thermoset matrix materials. An epoxy is a polymer that contains an epoxide group (one oxygen atom and two carbon atoms) in its chemical structure; see Figure 3.7. Diglycidyl ether of bisphenol A (DGEBA) is an example. DGEBA, containing two epoxide groups, is a low-molecular-weight organic liquid. Frequently, one uses various additives to modify the characteristics of epoxies. For example, diluents are used to

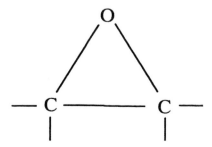

Fig. 3.7. Chemical structure of the epoxide group.

reduce the viscosity. Flexibilizing agents are used to make the epoxy flexible. Other agents are used for protection against ultraviolet radiation. Curing agents are organic amino or acid compounds, and cross-linking is obtained by introducing chemicals that react with the epoxy and hydroxy groups between adjacent chains. A common curing agent for DGEBA epoxy is di-ethylenetriamine (DETA). The extent of cross-linking is a function of the amount of curing agents. Generally, 10–15% by weight of amines or acid anhydrides is added, and they become part of the epoxy structure. An accelerator, if added, can speed up the curing process. In general, characteristics such as stiffness, strength, and glass transition temperature increase with increased crosslinking, but toughness decreases.

Epoxy resins are more expensive than polyesters, but they have better moisture resistance, lower shrinkage on curing (about 3%), a higher maximum use temperature, and good adhesion with glass fibers. A large number of proprietary formulations of epoxies is available, and a very large fraction of high-performance polymer matrix composites has thermosetting epoxies as matrices.

The curing reaction of an epoxy can be slowed by lowering the reaction temperature. An epoxy before it is fully crosslinked is said to be in stage B. In stage B, an epoxy has a characteristic *tackiness*. This B-stage resin is used to make a fiber prepreg and is shipped to a manufacturer where it can be fully cured into a hard solid. The type and extent of curing agent addition will control the total curing time (also called the *shelf life* or *pot life*). In order to increase the shelf life, such B-stage cured epoxies are transported and stored in refrigerated storage. Table 3.1 gives some important characteristics of epoxy. Epoxy matrix composites were originally formulated to withstand prolonged service at 180 °C (~ 350 °F). In the 1970s, it came to be recognized that they are prone to hygrothermal effects and their service temperature limitation is 120 °C (250 °F). Use temperature for DGEBA-based epoxies is about 150 °C. A detailed account of structure-property relationships of epoxies used as composite matrices is provided by Morgan (1985).

Polyester

An unsaturated polyester resin contains a number of C=C double bonds. A condensation reaction between a glycol (ethylene, propylene, or diethylene glycol) and an unsaturated dibasic acid (maleic or fumaric) results in a linear

Table 3.1. Some important characteristics of epoxy

Density, ρ (g cm^{-3})	Strength, σ (MPa)	Modulus, E (GPa)	Poisson's ratio, ν	CTE α (10^{-6} K^{-1})	Cure Shrinkage (%)	Use Temp. (°C)
1.2–1.3	50–125	2.5–4	0.2–0.33	50–100	1–5	150

Table 3.2. Some important characteristics of polyester

Density, ρ (g cm^{-3})	Strength, σ (MPa)	Modulus, E (GPa)	Poisson's ratio, ν	CTE α (10^{-6} K^{-1})	Cure Shrinkage (%)	Use Temp. (°C)
1.1–1.4	30–100	2–4	0.2–0.33	50–100	5–12	80

polyester that contains double bonds between certain carbon atoms. The term *unsaturated* means that there are reactive sites in the molecule. Diluents such as styrene are used to reduce the viscosity of polyester. Styrene contains C=C double bonds and crosslinks the adjacent polyester molecules at the unsaturation points. Hardening and curing agents and ultraviolet absorbents are usually added. Frequently, a catalyst such as an organic peroxide is added to initiate the curing action. One can hasten the curing process by raising the temperature; this increases the decomposition rate of the catalyst. This can also be accomplished by using an accelerator such as cobalt napthalate. Unsaturated polyester has adequate resistance to water and a variety of chemicals, weathering, aging, and, last but not least, it is very cheap. It can withstand temperatures up to about 80 °C and combines easily with glass fibers. Polyester resins shrink between 4 and 8% on curing. Table 3.2 gives some important properties of polyester.

Other Thermosets

Polyimides are thermosetting polymers that have a relatively high service temperature range, 250–300 °C. However, like other thermosetting resins, they are brittle. Their fracture energies are in the 15-to-70 J m^{-2} range. A major problem with polyimides is the elimination of water of condensation and solvents during processing.

Bismaleimides (BMI), thermosetting polymers, can have service temperatures between 180 and 200 °C. They have good resistance to hygrothermal effects. Because they are thermosets, they are quite brittle and must be cured at higher temperatures than conventional epoxies.

Curing

Figure 3.8 shows schematically the variation of viscosity of a thermoset as a function of time at two different temperatures ($T_1 > T_2$). There occurs a slight decrease in viscosity in the beginning because of the heat generated by the exothermic curing reaction. As the crosslinking progresses, the molecular mass of the thermoset increases; viscosity increases at first slowly and then very rapidly. At the time corresponding to a perceptible change in viscosity, t_{gel}, a gel-like lump appears. At times greater than t_{gel}, viscosity tends to infinity and the thermoset can treated as a solid. Figure 3.9 shows the variation of mechanical properties as a function of curing time. After a time

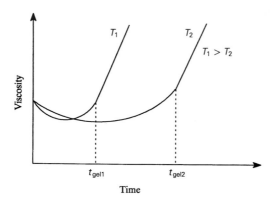

Fig. 3.8. Viscosity (η) vs. time (t) for a thermoset for two different temperatures, $T_1 > T_2$.

marked t_{cure}, the mechanical properties of a thermoset essentially do not change with time.

Electron Beam Curing

An electron beam (1–10 MeV) is a good source of ionizing radiation that can be used to initiate polymerization or the curing reaction in a polymer. Parts up to 25 mm thick can be cured. Higher energy beams are required for thicker parts. It turns out that one needs to modify the resin formulation for electron beam curing. Addition of 1 to 3 parts per hundred of a cationic photo-initiator can do the job. No hardener is required. However, other conventional additives, such as toughening agents or diluents, can be used to improve the processability of the resin. Among the advantages of electron beam curing, one may list

- Shorter cure times (minutes rather than the hours required by conventional thermal processes).
- Curing can be accomplished at a selected temperature (room temperature or near the service temperature).

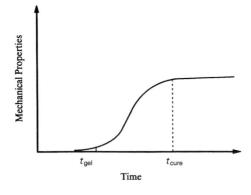

Fig. 3.9. Variation of mechanical properties with curing time.

- Lower energy consumption than in thermal curing.
- Removal of hardener means elimination of the undesirable volatile organic compounds. It also allows the use of thermoset resins with un-limited shelf life.

Among the limitations, one may list the following:

- Only a few electron beam curing facilities are available, especially for large and complex structures. This is especially so in the 5-to-15 MeV range.
- Shipping parts to outside accelerator facilities can be problematic.
- Long-term testing and experience in the field are missing; that is, customer confidence needs to be built.

Common Thermoplastic Matrix Materials

Thermoplastics are characterized by linear chain molecules and can be repeatedly melted or reprocessed. It is important to note that in this regard the cool-down time affects the degree of crystallinity of the thermoplastic. This is because the polymer needs time to get organized in the orderly pat-tern of the crystalline state; too quick a cooling rate will not allow crystal-lization to occur. Although repeated melting and processing is possible with thermoplastics, it should be recognized that thermal exposure (too high a temperature or too long a residence at a given temperature) can degrade the polymer properties such as, especially, impact properties.

Linear molecules in thermoplastic result in higher strain-to-failure values compared to those of thermosets, i.e., thermoplastics are tougher than the cross-linked thermosets. Thermoplastic matrix materials can have failure strains ranging from 30 to 100%, while the thermosets typically range from 1 to 3%. The large range of failure strains in thermoplastics stems from the rather large variations in the amount of crystallinity.

Common thermoplastic resins used as matrix materials in composites include some conventional thermoplastics such as nylon, thermoplastic poly-esters (PET, PBT), and polycarbonates. Some of the new thermplastic matrix materials include polyamide imide, polyphenylene sulfide (PPS), polyarylsulfone, and polyetherether ketone (PEEK). Figure 3.10 shows the chemical structure of some of these thermoplastics. PEEK is an attractive matrix material because of its toughness and impact properties, which are a function of its crystalline content and morphology. It should be pointed out that crystallization kinetics of a thermoplastic matrix can vary substantially because of the presence of fibers (Waddon et al., 1987). In order to make a thermoplastic matrix flow, heating must be done to a temperature above the melting point of the matrix. In the case of PEEK, the melting point of the crystalline component is 343 °C. In general, most thermoplastics are harder to flow vis à vis epoxy! Their viscosity decreases with increasing temperature, but the danger is decomposition of resin.

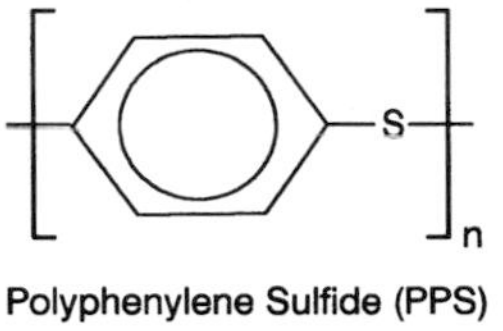

Polyphenylene Sulfide (PPS)

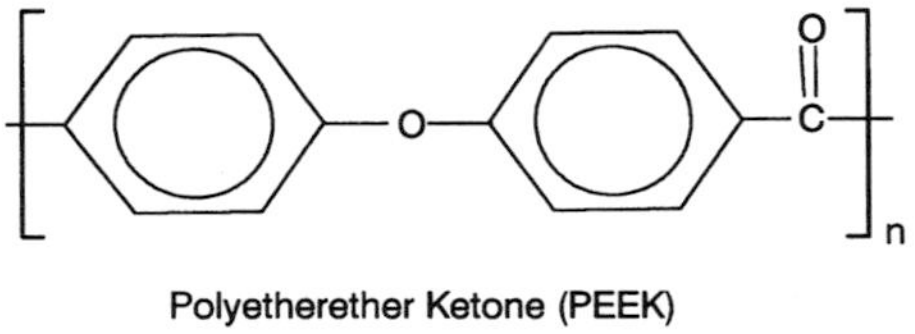

a

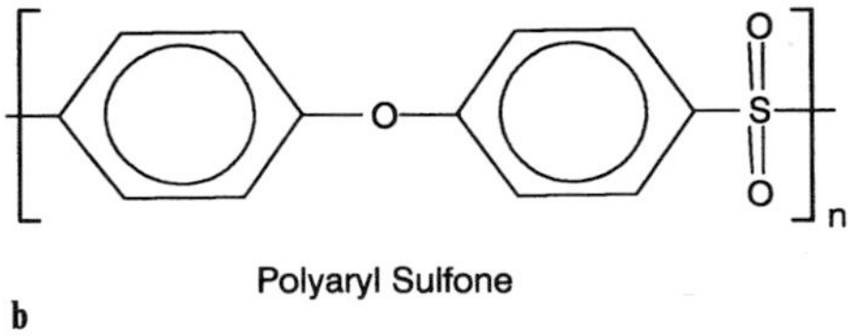

Fig. 3.10. Chemical structure of: **a** PPS, **b** Polyaryl Sulfone, and **c** Polyetherether Ketone PEEK.

Polyaryl Sulfone

b

Polyetherether Ketone (PEEK)

c

Polyphenylene sulfide (PPS) is a linear polymer of "modest" molecular weight ($\sim$150) and low mold shrinkage (0.1–0.5% range). It can be injection molded, and the scrap can be ground and reused without much effect on processability and performance of a part. Polyethersulfone (PES) belongs to the polysulfone group of thermoplastics. Its chemical structure has an aromatic sulfone unit, which imparts high thermal stability and mechanical strength. It is completely amorphous and can withstand loads for long periods up to 190 °C. Both PEEK and PES are wholly aromatic polymers suitable for high temperatures, and both can be processed as conventional thermoplastics. PEEK, however, is partially crystalline, while PES is amorphous. The deactivated nature of PEEK results in a high melting point (343 °C) and good oxidative stability. PEEK has a long-term use temperature of 250 °C and a short use temperature of 300 °C.

Polyimides are among the most temperature-resistant engineering polymers. They can be linear or crosslinked, aromatic polymers. Processing problems and high cost have limited their use. Avimid K is the tradename of Du Pont Co. for an amorphous, linear, thermoplastic polyimide resin that is used to make fibrous prepregs. These prepregs are processed by a vacuum bag lay-up technique in an autoclave. Avimid K resin is made by condensation. An aromatic diethyl ester diacid is reacted with an aromatic diamine in n-methyl pyrorolidone (NMP) solvent. The by-products of the reaction are water, ethanol, and the solvent.

Thermoplastic resins have the advantage that, to some extent, they can be recycled. Heat and pressure are applied to form and shape them. More often than not, short fibers are used with thermoplastic resins but in the late 1970s continuous fiber reinforced thermoplastics began to be produced. The disadvantages of thermoplastics include their rather large expansion and low viscosity characteristics.

An important problem with polymer matrices is that associated with environmental effects. Polymers can degrade at moderately high temperatures and through moisture absorption. Absorption of moisture from the environment causes swelling in the polymer as well as a reduction in its T_g. In the presence of fibers bonded to the matrix, these hygrothermal effects can lead to severe internal stresses in the composite. The presence of thermal stresses resulting from thermal mismatch between matrix and fiber is, of course, a general problem in all kinds of composite materials; it is much more so in polymer matrix composites because polymers have high thermal expansivities.

Typical properties of some common polymeric matrix materials are summarized in Table 3.3 (English, 1985).

Matrix Toughness

Thermosetting resins (e.g., polyesters, epoxies, and polyimides) are highly cross-linked and provide adequate modulus, strength, and creep resistance, but the same crosslinking of molecular chains causes extreme brittleness, that is, very low fracture toughness. By *fracture toughness*, we mean resistance to crack propagation. It came to be realized in the 1970s that matrix fracture characteristics (strain to failure, work of fracture, or fracture toughness) are as important as lightness, stiffness, and strength properties. Figure 3.11 compares some common materials in terms of their fracture toughness as measured by the fracture energy in J/m^2 (Ting, 1983). Note that thermosetting resins have values that are only slightly higher than those of glasses. Thermoplastic resins such as PMMA have fracture energies of about 1 kJ/m^2, while polysulfone thermoplastics have fracture energies of several kJ/m^2, almost approaching those of the 7075-T6 aluminum alloy. Amorphous thermoplastic polymers show higher fracture energy values because they have a large free volume available that absorbs the energy associated with crack propagation. Among the well-known modified thermoplastics are the acrylonitrile-butadiene-styrene (ABS) copolymer and high-impact polystyrene (HIPS). One class of thermosetting resins that comes close to polysulfones is the elastomer-modified epoxies. Elastomer-modified or rubber-modified thermosetting epoxies form multiphase systems, a kind of composite in their own right. Small (a few micrometers or less), soft, rubbery inclusions distributed in a hard, brittle epoxy matrix enhance its toughness by several orders of magnitude (Sultan and McGarry, 1973; Riew et al., 1976; Bascom and Cottington, 1976; St. Clair and St. Clair, 1981; Scott and Phillips, 1975).

Table 3.3. Representative properties of some polymeric matrix materials

Property	Epoxy	Polyimide	PEEK	Polyamideimide	Polyetherimide	Polysulfone	Polyphenylene Sulfide	Phenolics
Tensile strength (MPa)	35–85	120	92	95	105	75	70	50–55
Flexural modulus (GPa)	15–35	35	40	50	35	28	40	—
Density (g cm^{-3})	1.38	1.46	1.30	1.38	—	1.25	1.32	1.30
Continuous-service temperature (°C)	25–85	260–425	310	—	170	175–190	260	150–175
Coefficient of thermal expansion (10^{-5} °C^{-1})	8–11	9	—	6.3	5.6	9.4–10	9.9	4.5–11
Water absorption (24 h %)	0.1	0.3	0.1	0.3	0.25	0.2	0.2	0.1–0.2

Source: Adapted with permission from English (1985).

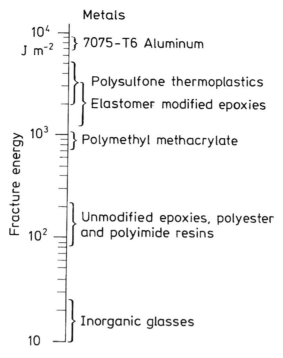

Fig. 3.11. Fracture energy for some common materials. [Adapted from Ting (1983), used with permission.]

Epoxy and polyester resins are commonly modified by introducing carboxyl-terminated butadiene-acrylonitrile copolymers (ctbn). Figure 3.12 shows the increase in fracture surface energy of an epoxy as a function of weight % of ctbn elastomer (Scott and Phillips, 1975). The methods of manufacture can be simple mechanical blending of the soft, rubbery particles and the resin or copolymerization of a mixture of the two.

Toughening of glassy polymers by elastomeric additions involves different mechanisms for different polymers. Among the mechanisms proposed for explaining this enhanced toughness are triaxial dilatation of rubber particles at the crack tip, particle elongation, and plastic flow of the epoxy. Ting (1983) studied such a rubber-modified epoxy containing glass or carbon fibers. He observed that the mechanical properties of rubber-modified composite improved more in flexure than in tension. Scott and Phillips (1975) obtained a large increase in matrix toughness by adding ctbn in unreinforced epoxy. But this large increase in toughness could be translated into only a modest increase in carbon fiber reinforced modified epoxy matrix composite. Introduction of a tough elastomeric phase, for example, a silicone rubber with good thermal resistance in a polyimide resin, produced a tough matrix

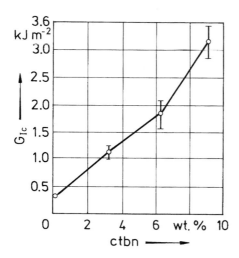

Fig. 3.12. Fracture surface energy of an epoxy as a function of weight % of carboxyl-terminated butadiene-acrylonitrile (ctbn). [Adapted from Scott and Phillips (1975), used with permission.]

material: a three- to five-fold gain in toughness G_{Ic} without a reduction in T_g (St. Clair and St. Clair, 1981).

Continuous fiber reinforced thermoplastics show superior toughness values owing to superior matrix toughness. PEEK is a semicrystalline aromatic thermoplastic (Hartness, 1983; Cogswell, 1983; Blundell et al., 1985) that is quite tough. PEEK can have 20 to 40% crystalline phase. At 35% crystallinity, the spherulite size is about 2 μm (Cogswell, 1983). Its glass transition temperature T_g is about 150 °C, and the crystalline phase melts at about 350 °C. It has an elastic modulus of about 4 GPa, a yield stress of 100 MPa, and a relatively high fracture energy of about 500 J/m^2. In addition to PEEK, other tough thermoplastic resins are available, for example, thermoplastic polyimides and polyphenylene sulfide (PPS), which is a semicrystalline aromatic sulfide. PPS is the simplest member of a family of polyarlene sulfides (O'Connor et al., 1986). PPS (tradename Ryton), a semicrystalline polymer, has been reinforced by chopped carbon fibers and prepregged with continuous carbon fibers (O'Connor et al., 1986).

3.2 Metals

Metals are very versatile engineering materials. They are strong and tough. They can be plastically deformed, and they can be strengthened by a wide variety of methods, mostly involving obstruction of movement of linear defects called *dislocations*.

3.2.1 Structure

Metals, with the exception of metallic glasses, are crystalline materials. Most metals exists in one of the following three crystalline forms:

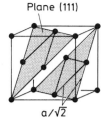
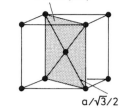
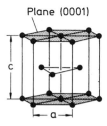

Plane (111)   Plane (110)   Plane (0001)

$a/\sqrt{2}$ $a/\sqrt{3}/2$ a

a Face-centered cubic   **b** Body-centered cubic   **c** Hexagonal close packed

Fig. 3.13a–c. Three crystalline forms of metals: **a** Face-centered cubic, **b** body-centered cubic, **c** hexagonal close packed.

- Face-centered cubic (fcc)
- Body-centered cubic (bcc)
- Hexagonal close packed (hcp)

Figure 3.13 shows these three structures. The black dots mark the centers of the atomic positions, and some of the atomic planes are shown shaded. In fact, the atoms touch each other and all the space is filled up. Some important metals with their respective crystalline structures are listed in Table 3.4. Metals are crystalline materials; however, the crystalline structure is never perfect. Metals contain a variety of crystal imperfections. We can classify these as follows:

1. Point defects (zero dimensional)
2. Line defects (unidimensional)
3. Planar or interfacial defects (bidimensional)
4. Volume defects (tridimensional)

Point defects can be of three types. A *vacancy* is created when an atomic position in the crystal lattice is vacant. An *interstitial* is produced when

Table 3.4. Crystal structure of some important metals

fcc	bcc	hcp
Iron (910–1390 °C)	Iron ($T < 910$ °C and $T > 1390$ °C)	Titanium[a]
Nickel	Beryllium ($T > 1250$ °C)	Beryllium ($T < 1250$ °C)
Copper	Cobalt ($T > 427$ °C)	Cobalt ($T < 427$ °C)
Aluminum	Tungsten	Cerium (-150 °C $< T < -10$ °C)
Gold	Molybdenum	Zinc
Lead	Chromium	Magnesium
Platinum	Vanadium	Zirconium[a]
Silver	Niobium	Hafnium[a] ($T < 1950$ °C)

[a] Undergoes bcc $\rightleftharpoons$ hcp transformation at different temperatures.

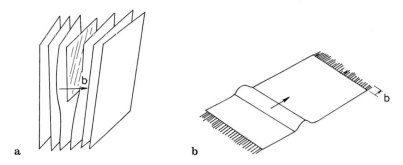

Fig. 3.14. a Edge dislocation, **b** dislocation in a carpet.

an atom of material or a foreign atom occupies an interstitial or nonlattice position. A *substitutional point defect* comes into being when a regular atomic position is occupied by a foreign atom. *Intrinsic point defects* (vacancies and self-interstitials) in metals exist at a given temperature in equilibrium concentrations. Increased concentrations of these defects can be produced by quenching from high temperatures, bombarding with energetic particles such as neutrons, and plastic deformation. Point defects can have a marked effect on the mechanical properties.

Line imperfections, called dislocations, represent a critically important structural imperfection that plays a very important role in the area of physical and mechanical metallurgy, diffusion, and corrosion. A dislocation is defined by two vectors: a dislocation line vector (tangent to the line) and its Burgers vector, which gives the direction of atomic displacement associated with the dislocation. The dislocation has a kind of lever effect because its movement allows one part of metal to be sheared over the other without the need for simultaneous movement of atoms across a plane. It is the presence of these line imperfections that makes it easy to deform metals plastically. Under normal circumstances then, the plastic deformation of metals is accomplished by the movement of these dislocations. Figure 3.14a shows an edge dislocation. The two vectors, defining the dislocation line and the Burgers vector, are designated as t (not shown) and b, respectively (see Figure 3.14). The dislocation permits shear in metals at stresses much below those required for simultaneous shear across a plane. Figure 3.14b shows this in an analogy. A carpet, of course, can be moved by pushing or pulling. However, a much lower force is required to move the carpet by a distance b if a defect is introduced into it and made to move the whole extension of the carpet. Figure 3.15 shows dislocations as seen by transmission electron microscopy in a thin foil of steel. This is a dark field electron micrograph; dislocation lines appear as white lines. Dislocations become visible in the transmission electron microscope because of the distortion of the atomic planes owing to the presence of dislocations. Also visible in this micrograph are equiaxial precipitate particles pinning the dislocations at various points.

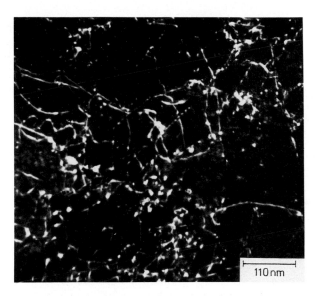

Fig. 3.15. Dislocations (*white lines*) in a steel sample: dark field transmission electron micrograph. Also visible are equiaxial precipitate particles.

The interfacial or planar defects occupy an area or surface of the crystal, for example, grain boundaries, twin boundaries, domains, or antiphase boundaries. Grain boundaries are, by far, the most important of these planar defects from the mechanical metallurgy point of view. Among the volumetric or tridimensional defects we can include large inclusions and gas porosity existing cracks.

3.2.2   Conventional Strengthening Methods

Experimental results show that work hardening (or strain hardening), which is the ability of a metal to become more resistant to deformation as it is deformed, is related in a singular way to the dislocation density (ρ) after deformation. There exists a linear relationship between the flow stress τ and $\sqrt{\rho}$ (Wiedersich, 1964):

$$\tau = \tau_0 + \alpha G b \sqrt{\rho} \qquad (3.1)$$

where G is the shear modulus, b is the Burgers vector, α is an adjustable parameter, and τ_0 is the shear stress required to move a dislocation in the absence of any other dislocations. Basically, work hardening results from the interactions among dislocations moving on different slip planes. A tangled dislocation network results after a small plastic deformation, which impedes the motion of other dislocations. This, in turn, requires higher loads for further plastic deformation. Various theories (Seeger, 1957; Kuhlmann-Wilsdorf, 1977) explain the interactions of dislocations with different kinds of barrier (e.g., dislocations, grain boundaries, solute atoms, and precipitates) that result in characteristic strain hardening of metals. All these theories arrive at the relationship between τ and ρ given in Equation (3.1), indicating that a particular dislocation distribution is not crucial and that strain hardening in practice is a statistical result of some factor that remains the same for various distributions. Cold working of metals, which leads to the strengthening of metals as a result of work hardening, is a routinely used strengthening technique.

A similar relationship exists between the flow stress τ and the mean grain size (or dislocation cell size) to an undetermined level

$$\tau = \tau_0 + \frac{\alpha' Gb}{D^{1/2}} \tag{3.2}$$

where D is the mean grain diameter. This relationship is known as the *Hall-Petch relationship* after the two researchers who first postulated it (Hall, 1951; Petch, 1953). Again, various models have been proposed to explain this square root dependence on grain size. Earlier explanations involved a dislocation pile-up bursting through the boundary owing to stress concentrations at the pileup tip and activation of dislocation sources in adjacent grains (Cottrell, 1958). Later theories involved the activation of grain boundary dislocations into grain interiors, elastic incompatibility stresses between adjacent grains leading to localized plastic flow at the grain boundaries, and so on (Li, 1963; Meyers and Ashworth, 1982). An important aspect of strengthening by grain refinement is that, unlike other strengthening mechanisms, it results in an improvement in toughness concurrent with that in strength (again to an undefined lower grain size). Another easy way of strengthening metals by impeding dislocation motion is that of introducing heterogeneities such as solute atoms or precipitates or hard particles in a ductile matrix. When we introduce solute atoms (say, carbon, nitrogen, or manganese in iron) we obtain solid solution hardening. Interstitial solutes such as carbon and nitrogen are much more efficient strengthening agents than substitutional solutes such as manganese and silicon. This is because the interstitials cause a tetragonal distortion in the lattice and thus interact with both screw and edge dislocations, while the substitutional atoms cause a spherical distortion that interacts only with edge dislocations, because the screw dislocations have a pure shear stress field and no hydrostatic component. Precipitation hardening of a metal is obtained by decomposing a supersaturated solid solution to form a finely distributed second phase. Classical examples of precipitation strengthening are those of Al-Cu and Al-Zn-Mg alloys, which are used in the aircraft industry (Fine, 1964). Oxide dispersion strengthening involves artificially dispersing rather small volume fractions (0.5–3 vol. %) of strong and inert oxide particles (e.g., Al_2O_3, Y_2O_3, and ThO_2) in a ductile matrix by internal oxidation or powder metallurgy blending techniques (Ansell, 1968). Both the second-phase precipitates and dispersoids act as barriers to dislocation motion in the ductile matrix, thus making the matrix more deformation-resistant. Dispersion-hardened systems (e.g., $Al + Al_2O_3$) show high strength levels at elevated temperatures while precipitates (say, $CuAl_2$ in aluminum) tend to dissolve at those temperatures. Precipitation hardening systems, however, have the advantage of enabling one to process the alloy in a soft condition and to give the precipitation treatment to the finished part. The precipitation process carried out for long periods of time can also lead to overaging and solution, that is, a weakening effect.

Quenching a steel to produce a martensitic phase has been a time-honored strengthening mechanism for steels. The strength of the martensite phase in steel depends on a variety of factors, the most important being the amount of carbon. The chemical composition of martensite is the same as that of the parent austenite phase from which it formed, but it is supersaturated with carbon (Roberts and Owen, 1968). Carbon saturation and the lattice distortion that accompanies the transformation lead to the high hardness and strength of martensite.

Another approach to obtaining enhanced mechanical performance is rapid solidification processing (Grant, 1985). By cooling metals at rates in the 10^4–10^9 Ks^{-1} range, it is possible to produce unique microstructures. Very fine powders or ribbons of rapidly solidified materials are processed into bulk materials by hot pressing, hot isostatic pressing, or hot extrusion. The rapidly solidified materials can be amorphous (noncrystalline), nano-crystalline (grain size in the nm range), or microdendritic solid solutions containing solute concentrations vastly superior to those of conventionally processed materials. Effectively, massive second-phase particles are eliminated. These unique microstructures lead to very favorable mechanical properties.

3.2.3 Properties of Metals

Typical values of elastic modulus, yield strength, and ultimate strength in tension, as well as those of fracture toughness of some common metals and their alloys, are listed in Table 3.5, while typical engineering stress-strain curves in tension are shown in Figure 3.16. Note the large plastic strain range.

3.2.4 Why Fiber Reinforcement of Metals?

Precipitation or dispersion hardening of a metal can result in a dramatic increase in the yield stress and/or the work hardening rate. The influence of these obstacles on the elastic modulus is negligible. This is because the intrinsic properties of the strong particles (e.g., the high elastic modulus) are not used. Their only function is to impede dislocation movements in the metal. The improvement in stiffness can be profitably obtained by incorporating so-called advanced high-modulus fibers in a metal matrix. It turns out

Table 3.5. Mechanical properties of some common metals and alloys

Property	E (GPa)	σ_y (MPa)	σ_{max} (MPa)	K_{Ic} (MPa m$^{1/2}$)
Pure (ductile) metals				
Aluminum	70	40	200	100
Copper	120	60	400	?
Nickel	210	70	400	350
Ti-6Al-4V	110	900	1000	120
Aluminum alloys				
(high strength–				
low strength)	70	100–380	250–480	23–40
Plain carbon steel	210	250	420	140
Stainless steel (304)	195	240	365	200

that most of these high-modulus fibers are also lighter than the metallic matrix materials; the only exception is tungsten, which has a high modulus and is very heavy. Table 3.6 lists some common metals and their densities, ρ. The densities of various fibers were given in Chapter 2.

Although one generally thinks of a high Young's modulus as something very desirable from a structural point of view, it turns out that for structural applications involving compression or flexural loading, for example, of beams (say, in a plane, rocket, or truck), it is the E/p^2 value that should be maximized. Consider a simple square section cantilever beam, of length l and thickness t under an applied force P. The elastic deflection of this beam, ignoring self-weight, is given by (Fitzgerald, 1982):

$$\delta = \frac{Pl^3}{3EI}$$

where I is the moment of inertia; in this case it is equal to $t^4/12$.

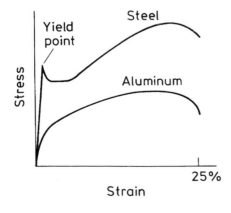

Fig. 3.16. Stress-strain curves of two common metals. Note the large plastic strain range.

Table 3.6. Density of some common metals

Metal	Density (g cm^{-3})	Metal	Density (g cm^{-3})
Aluminum	2.7	Lead	11.0
Beryllium	1.8	Nickel	8.9
Copper	8.9	Silver	10.5
Gold	19.3	Titanium	4.5
Iron	7.9	Tungsten	19.3

Therefore,

$$\delta = \frac{4l^3 P}{Et^4} \tag{3.3}$$

The mass of this beam is

$$M = \text{volume} \times \text{density} = lt^2\rho$$

or

$$t = \left(\frac{M}{l\rho}\right)^{1/2} \tag{3.4}$$

From Equations (3.3) and (3.4), we have

$$\delta = \left(\frac{4l^3 P}{E}\right)\left(\frac{l^2\rho^2}{M^2}\right)$$

or

$$M = \left(\frac{4l^5 P}{\delta}\right)\left(\frac{\rho^2}{E}\right)^{1/2}$$

Thus, for a given rigidity or stiffness P/δ, we have a minimum of mass when the parameter E/ρ^2 is a maximum. What this simple analysis shows is that it makes good sense to use high-modulus fibers to reinforce metals in a structural application and it makes eminently more sense to use fibers that are not only stiffer than metallic matrices but also lighter.

3.3 Ceramic Matrix Materials

Ceramic materials are very hard and brittle. Generally, they consist of one or more metals combined with a nonmetal such as oxygen, carbon, or nitrogen. They have strong covalent and ionic bonds and very few slip systems available compared to metals. Thus, characteristically, ceramics have low failure strains and low toughness or fracture energies. In addition to being brittle, they lack uniformity in properties, have low thermal and mechanical shock resistance, and have low tensile strength. On the other hand, ceramic materials have very high elastic moduli, low densities, and can withstand very high temperatures. The last item is very important and is the real driving force behind the effort to produce tough ceramics. Consider the fact that metallic superalloys, used in jet engines, can easily withstand temperatures up to 800 °C and can go up to 1100 °C with oxidation-resistant coatings. Beyond this temperature, one must use ceramic materials.

By far, the major disadvantage of ceramics is their extreme brittleness. Even the minutest of surface flaws (scratches or nicks) or internal flaws (inclusions, pores, or micro cracks) can have disastrous results. One important approach to toughen ceramics involves fiber reinforcement of brittle ceramics. We shall describe the ceramic matrix composites in Chapter 7. Here we make a brief survey of ceramic materials, emphasizing the ones that are commonly used as matrices.

3.3.1 Bonding and Structure

Ceramic materials, with the exception of glasses, are crystalline, as are metals. Unlike metals, however, ceramic materials have mostly ionic bonding and some covalent bonding. Ionic bonding involves electron transfer between atomic species constituting the ceramic compound; that is, one atom gives up an electron(s) while another accepts an electron(s). Electrical neutrality is maintained; that is, positively charged ions (*cations*) balance the negatively charged ions (*anions*). Generally, ceramic compounds are stoichiometric; that is, there exists a fixed ratio of cations to anions. Examples are alumina (Al_2O_3), beryllia (BeO), spinels ($MgAl_2O_4$), silicon carbide (SiC), and silicon nitride (Si_3N_4). It is not uncommon, however, to have nonstoichiometric ceramic compounds, for example, $Fe_{0.96}O$. The oxygen ion (anion) is generally very large compared to the metal ion (cation). Thus, cations occupy interstitial positions in a crystalline array of anions.

Crystalline ceramics generally exhibit close-packed cubic and hexagonal close-packed structures. The simple cubic structure is also called the cesium chloride structure. It is, however, not very common. CsCl, CsBr, and CsI show this structure. The two species form an interpenetrating cubic array, with anions occupying the cube corner positions while cations go to the interstitial sites. Cubic close packed is really a variation of the fcc structure described in Section 3.2. Oxygen ions (anions) make the proper fcc structure with metal ions (cations) in the interstices. Many ceramic materials show this structure, also called the NaCl or rock salt–type structure. Examples include MgO, CaO, FeO, NiO, MnO, and BaO. There are other variations of fcc close-packed structures, for example, zinc blende types (ZnS) and fluorite types (CaF). The hexagonal close-packed structure is also observed in ceramics. ZnS, for example, also crystallizes in the hcp form. Other examples are nickel arsenide (NiAs) and corundum (Al_2O_3). Figure 3.17 shows the hcp crystal structure of α-Al_2O_3. A and B layers consist of oxygen atoms

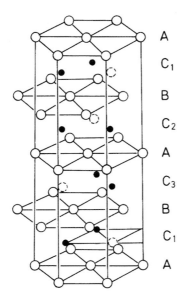

O Oxygen
• Aluminum
◌ Vacant

Fig. 3.17. Hexagonal closed-packed structure of α-alumina. A and B layers contain oxygen atoms. C_1, C_2, and C_3 contain aluminum atoms. The C layers are only two-thirds full.

while C_1, C_2, and C_3 layers contain aluminum atoms. The C layers are only two-thirds full.

Glass-ceramic materials form yet another important category of ceramics. They form a sort of composite material because they consist of 95–98% by volume of crystalline phase and the rest glassy phase. The crystalline phase is very fine (grain size less than 1 μm in diameter). Such a fine grain size is obtained by adding nucleating agents (commonly TiO_2 and ZrO_2) during the melting operation, followed by controlled crystallization. Important examples of glass-ceramic systems include:

1. Li_2O-Al_2O_3-SiO_2: This has a very low thermal expansion and is therefore very resistant to thermal shock. Corningware is a well-known tradename of this class of glass-ceramic.
2. MgO-Al_2O_3-SiO_2: This has high electrical resistance coupled with high mechanical strength.

Ceramic materials can also form solid solutions. Unlike metals, however, interstitial solid solutions are less likely in ceramics because the normal interstitial sites are already filled. Introduction of solute ions disrupts the charge neutrality. Vacancies accommodate the unbalanced charge. For example, FeO has a NaCl-type structure with an equal number of Fe^{2+} and O^{2-} ions. If, however, two Fe^{3+} ions were to replace three Fe^{2+} ions we would have a vacancy where an iron ion would form.

Glasses, the traditional silicate ceramic materials, are inorganic solid-like materials that do not crystallize when cooled from the liquid state. Their structure (see Fig. 2.5) is not crystalline but that of a supercooled liquid. In this case we have a specific volume versus temperature curve similar to the one for polymers (Fig. 3.2) and a characteristic glass transition temperature T_g. Under certain conditions, crystallization of glass can occur with an accompanying abrupt decrease in volume at the melting point because the atoms take up ordered positions.

3.3.2   Effect of Flaws on Strength

As in metals, imperfections in crystal packing of ceramics do exist and reduce their strength. The difference is that important defects in ceramic materials are surface flaws and vacancies. Dislocations do exist but are relatively immobile. Grain boundaries and free surfaces are important planar defects. As in metals, small grain size improves the mechanical properties of ceramics at low- to medium-temperatures.

Surface flaws and internal pores (Griffith flaws) are particularly dangerous for strength and fracture toughness of ceramics. Fracture stress for an elastic material having an internal crack of length $2a$ is given by the Griffith relationship:

$$\sigma_f = \left(\frac{2E\gamma}{a}\right)^{1/2}$$

where E is the Young's modulus and γ is the surface energy of the crack surface. Linear elastic fracture mechanics treats this problem of brittle fracture in terms of a parameter called the *stress intensity factor*, K. The stresses at the crack tip are given by

$$\sigma_{ij} = \left[\frac{K}{(2\pi r)^{1/2}}\right] f_{ij}(\theta)$$

where $f_{ij}(\theta)$ is a function of the angle θ. Fracture occurs when K attains a critical value K_{Ic}. Yet another approach is based on the energy viewpoint, a modification of the Griffith idea. Fracture occurs, according to this approach, when the crack extension force G reaches a critical value G_{Ic}. For ceramic materials, $G_{Ic} = 2\gamma$. It can also be shown that $K^2 = EG$ for opening failure mode and plane stress; that is, the stress intensity factor and the energy approaches are equivalent.

3.3.3   Common Ceramic Matrix Materials

Silicon carbide has excellent high-temperature resistance. The major problem is that it is quite brittle up to very high temperatures and in all environments. Silicon nitride is also an important nonoxide ceramic matrix material.

Table 3.7. Properties of some ceramic matrix materials

Material	Young's modulus (GPa)	Tensile strength (MPa)	Coefficient of thermal expansion (10^{-6} K^{-1})	Density (g cm^{-3})
Borosilicate glass	60	100	3.5	2.3
Soda glass	60	100	8.9	2.5
Lithium aluminosilicate glass-ceramic	100	100–150	1.5	2.0
Magnesium aluminosilicate glass-ceramic	120	110–170	2.5–5.5	2.6–2.8
Mullite	143	83	5.3	
MgO	210–300	97–130	13.8	3.6
Si_3N_4	310	410	2.25–2.87	3.2
Al_2O_3	360–400	250–300	8.5	3.9–4.0
SiC	400–440	310	4.8	3.2

Source: Adapted with permission from Phillips (1983).

Among the oxide ceramics, alumina and mullite are quite promising. Silica-based glasses and glass-ceramics are other ceramic matrices. With glass-ceramics one can densify the matrix in a glassy state with fibers, followed by crystallization of the matrix to obtain high-temperature stability.

Ceramic matrices are used in fiber reinforced composites to achieve, in addition to high strength and stiffness, high-temperature stability and adequate fracture toughness. Table 3.7 summarizes some of the important characteristics of common ceramic matrix materials (Phillips, 1983).

References

G.S. Ansell (Ed.) (1968). *Oxide Dispersion Strengthening*, Gordon & Breach, New York.

W.D. Bascom and R.L. Cottington (1976). *J. Adhesion*, **7**, 333.

D.J. Blundell, J.M. Chalmers, M.W. Mackenzie, and W.F. Gaskin (July 1985). *Sampe Q.*, **16**, 22.

F.N. Cogswell (July 1983). *Sampe Q.*, **14**, 33.

A.H. Cottrell (1958). *Trans. TMS-AIME*, **212**, 192.

L.K. English (Sept. 1985). *Mater. Eng.*, **102**, 32.

M.E. Fine (1964). *Phase Transformations in Condensed Systems*, Macmillan, New York.

R.W. Fitzgerald (1982). *Mechanics of Materials*, second ed., Addison-Wesley, Reading, MA, p. 205.

N.J. Grant (1985). In *Frontiers in Materials Technologies*, Elsevier, New York, p. 125.

E.O. Hall (1951). *Proc. R. Soc. London*, **B64**, 474.

J.T. Hartness (Jan. 1983). *Sampe Q.*, **14**, 33.

D. Kuhlmann-Wilsdorf (1977). In *Work Haraening in Tension and Fatigue*, TMS-AIME, New York, p. 1.

J.C.M. Li (1963). *Trans. TMS-AIME*, **227**, 239.

L. Mandelkern (1983). *An Introduction to Macromolecules*, second ed., Springer-Verlag, New York, p. 1.

M.A. Meyers and E. Ashworth (1982). *Philos. Mag.*, **46**, 737.

R.J. Morgan (1985). In *Epoxy Resins and Composites*, Springer-Verlag, Berlin, p. 1.

J.E. O'Connor, W.R. Beever, and J.F. Geibel (1986). *Proc. Sampe Materials Symp.*, **31**, 1313.

N.J. Petch (1953). *J. Iron Steel Inst.*, **174**, 25.

D.C. Phillips (1983). In *Fabrication of Composites*, North-Holland, Amsterdam, p. 373.

C.K. Riew, E.H. Rowe, and A.R. Siehert (1976). In *Toughness and Brittleness of Plastics*, American Chemical Society, Advances in Chemistry series, vol. 154, p. 326.

M.J. Roberts and W.S. Owen (1968). *J. Iron Steel Inst.*, **206**, 375.

J.M. Scott and D.C. Phillips (1975). *J. Mater. Sci.*, **10**, 551.

A. Seeger (1957). In *Dislocations and Mechanical Properties of Crystals*, John Wiley & Sons, New York, p. 23.

A.K. St. Clair and T.L. St. Clair (1981). *J. Adhesion and Adhesives*, **1**, 249.

J.N. Sultan and F.J. McGarry (1973). *Polym. Eng. Sci.*, **13**, 29.

R.Y. Ting (1983). In *The Role of Polymeric Matrix in the Processing and Structural Properties of Composite Materials*, Plenum Press, New York, p. 171.

A.J Waddon, M.J. Hill, A. Keller, and D.J. Blundell (1987). *J. Mater Sci.*, **22**, 1773.

H. Wiedersich (1964). *J. Met.*, **10**, 425.

Suggested Reading

C.S. Barrett and T.B. Massalski (1980). *Structure of Metals*, third ed. Pergamon Press, Oxford.

K. Dusek (Ed.) (1985). *Epoxy Resins and Composites I* (Advances in Polymer Science, vol. 72), Springer-Verlag, Berlin.

W.D. Kingery, H.K. Bowen, and D.R. Uhlmann (1976). *Introduction to Ceramics*, second ed , John Wiley, New York.

L. Mandelkern (1983). *An Introduction to Macromolecules*, second ed., Springer-Verlag, New York.

M.A. Meyers and K.K. Chawla (1984). *Mechanical Metallurgy: Principles and Applications*, Prentice-Hall, Englewood Cliffs, NJ.

M.A. Meyers and K.K. Chawla (1998). *Mechanical Behavior of Materials*, Prentice-Hall, Upper Saddle River, NJ.

R.E. Reed-Hill and R. Abbaschian (1992). *Physical Metallurgy Principles*, third ed., PWS-Kent, Boston.

J.B. Wachtman (1996). *Mechanical Properties of Ceramics*, John Wiley & Sons, New York.

CHAPTER 4

Interfaces

We can define an interface between a reinforcement and a matrix as the bounding surface between the two across which a discontinuity in some parameter occurs. The discontinuity across the interface may be sharp or gradual. Mathematically, interface is a bidimensional region. In practice, we have an interfacial region with a finite thickness. In any event, an interface is the region through which material parameters, such as concentration of an element, crystal structure, atomic registry, elastic modulus, density, coefficient of thermal expansion, etc. change from one side to another. Clearly, a given interface may involve one or more of these items.

The behavior of a composite material is a result of the combined behavior of the following three entities:

- Fiber or the reinforcing element
- Matrix
- Fiber/matrix interface

The reason the interface in a composite is of great importance is that the internal surface area occupied by the interface is quite extensive. It can easily go as high as 3000 cm^2/cm^3 in a composite containing a reasonable fiber volume fraction. We can demonstrate this very easily for a cylindrical fiber in a matrix. The fiber surface area is essentially the same as the interfacial area. Ignoring the fiber ends, one can write the surface-to-volume ratio (S/V) of the fiber as

$$S/V = 2\pi r l / \pi r^2 l = 2/r \tag{4.1}$$

where r and l are the fiber radius and length of the fiber, respectively. Thus, the surface area of a fiber or the interfacial area per unit volume increases as r decreases. Clearly, it is important that the fibers not be weakened by flaws because of an adverse interfacial reaction. Also, the applied load should be effectively transferred from the matrix to the fibers via the interface. Thus, it becomes extremely important to understand the nature of the interface region of any given composite system under a given set of conditions. Specifically, in the case of a fiber reinforced composite material, the interface, or more precisely the interfacial zone, consists of near-surface layers of fiber and matrix and any layer(s) of material existing between these surfaces. Wettability of the fiber by the matrix and the type of bonding between the two components constitute the primary considerations. Additionally, one should determine the characteristics of the interface and how they are affected by temperature, diffusion, residual stresses, and so on. We will discuss some of the interfacial characteristics and the associated problems in composites in a general way. The details regarding interfaces in polymer matrix, metal matrix, and ceramic matrix composites are given in specific chapters devoted to those composite types.

4.1 Wettability

Various mechanisms can assist or impede adhesion (Baier et al., 1968). A key concept in this regard is that of wettability. Wettability tells us about the ability of a liquid to spread on a solid surface. We can measure the wettability of a given solid by a liquid by considering the equilibrium of forces in a system consisting of a drop of liquid resting on a plane solid surface in the appropriate atmosphere. Figure 4.1 shows the situation schematically. The liquid drop will spread and wet the surface completely only if this results in a net reduction of the system free energy. Note that a portion of the solid/vapor interface is substituted by the solid/liquid interface. Contact angle, θ, of a liquid on the solid surface fiber is a convenient and important parameter to characterize wettability. Commonly, the contact angle is measured by putting a sessile drop of the liquid on the flat surface of a solid substrate. The contact angle is obtained from the tangents along three interfaces: solid/liquid, liquid/vapor, and solid/vapor. The contact angle, θ can be measured directly by a goniometer or calculated by using simple trigonometric relationships involving drop dimensions. In theory, one can use the following expression, called Young's equation,

$$\gamma_{SV} = \gamma_{SL} + \gamma_{LV} \cos \theta \qquad (4.2)$$

where γ is the specific surface energy, and the subscripts SV, LS, and LV represent solid/vapor, liquid/solid, and liquid/vapor interfaces, respectively. If this process of substitution of the solid/vapor interface involves an increase in the free energy of the system, then complete spontaneous wetting will not result. Under such conditions, the liquid will spread until a balance of forces acting on the surface is attained; that is, we shall have partial wetting. A small θ implies good wetting. The extreme cases being $\theta = 0°$, corresponding to perfect wetting, and $\theta = 180°$, corresponding to no wetting. In practice, it is rarely possible to obtain a unique equilibrium value of θ. Also, there exists a range of contact angles between the maximum or advancing angle, θ_a, and the minimum or receding angle, θ_r. This phenomenon, called the *contact-angle hysteresis*, is generally observed in polymeric systems. Among the

Fig. 4.1. Three different conditions of wetting: complete wetting, no wetting, and partial wetting. The terms γ_{SV}, γ_{LS}, and γ_{LV} denote the surface energies of solid/vapor, liquid/solid, and liquid/vapor interfaces, respectively.

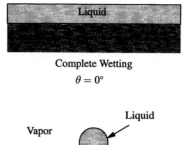

Complete Wetting
$\theta = 0°$

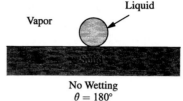

No Wetting
$\theta = 180°$

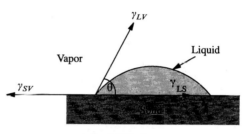

Partial Wetting

sources of this hysteresis are: chemical attack, dissolution, inhomogeneity of chemical composition of solid surface, surface roughness, and local adsorption.

It is important to realize that wettability and bonding are not synonymous terms. Wettability describes the extent of intimate contact between a liquid and a solid; it does not necessarily mean a strong bond at the interface. One can have excellent wettability and a weak van der Waals–type low-energy bond. A low contact angle, meaning good wettability, is a necessary but not sufficient condition for strong bonding. Consider again a liquid droplet lying on a solid surface. In such a case, Young's equation, Equation 4.1, is commonly used to express the equilibrium among surface tensions in the horizontal directions. What is normally neglected in such an analysis is that there is also a vertical force $\gamma_{LV} \sin \theta$, which must be balanced by a stress in the solid acting perpendicular to the interface. This was first pointed out by Bikerman and Zisman in their discussion of the proof of Young's equation by Johnson (1959). The effect of internal stress in the solid for this configuration was discussed by Cahn et al. (1964, 1979). In general, Young's equation has been applied to void formation in solids without regard to the precise state of internal stress. Fine et al. (1993) analyzed the conditions for occurrence of these internal stresses and their effect on determining work of adhesion in particle reinforced composites.

Wettability is very important in PMCs because in the PMC manufacturing the liquid matrix must penetrate and wet fiber tows. Among polymeric resins that are commonly used as matrix materials, thermoset resins have a viscosity in the 1–10 Pa s range. The melt viscosities of thermoplastics are two to three orders of magnitude higher than those of thermosets and they show, comparatively, poorer fiber wetting characteristics and poorer composites. Although the contact angle is a measure of wettability, the reader should realize that its magnitude will depend on the following important variables: time and temperature of contact; interfacial reactions; stoichiometry, surface roughness and geometry; heat of formation; and electronic configuration.

Example 4.1

Consider a laminated composite made by laminating sheets of two materials (1 and 2) in alternate sequence. Let the thickness of the laminae of the two materials be t_1 and t_2, and the number of sheets of each be N_1 and N_2, respectively. For a given volume fraction of component 1, V_1 (remember that $V_1 + V_2 = 1$),

derive an expression for the interfacial area as a function of t_1 and t_2.

Solution

Let the area of cross section of the laminate be A. Let v, V, N, and t represent the volume, volume fraction, number, and thickness of the laminae, respectively, and let subscripts $_1$ and $_2$ denote the two components. Then, we can write

$$V_1 = \text{volume of component 1/total volume} = AN_1t_1/v$$

$$V_2 = \text{volume of component 2/total volume} = AN_2t_2/v$$

$$V_1 + V_2 = 1$$

$$A(N_1t_1/v + N_2t_2/v) = 1$$

or

$$A/v = 1/(N_1t_1 + N_2t_2) \tag{4.3}$$

$$\text{Total number of interfaces} = (N_1 + N_2 - 1)$$

$$\text{Total interfacial area per unit volume, } I_a = (N_1 + N_2 - 1)A/v \tag{4.4}$$

From Eqs. (4.3) and (4.4), we have

$$I_a = (N_1 + N_2 - 1)/(N_1t_1 + N_2t_2)$$

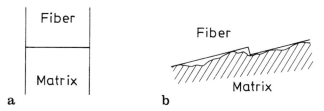

a   b

Fig. 4.2. a An ideal planar interface between reinforcement and matrix. **b** A more likely jagged interface between fiber and reinforcement.

Taking $t_1 = t_2 = t$, we get

$$I_a = (N_1 + N_2 - 1)/(N_1 + N_2)t = k/t$$

where $k = (N_1 + N_2 - 1)/(N_1 + N_2) = $ a constant. The constant k will be approximately equal to 1 when N_1 and N_2 are very large compared to unity. Thus, the interfacial area is inversely proportional to the thickness of the laminae.

Effect of Surface Roughness

In the earlier discussion, it is implicitly assumed that the substrate is perfectly smooth. This, however, is far from true in practice. More often than not, the interface between fiber and matrix is rather rough instead of the ideal planar interface; see Figure 4.2. Most fibers or reinforcements show some degree of roughness (Chawla, 1998). Surface roughness profiles of the fiber surface obtained by atomic force microscopy (AFM) can provide detailed, quantitative information on the surface morphology and roughness of the fibers. It would appear that AFM can be a useful tool in characterizing the fiber surface roughness (Chawla et al., 1993; Chawla et al., 1995; Jangehud et al., 1993; Chawla and Xu, 1994). Figure 4.3 shows an example of roughness characterization by AFM of the surface of a polycrystalline alumina fiber (Nextel 610).

Generally, the fiber/matrix interface will assume the same roughness profile as that of the fiber. In the case of polymer matrix composites, an intimate contact at the molecular level between the fiber and the matrix brings intermolecular forces into play with or without causing a chemical linkage between the components. This intimate contact between the fiber and the matrix requires that the latter in liquid form must wet the former. Coupling agents are frequently used to improve the wettability between the components. At times, other approaches, such as modifying the matrix composition, are used. Wenzel (1936) discussed the effect of surface roughness on wettability and pointed out that "within a measured unit on a rough surface, there is actually more surface, and in a sense therefore a greater surface energy, than in the same measured unit area on a smooth surface." Follow-

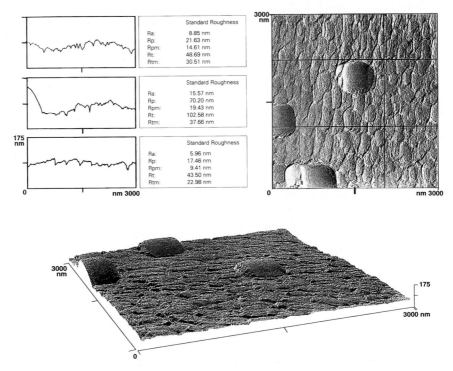

Fig. 4.3. Characterization of surface roughness of a polycrystalline alumina fiber (Nextel 610) by atomic force microscopy. The three scans on the left hand side correspond to the two horizonal and one vertical lines on the right hand side figure.

ing Wenzel, the effect of surface roughness on wettability can be described in terms of r, the ratio of real area, A_{real}, to projected area, A_{proj}, of the interface. Thus,

$$\cos \theta_0 = r \frac{\gamma_{SV} - \gamma_{LS}}{\gamma_{LV}} \tag{4.5}$$

where $r = A_{real}/A_{proj}$.

If $\theta_o < 90°$, wettability is enhanced by roughness, and if $\theta_0 > 90°$, wettability is reduced by roughness. If wetting is poor ($\theta_0 > 90°$), surface roughness can reduce bonded area and lead to void formation and possible stress concentrations.

4.2   Crystallographic Nature of Interface

Most of the physical, chemical, and mechanical discontinuities at the interface mentioned previously are self-explanatory. The concept of atomic registry or the crystallographic nature of an interface needs some elaboration. In terms of the types of atomic registry, we can have a coherent, semicoherent, or incoherent interface. A *coherent* interface is one where atoms at the interface form part of both the crystal lattices; that is, there exists a one-to-one correspondence between lattice planes on the two sides of the interface. A coherent interface thus has some coherency strains associated with it because of the straining of the lattice planes in the two phases to provide the continuity at the interface atomic sites on the two sides of the interface. In general, a perfect atomic registry does not occur between unconstrained crystals. Rather, coherency at the interface invariably involves an elastic deformation of the crystals. A coherent interface, however, has a lower energy than an incoherent one. A classic example of a coherent interface is the interface between Guinier-Preston (G-P) zones and the aluminum matrix. These G-P zones are precursors to the precipitates in aluminum matrix. With increasing size of the crystals, the elastic strain energy becomes more than the interfacial energy, leading to a lowering of the free energy of the system by introducing dislocations at the interface. Such an interface, containing dislocations to accommodate the large interfacial strains and thus having only a partial atomic registry, is called a *semicoherent interface.* Thus, a semicoherent interface is one that does not have a very large lattice mismatch between the phases, and the small mismatch is accommodated by the introduction of dislocations at the interface. As examples, we cite interfaces between a precipitate and a matrix as well as interfaces in some eutectic composites such as NiAl-Cr system (Walter et al., 1969), which has semicoherent interfaces between phases. With still further increases in crystal sizes, the dislocation density at the interface increases, and eventually the dislocations lose their distinct identity; that is, it is no longer possible to specify individual atomic positions at the interface. Such an interface is called an *incoherent interface.* An incoherent interface consists of such severe atomic disorder that no matching of lattice planes occurs across the boundary, i.e., no continuity of lattice planes is maintained across the interface. This eliminates coherency strains, but the energy associated with the boundary increases because of severe atomic disorder at the grain boundary. The atoms located at such an interface do not correspond to the structure of either of the two crystals or grains. Crystallographically, most of the interfaces that one encounters in fiber, whisker, or particle reinforced composites are incoherent.

4.3 Interactions at the Interface

We mentioned earlier that interfaces are bidimensional regions. An initially planar interface, however, can become an interfacial zone with multiple interfaces resulting from the formation of different intermetallic compounds, interdiffusion, and so on. In such a case, in addition to the compositional parameter, we need other parameters to characterize the interfacial zone: for example, geometry and dimensions; microstructure and morphology; and mechanical, physical, chemical, and thermal characteristics of different phases present in the interfacial zone. It commonly occurs that initially the components of a composite system are chosen on the basis of their mechanical and physical characteristics in isolation. It is important to remember, however, that when one puts together two components to make a composite, the composite will rarely be a system in thermodynamic equilibrium. More often than not, there will be a driving force for some kind of interfacial reaction(s) between the two components, leading to a state of thermodynamic equilibrium for the composite system. Of course, thermodynamic information, such as phase diagrams, can help predict the final equilibrium state of the composite. Data regarding reaction kinetics, for example, diffusivities of one constituent in another, can provide information about the rate at which the system would tend to attain the equilibrium state. In the absence of thermodynamic and kinetic data, experimental studies would have to be done to determine the compatibility of the components. Quite frequently, the very process of fabrication of a composite can involve interfacial interactions that can cause changes in the constituent properties and/or interface structure. For example, if the fabrication process involves cooling from high temperatures to ambient temperature, the difference in the expansion coefficients of the two components can give rise to thermal stresses of such a magnitude that the softer component (generally the matrix) will deform plastically. Chawla and Metzger (1972) observed in a tungsten-reinforced copper matrix (nonreacting components) that liquid copper infiltration of tungsten fibers at about 1100 °C followed by cooling to room temperature resulted in a dislocation density in the copper matrix that was much higher in the vicinity of the interface than away from the interface. The high dislocation density in the matrix near the interface occurred because of plastic deformation of the matrix caused by high thermal stresses near the interface. Arsenault and coworkers (Arsenault and Fisher, 1983; Vogelsang et al., 1986) found similar results in SiC whiskers/aluminum matrix composite. Many other researchers have observed dislocation generation in the vicinity of reinforcements/matrix interface due to the thermal mismatch between the reinforcement and metal matrix. In PMCs and CMCs, the matrix is unlikely to deform plastically in response to the thermal stresses. It is more likely to relieve those stresses by matrix microcracking. In powder processing techniques, the nature of the powder surface will influence the interfacial interactions. For example, an oxide film, which is invariably present on the surface of powder particles, will affect the chemical nature of the powder. Topographic characteristics of the components can also affect the degree of atomic contact that can be obtained between the components. This can result in geometrical irregularities (e.g., asperities and voids) at the interface, which can be a source of stress concentrations.

Example 4.2

Distinguish between the terms *surface energy* and *surface tension*.

Answer

Surfaces in solids and liquids have special characteristics because surfaces represent the termination of the phase. Consider an atom or a molecule in the interior of an infinite solid. It will be bonded in all directions, and this balanced bonding results in a reduced potential energy of the system. At a free surface, atoms or molecules are not surrounded by other atoms or molecules; they have bonds or neighbors on only one side. Thus there exists an imbalance of forces at the surface (it is true at any interface, really) that results in a rearrangement of atoms or molecules at the surface. We say that a surface has an extra energy called *surface energy*, i.e., surface energy is the excess energy per unit area associated with the surface because of the unsatisfied bonds at the surface. The units of surface energy are J m^{-2}. We can also define it as the energy needed to create a unit surface area. Surface energy depends on the crystallographic orientation. For example, if we hold a single crystal at high temperature, it will assume a shape bounded by low-energy crystallographic planes of minimum surface energy. *Surface tension* is the tendency to minimize the total surface energy by minimizing the surface area. Surface energy and surface tension are numerically equal for isotropic materials. Surface tension is generally given in units of N m^{-1}, which is the same as J m^{-2}. This is not true for anisotropic solids. Consider a surface of area A with a surface energy of γ. If we increase the surface area by a small amount, the work done per unit increase of area can be written as

$$d(A\gamma)/dA = \gamma + \partial\gamma/\partial A$$

The term $\partial\gamma/\partial A$ is zero for liquids because of atomic or molecular mobility in the liquid state. The structure of a liquid surface is unchanged when we increase its surface area. Actually, for any material that is incapable of withstanding shear, $\partial\gamma/\partial A = 0$. In general, such materials include liquids and solids at high temperatures. Thus, for a liquid, the surface tension and surface energy are equal. Because, thermodynamically, the most stable state is the one with a minimum of free energy, isotropic liquids tend toward a minimum area/unit volume, i.e., a sphere. Such an equality does not hold for solids that can withstand shear. If we stretch the surface of a solid, the atoms or molecules at the surface are pulled apart, γ decreases, and the quantity $\partial\gamma/\partial A$ becomes negative. Thus, for solids, the surface tension is not equal to surface energy. That is why a piece of solid metal does not assume a spherical shape when left to stand at room temperature. In fact, the surface energy of a crystal varies with crystallographic orientation. Generally, the more densely packed planes have a lower surface energy, and they end up forming the stable planes on the surface.

4.4   Types of Bonding at the Interface

It is important to be able to control the degree of bonding between the matrix and the reinforcement. To do so, it is necessary to understand all the

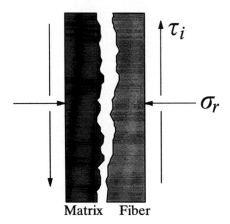

Fig. 4.4. Mechanical gripping due to radial shrinkage of a matrix in a composite more than the fiber on cooling from a high temperature.

Matrix   Fiber

different possible bonding types, one or more of which may be acting at any given instant. We can conveniently classify the important types of interfacial bonding as follows:

- Mechanical bonding
- Physical bonding
- Chemical bonding
  - Dissolution bonding
  - Reaction bonding

Mechanical Bonding

Simple mechanical keying or interlocking effects between two surfaces can lead to a considerable degree of bonding. Any contraction of the matrix onto a central fiber would result in a gripping of the fiber by the matrix. Imagine, for example, a situation in which the matrix in a composite radially shrinks more than the fiber on cooling from a high temperature. This would lead to a gripping of the fiber by the matrix even in the absence of any chemical bonding (Fig. 4.4). The matrix penetrating the crevices on the fiber surface, by liquid or viscous flow or high-temperature diffusion, can also lead to some mechanical bonding. In Figure 4.4, we have a radial gripping stress, σ_r. This is related to the interfacial shear stress, τ_i, as

$$\tau_i = \mu\sigma_r \tag{4.6}$$

where μ is the coefficient of friction, generally between 0.1 and 0.6.

In general, mechanical bonding is a low-energy bond vis à vis a chemical bond, i.e., the strength of a mechanical bond is lower than that of a chemical bond. There has been some work (Vennett et al., 1970; Schoene and Scala, 1970) on metallic wires in metal matrices that indicates that in the presence of internal compressive forces, a wetting or metallurgical bond is not quite necessary because the mechanical gripping of the fibers by the matrix is

Fig. 4.5. a. Good mechanical bond b. Lack of wettability can make a liquid polymer or metal unable to penetrate the asperities on the fiber surface, leading to interfacial voids.

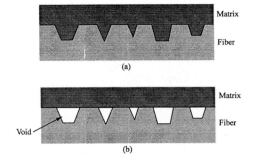

sufficient to cause an effective reinforcement, as indicated by the occurrence of multiple necking in fibers. Hill et al. (1969) confirmed the mechanical bonding effects in tungsten filament/aluminum matrix composites. Chawla and Metzger (1978) studied bonding between an aluminum substrate and anodized alumina (Al_2O_3) films and found that with a rough interface a more efficient load transfer from the aluminum matrix to the alumina occurred. Pure mechanical bonding alone is not enough in most cases. However, mechanical bonding could add, in the presence of reaction bonding, to the overall bonding. Also, mechanical bonding is efficient in load transfer when the applied force is parallel to the interface. In the case of mechanical bonding, the matrix must fill the pores and surface roughness of the reinforcement. Rugosity, or surface roughness, can contribute to bond strength only if the liquid matrix can wet the reinforcement surface. If the matrix, for example, liquid polymer or molten metal, is unable to penetrate the asperities on the fiber surface, then the matrix will solidify and leave interfacial voids, as shown in Figure 4.5. Examples of surface roughness contributing to interfacial strength include:

1. Surface treatments of carbon fibers, e.g., nitric acid oxidation of carbon fibers, which increase specific surface area and lead to good wetting in PMCs, consequently an improved interlaminar shear strength (ILSS) of the composite (see Chap. 8)
2. Most metal matrix composites will have some roughness-induced mechanical bonding between the ceramic reinforcement and the metal matrix (see Chap. 6).
3. Most CMC systems also show a mechanical gripping between the fiber and the matrix (see Chap. 7).

We can make some qualitative remarks about general interfacial characteristics that are desirable in different composites. In PMCs and MMCs, one would like to have mechanical bonding in addition to chemical bonding. In CMCs, on the other hand, it would be desirable to have mechanical bonding in lieu of chemical bonding. In any ceramic matrix composite, roughness-induced gripping at the interface is quite important. Specifically, in fiber reinforced ceramic matrix composites, interfacial roughness-induced radial stress will affect the interface debonding, the sliding friction of debonded fibers, and the fiber pullout length.

Physical Bonding

Any bonding involving weak, secondary or van der Waals forces, dipolar interactions, and hydrogen bonding can be classified as *physical bonding*. The bond energy in such physical bonding is approximately 8–16 kJ/mol.

Chemical Bonding

Atomic or molecular transport, by diffusional processes, is involved in chemical bonding. Solid solution and compound formation may occur at the interface, resulting in a reinforcement/matrix interfacial reaction zone with a certain thickness. This encompasses all types of covalent, ionic, and metallic bonding. Chemical bonding involves primary forces and the bond energy in the range of approximately 40–400 kJ/mol.

There are two main types chemical bonding:

1. *Dissolution bonding*: In this case, interaction between components occurs at an electronic scale. Because these interactions are of rather short range, it is important that the components come into intimate contact on an atomic scale. This implies that surfaces should be appropriately treated to remove any impurities. Any contamination of fiber surfaces, or entrapped air or gas bubbles at the interface, will hinder the required intimate contact between the components.
2. *Reaction bonding*: In this case, a transport of molecules, atoms, or ions occurs from one or both of the components to the reaction site, that is, the interface. This atomic transport is controlled by diffusional processes. Such a bonding can exist at a variety of interfaces, e.g., glass/polymer, metal/metal, metal/ceramic, or ceramic/ceramic.

Two polymer surfaces may form a bond owing to the diffusion of matrix molecules to the molecular network of the fiber, thus forming tangled molecular bonds at the interface. Coupling agents (silanes are the most common ones) are used for glass fibers in resin matrices; see Chapter 5 for details. Surface treatments (oxidative or nonoxidative) are given to carbon fibers to be used in polymeric materials; we describe these in Chapter 8. In metallic systems, one commonly finds solid solution and intermetallic compound formation at the interface. A schematic of the diffusion phenomenon between a fiber and a matrix resulting in solid solution as well as a layer of an intermetallic compound, M_xF_y is shown in Figure 4.6. The plateau region, which has a constant proportion of the two atomic species, is the region of intermetallic compound formation. The reaction products and the reaction rates can vary, depending on the matrix composition, reaction time,

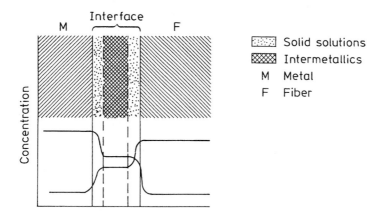

Fig. 4.6. Interface zone in a metal matrix composite showing solid solution and intermetallic compound formation.

and temperature. Generally, one tries to fit such data to an expression of the form

$$x^2 \simeq Dt \tag{4.7}$$

where x is the thickness of the reaction zone, D is the diffusivity, and t is the time. This expression follows from the theory of diffusion in solids. The diffusivity, D, depends on the temperature in an exponential manner

$$D = A \exp(-\Delta Q/kT) \tag{4.8}$$

where A is a preexponential constant, ΔQ is the activation energy for the rate-controlling process, k is Boltzmann's constant, and T is the temperature in kelvin. This relationship comes from the diffusion-controlled growth in an infinite diffusion couple with planar interface. For a composite containing cylindrical fibers of a small diameter, the diffusion distance is very small, i.e., the condition of an infinite diffusion couple with planar interface is not likely to be valid. However, to a first approximation, we can write

$$x^2 = ct \tag{4.9}$$

where c is a pseudo-diffusivity and has the dimensions of diffusivity (m^2 s^{-1}). The reader should bear in mind that this approximate relationship can be expected to work for composites in which the reaction thickness is small compared to the interfiber spacing. Under these conditions, one can use an Arrhenius-type relationship, $c = A \exp(-\Delta Q/kT)$, where A is a preexponential constant. A plot of $\ln c$ vs. $1/T$ can be used to obtain the activation energy, ΔQ, for a fiber/matrix reaction in a given temperature range. The preexponential constant A depends on the matrix composition, fiber, and the environment.

4.5   Optimum Interfacial Bond Strength

Two general ways of obtaining an optimum interfacial bond involve fiber or reinforcement surface treatments or modification of matrix composition. It should be emphasized that maximizing the bond strength is not always the goal. In brittle matrix composites, too strong a bond would cause embrittlement. We can illustrate the situation by examining the following three cases.

Very Weak Interface or Fiber Bundle (No Matrix)

This extreme situation will prevail when we have no matrix and the composite consists of only a fiber bundle. The bond strength in such a composite will only be due to interfiber friction. A statistical treatment of fiber bundle strength (see Chap. 12) shows that the fiber bundle strength is about 70 to 80% of average single-fiber strength.

Very Strong Interface

The other extreme in interfacial strength is when the interface is as strong or stronger than the higher-strength component of the composite, generally the reinforcement. In this case, of the three components—reinforcement, matrix, and interface—the interface will have the lowest strain to failure. The composite will fail when any weak cracking occurs at a weak spot along the brittle interface. Typically, in such a case, a catastrophic failure will occur, and we have a composite with very low toughness.

Optimum Interfacial Bond Strength

An interface with an optimum interfacial bond strength will result in a composite with an enhanced toughness, but without a severe penalty on the strength parameters. Such a composite will have multiple failure sites, most likely spread over the interfacial area, which will result in a diffused or global spread of damage, rather than a very local damage.

4.6   Tests for Measuring Interfacial Strength

Numerous tests have been devised for characterizing the fiber/matrix interface strength. We now briefly describe some of these.

4.6.1   Flexural Tests

Flexural or bend tests are very easy to do and can be used to get a semi-qualitative idea of the fiber/matrix interfacial strength of a composite. There are two basic governing equations for a simple beam elastically stressed in bending:

$$\frac{M}{I} = \frac{E}{R}$$

(4.10)

and

$$\frac{M}{I} = \frac{\sigma}{y}$$

(4.11)

where M is the applied bending moment, I is the second moment of area of the beam section about the neutral plane, E is the Young's modulus of elasticity of the material, R is the radius of curvature of the bent beam, and σ is the tensile or compressive stress on a plane distance y from the neutral plane. For a uniform, circular section beam

$$I = \frac{\pi d^4}{64}$$

(4.12)

where d is the diameter of the circular section beam. For a beam of a uniform, rectangular section, we have

$$I = \frac{bh^3}{12}$$

(4.13)

where b is the beam width and h is the height of the beam. Bending takes place in the direction of the depth, i.e., h and y are measured in the same direction. Also, for a beam with symmetrical section with respect to the neutral plane, replacing y in Equation (4.11) by $h/2$ gives the stress at the beam surface. When an elastic beam is bent, the stress and strain vary linearly with thickness, y across the section, with the neutral plane representing the zero level. The material on the outside or above the neutral plane of the bent beam is stressed in tension while that on the inside or below the neutral plane is stressed in compression. In the elastic regime, the stress and strain are related by

$$\sigma = E\varepsilon$$

(4.14)

From Equations (4.10), (4.11), and (4.14) we can obtain the following simple relation valid in the elastic regime:

$$\varepsilon = \frac{y}{R}$$

(4.15)

Thus, the strain ε in a beam bent to a radius of curvature R varies linearly with distance y from the neutral plane across the beam thickness. There are many variants of this type of test.

Three-Point Bending

The bending moment in three-point bending is given by

$$M = \frac{P}{2} \cdot \frac{S}{2} = \frac{PS}{4} \qquad (4.16)$$

where P is the load and S is the span. The important point to note is that the bending moment in a three-point bend test increases from the two extremities of the beam to a maximum value at the midpoint, i.e., the maximum stress is reached along a line at the center of the beam. From Equations (4.11), (4.13) and (4.16), and taking $y = h/2$, we get the following expression for the maximum stress for a rectangular beam in three-point bending:

$$\sigma = \frac{6PS}{4bh^2} = \frac{3PS}{2bh^2} \qquad (4.17)$$

We can have the fibers running parallel or perpendicular to the specimen length. When the fibers are running perpendicular to the specimen length, we obtain a measure of transverse strength of the fiber/matrix interface.

The shear stress in a three-point bend test is constant. The maximum in shear stress, τ_{max} will correspond to the maximum in load P_{max} and is given by

$$\tau_{max} = \frac{3P_{max}}{4bh} \qquad (4.18)$$

Four-Point Bending

This is also called *pure bending* because there are no transverse shear stresses on the cross sections of the beam between the two inner loading points. For an elastic beam bent in four-point, the bending moment increases from zero at the two extremities to a constant value over the inner span length. This bending moment in four-point is given by

$$M = \frac{P}{2} \cdot \frac{S}{4} = \frac{PS}{8} \qquad (4.19)$$

where S is the outer span and the stress can be written as

$$\sigma = \frac{6PS}{8bh^2} = \frac{3PS}{4bh^2} \qquad (4.20)$$

Short-Beam Shear Test (Interlaminar Shear Stress Test)

This test is a special longitudinal three-point bend test with fibers parallel to the length of the bend bar and the length of the bar being very small. It is also known as the *interlaminar shear strength* (ILSS) test. The maximum shear stress, τ, occurs at the midplane and is given by Equation (4.18). The maximum tensile stress occurs at the outermost surface and is given by Equation 4.17. Dividing Equation 4.18 by Equation 4.17, we get

$$\frac{\tau}{\sigma} = \frac{h}{2S} \tag{4.21}$$

Equation 4.21 says that if we make the load span, S, very small, we can maximize the shear stress, τ, so that the specimen fails under shear with a crack running along the midplane.

The reader should bear in mind that the interpretation of this test is not straightforward. Clearly, the test becomes invalid if the fibers fail in tension before shear-induced failure occurs. The test will also be invalid if shear and tensile failure occur simultaneously. It is advisable to make an examination of the fracture surface after the test and ensure that the crack is along the interface and not through the matrix. This test is standardized by ASTM (D2344). Among the advantages of this test are the following:

- Simple test, short span ($S = 5$ h)
- Easy specimen preparation
- Good for quality assessment, interfacial coatings, and so on.

The main disadvantages of this test are the following:

- Meaningful quantitative results on the fiber/matrix interface strength are difficult to obtain.
- It is difficult to ensure a pure shear failure along the interface.

Iosipescu Shear Test

This is a special test devised for measuring interfacial shear strength (Iosipescu, 1967). In this test, a double-edged notched specimen is subjected to two opposing force couples. This is a special type of four-point bend test in which the rollers are offset, as shown in Figure 4.7, to accentuate the shear deformation. A state of almost pure and constant shear stress is obtained across the section between the notches by selecting a proper notch angle and notch depth (90° and 22% of full width). The average shear stress in this configuration is given by

$$\tau = \frac{P}{bh} \tag{4.22}$$

The main advantage of this test is that a large region of uniform shear is obtained vis à vis other tests. However, there can be a substantial stress concentration near the notch tip in orthotropic materials (not so in isotropic materials) such as fiber reinforced composites. The stress concentration is proportional to the fiber orientation and the fiber volume fraction.

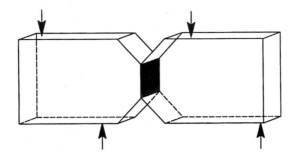

Fig. 4.7. A special type of four-point bend test, called the Iosipescu test, in which the rollers are offset to accentuate the shear deformation.

4.6.2   Single Fiber Pullout Tests

Single fiber pullout and pushout tests have been devised to measure interfacial characteristics. They frequently result in a peak load corresponding to fiber/matrix debonding and a frictional load corresponding to the fiber pullout from the matrix. The mechanics and interpretation of these tests are rather involved, and knowledge of the underlying assumptions is important in order to get useful information from such tests.

One averages the load values over the entire interfacial surface area to get the interface debond strength and/or frictional strength. Analytical and finite element analyses show that the shear stress is a maximum close to the surface and falls rapidly within a distance of a few fiber diameters. Thus, one would expect the interface debonding to start near the surface and progressively propagate along the embedded length. We describe the salient features of these tests.

Single Fiber Pullout Tests

These tests can provide useful information about the interface strength in model composite systems. They are not very helpful in the case of commercially available composites. One must also carefully avoid any fiber misalignment and introduction of bending moments. The mechanics of the single-fiber pullout test are rather complicated (Chamis, 1974; Penn and Lee, 1989; Kerans et al., 1989; Marshall et al., 1992; Kerans and Parthasarathy, 1991).

The fabrication of the single-fiber pullout test sample is often the most difficult part; it entails embedding a part of the single fiber in the matrix. A modified variation of this simple method is to embed both ends of the single fiber in the matrix material, leaving the center region of the fiber uncovered. In all of these methods, the fiber is pulled out of the matrix in a tensile testing machine and a load vs. displacement record is obtained.

The peak load corresponds to the initial debonding of the interface. This is followed by frictional sliding at the interface, and finally by the fiber pull-out from the matrix, during which a steady decrease in the load with displacement is observed. The steady decrease in the load is attributed to the decreasing area of the interface as the fiber is pulled out. Thus, the test simulates the fiber pullout that may occur in the actual composite, and more importantly, provides the bond strength and frictional stress values.

The effect of different Poisson's contractions of fiber and matrix can result in a radial tensile stress at the interface (see Chap. 10). The radial tensile stress will no doubt aid the fiber/matrix debonding process. The effect of Poisson's contraction, together with the fact that the imposed shear stress is not constant along the interface, complicates the analysis of the fiber pullout test. Lawrence (1972) provides the following expressions for the debonding load P_d and the frictional load, P_f

$$P_d = \frac{2\pi r l \tau_d}{\alpha} \tan h(\alpha l) \qquad (4.23)$$

$$P_f = \frac{\pi r^2 \sigma_0}{K} \left[1 - \exp\left(-\frac{2\mu K l}{r} \right) \right] \qquad (4.24)$$

where l is the embedded length of the fiber, $2r$ is the fiber diameter, α is a shear lag parameter that is dependent on elastic constants, τ_d is the debonding stress, μ is the coefficient of friction, σ_0 is the residual compressive stress on the interface, and K is a parameter that is dependent on elastic constants. When the embedded length, l, is small compared to the diameter of the fiber, $2r$, Eqs. (4.23) and (4.24) simplify to (Lawrence, 1972):

$$P_d = 2\pi r l \tau_d \qquad (4.25)$$

$$P_f = 2\pi r l \tau_i \qquad (4.26)$$

These tests can provide useful information about the interface strength in model composite systems. One must also carefully avoid any fiber misalignment and introduction of bending moments. Figure 4.8a shows the experimental setup for such a test. A portion of fiber, length l, is embedded in a matrix and a pulling tensile force is applied as shown. If we measure the stress required to pull the fiber out of the matrix as a function of the embedded fiber length, we get the plot shown in Figure 4.8b. The stress required to pull the fiber out without breaking it increases linearly with the embedded fiber length, up to a critical length, l_c. At embedded fiber lengths greater than or equal to l_c, the fiber will fracture under the action of the tensile stress, σ, acting on the fiber. Consider Figure 4.8a again. The tensile stress, σ, acting on the fiber results in a shear stress, τ, at the fiber/matrix interface. A simple force balance along the fiber length gives

$$\sigma \pi r^2 = \tau 2 \pi r l$$

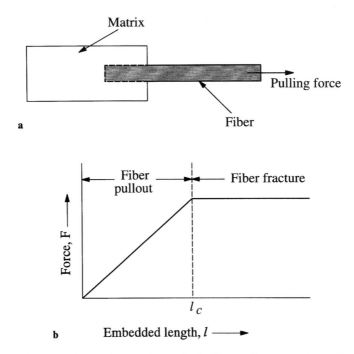

Fig. 4.8. a An experimental setup for a single-fiber pullout test. A portion of fiber, length *l*, is embedded in a matrix and a pulling tensile force is applied as shown. **b** The stress required to pull the fiber out of the matrix as a function of the embedded fiber length.

For $l < l_c$, the fiber is pulled out and the interfacial shear strength is given by

$$\tau = \frac{\sigma r}{2l} \tag{4.27}$$

One measures the load required to debond as a function of the embedded fiber length. Then we can write

$$P = 2\pi r l \tau \tag{4.28}$$

and the interfacial shear strength, τ, can be calculated from the slope of the *P* vs. *l*. There is an implicit assumption in this analysis, viz., the shear stress acting along the fiber/matrix interface is a constant. At $l > l_c$, fiber failure rather than pullout occurs.

The interfacial shear strength is a function of the coefficient of friction, μ, and any normal compressive stress at the interface, σ_r. The source of radial compressive stress is the shrinkage of the matrix during cooling from the processing temperature.

4.6.3   Curved Neck Specimen Test

This technique was devised for PMCs. A special mold is used to prepare a curved neck specimen of the composite containing a single fiber along its central axis. The specimen is compressed and the fiber/matrix debonding is observed visually. The curved neck shape of the specimen enhances the transverse tensile stress at the fiber/matrix interface. The transverse tensile stress leading to interface debonding results from the fact that the matrix and the fiber have different Poisson's ratios. If the matrix Poisson ratio, v_m, is greater than that of the fiber, v_f, then on compression, there will result a transverse tensile stress at the center of the neck and perpendicular to the interface whose magnitude is given by (Broutman, 1969):

$$\sigma_i = \frac{\sigma(v_m - v_f)E_f}{[(1 + v_m)E_f + (1 - v_f - 2v_f^2)E_m]} \tag{4.29}$$

where σ is the net section compressive stress (i.e., load/minimum area), E is the Young's modulus, v is the Poisson's ratio, and the subscripts f and m denote fiber and matrix, respectively.

One can measure the net section stress, σ, corresponding to interface debonding and compute the interfacial tensile strength from Eq. (4.29). There are some important points that should be taken into account before using this test. One needs a special mold to prepare the specimen, and a very precise alignment of the fiber along the central axis is a must. Finally, a visual examination is required to determine the interface debonding point. This would limit the technique to transparent matrix materials. Acoustic emission detection techniques may be used to avoid visual examination.

4.6.4   Instrumented Indentation Tests

Many instrumented indentation tests have been developed that allow extremely small forces and displacements to be measured. Indentation instruments have been in use for hardness measurement for quite a long time, but depth sensing instruments with high resolution became available in the 1980s (Doerner and Nix, 1986; Ferber et al., 1993, 1995; Janczak et al., 1997). Such instruments allow very small volumes of a material to be studied, and a very local characterization of microstructural variations is possible by mechanical means. Hence, the name *mechanical microprobe* is often given to such an instrument (cf. electron microprobe). Such an instrument records the total penetration of an indenter into the sample. The indenter position is determined by a capacitance displacement gage. Pointed or conical and rounded indenters can be used to displace a fiber aligned perpendicular to the composite surface. Figure 4.9 shows three different types of indenters available commercially: cylindrical with a flat end, conical with a flat end, and pointed with a flat end. An example of fiber pushin is shown in a series of four photographs in Figure 4.10. By measuring the applied force and the displacement, interfacial stress can be obtained. The indenter can be moved toward the sample or away from the sample by means of a magnetic coil assembly. Such instruments are available commercially. One special instrument (Touchstone Res. Lab., Tridelphia, WV) combines an indentation system within the chamber of an SEM. Such an instrument combines the materials characterization ability of an SEM with a fiber pushout apparatus. Commonly, some assumptions are made in making an interpretation of an indentation test to determine the strength characteristics of the interface region. For example:

1. Any elastic depression of the matrix adjacent to the fiber is negligible.
2. There are no surface stress concentrations.
3. There is no change in the fiber diameter due to the Poisson expansion during compression of the fiber.
4. There are no residual stresses.

The specimen thickness must be large compared to the fiber diameter for these assumptions to be valid.

Many methods involving the pressing of an indenter on a fiber cross section have been devised for measuring the interfacial bond strength in a fiber-reinforced composite. The pushout test uses a thin specimen (1–3 mm). In the pushout test, if one plots force squared vs. displacement, a three-region curve is obtained (Fig. 4.11) (Cranmer, 1991). In the first region, the indenter is in contact with the fiber and the fiber sliding length l is less then the specimen thickness t. This is followed by a horizontal region in which the fiber sliding length is greater than or equal to the sample thickness. In the third region, the indenter comes in contact with the matrix.

In the first region, we can determine the interfacial frictional shear stress, τ, from the following expression due to Marshall (1984):

$$\tau = F^2 / 4\pi^2 u r^3 E_f \qquad (4.30)$$

where F, the force $= 2a^2 H_f$ and u, the fiber displacement $= (b - a) \cot 74°$, a is the half-diagonal length of the indentation on the fiber, b is the half-diagonal length on the matrix surrounding the fiber; and r, E_f, and H_f are the fiber radius, Young's modulus, and hardness, respectively. One can measure the parameters a, b, and r in a suitable microscope. Equation (4.30) is based on the assumption of a constant shear stress at the fiber/matrix interface. The specimen thickness should be much greater than the fiber diameter for this relationship to be valid.

In the horizontal region, the interfacial shear stress is given by:

$$\tau = F / 2\pi r t \qquad (4.31)$$

where t is the specimen thickness. In the third region of Figure 4.11, the value of the interfacial shear stress cannot be determined because the indenter comes in contact with the matrix.

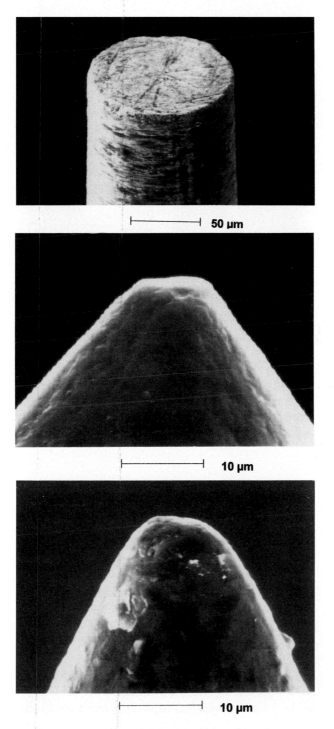

50 μm

10 μm

10 μm

Fig. 4.9. Three different types of indenters: cylindrical with a flat end, conical with a flat end, and pointed with a flat end. (Courtesy of J. Janczak-Ruseh and L. Rohr.)

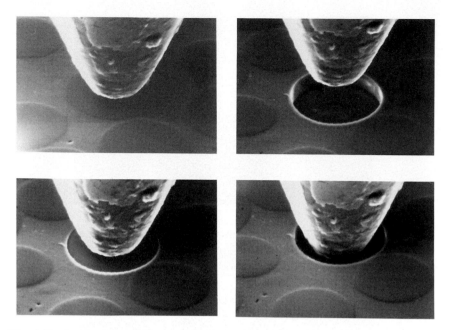

Fig. 4.10. An example of fiber pushin shown in a series of four photographs. (Courtesy of J. Janczak-Ruseh and L. Rohr.)

The capacitance gage can detect displacement changes smaller than 1 nm while the applied force can be detected to less than 1 µN. The hardness, H, is given by

$$H = P/A$$

where P is the load and A is the area of indentation.

The area of indentation A is calculated by means of the following expression:

$$A = a + bh_i^{1/2} + ch_i + dh_i^{3/2} + 24.56h_i^2$$

where h_i is the plastic depth of the indentation and a, b, c, and d are adjustable coefficients. For a perfect tip, $a = b = c = d = 0$, and the only coefficient is 24.56.

It has been observed that the fiber will slide along the interface over a distance that is dependent on the load applied by the indenter. In this model, the load on the indenter is assumed to be balanced by the frictional stress at the interface, and the effect of radial expansion during indentation is neglected. The fiber is elastically compressed by the indenter load over the debonded length, which is assumed to be dependent on the interfacial fric-

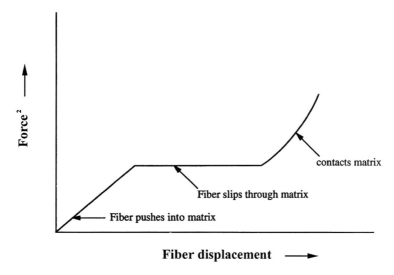

Fig. 4.11. Force squared vs. displacement showing a three-region curve obtained in an indentation test. [After Cranmer (1991)].

tion. There are two analytical models due to Kerans and Parthasarthy (1991) and Hsueh (1992) that take into account progressive debonding and sliding during fiber pushin and pushout. Radial and axial stresses are taken into account in both models and Coulombic friction is assumed at the fiber/ matrix interface. Lara-Curzio and Ferber (1994) found that the two models gave almost identical results in Nicalon fiber/calcium aluminosilicate matrix composites.

An instrumented indentation test to be used at high temperature was developed by Eldridge (1995). Such a test is very useful for CMCs. Such a test is also useful to analyze the effect of residual stresses.

4.6.5 Fragmentation Test

Drzal et al. (1983, 1994, 1997) and others have used this technique extensively to characterize carbon fiber reinforced polymer matrix composites. A single fiber is embedded in a dog-bone-type tensile sample of matrix. When a tensile load is applied to such a sample, the load is transferred to the fiber via shear strains and stresses produced on planes parallel to the fiber/matrix interface. We discuss the subject of load transfer in fiber reinforced composites in Chapter 10. When the tensile stress in the fiber reaches its ultimate strength, it fragments into two parts. If we continue loading, this process of fiber fragmentation continues, i.e., the single fiber continues to fragment into even smaller pieces until the fiber fragment length becomes too small to

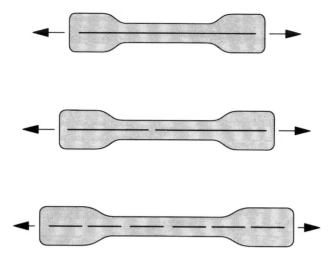

Fig. 4.12. A single fiber fragmentation test. On loading the composite, the load is transferred to the fiber and on continued loading it fragments into smaller pieces until the fiber fragment length becomes too small to enable loading it to fracture.

enable loading it to fracture. This is shown schematically in Figure 4.12 This fiber length is called the *critical length*, l_c. From a consideration of equilibrium of forces over an element dx of the fiber, we can write

$$\pi r^2\, d\sigma = 2\pi r\, dx\, \tau$$

$$d\sigma/dx = 2\tau/r \tag{4.32}$$

From Eq. 4.32, we obtain the critical length as follows. For simplicity, we consider that the matrix is perfectly plastic, i.e., we ignore any strain hardening effects and that the matrix yields in shear at shear stress of τ_y. We also assume that the shear stress along the fiber/matrix interface is a constant over the length of the fiber fragment, l_c. Integrating Eq. 4.32, we get

$$\int_0^{\sigma_{\max}} d\sigma = \int_0^{l_c/2} 2\tau\, dx/r$$

$$\sigma^{\max} = \tau l_c/r$$

or

$$\tau = \sigma^{\max} r/l_c = \sigma^{\max} d/2l_c. \tag{4.33}$$

The fiber fragmentation technique is a simple technique that gives us a qualitative measure of the fiber/matrix interfacial strength. Clearly, a transparent matrix is required. The technique will work only if the fiber failure strain is less than that of the matrix. The major shortcoming or the doubtful assumption is that the interfacial shear stress is constant over the fiber length. In addition, the real material is rarely perfectly plastic.

4.6.6   Laser Spallation Technique

Gupta et al. (1990, 1992) devised a laser spallation technique to determine the tensile strength of a planar interface between a coating (thickness $> 0.5\,\mu m$) and a substrate. Figure 4.13 shows their experimental setup. A collimated laser pulse impinges on a thin film sandwiched between the substrate and a confining plate. This plate is made of fused quartz, which is transparent to Nd:YAG laser (wavelength $= 1.06\,\mu m$). An aluminum film is used as the laser-absorbing medium. Absorption of the laser energy in the constrained aluminum film causes a sudden expansion of the film, which

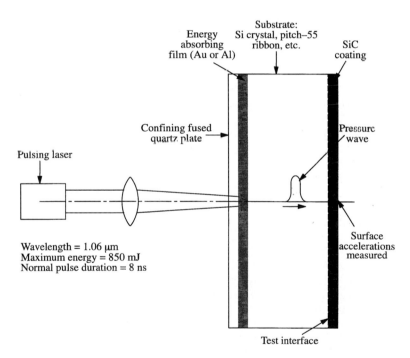

Fig. 4.13. Experimental setup for a laser spallation test. A collimated laser pulse impinges on a thin film sandwiched between the substrate and a confining plate. Absorption of the laser energy in the constrained aluminum film causes a sudden expansion of the film, which produces a compressive shock wave in the substrate that moves toward the coating/substrate interface. This compressive pulse is reflected into a tensile pulse at the free surface of the coating. If this tensile pulse is of a sufficient magnitude, it will remove the coating from the substrate.

produces a compressive shock wave in the substrate that moves toward the coating/substrate interface. When the compression pulse hits the interface, part of it is transmitted into the coating. This compressive pulse is reflected into a tensile pulse at the free surface of the coating. If this tensile pulse is of a sufficient magnitude, it will remove the coating from the substrate. A laser Doppler displacement interferometer is used to record the time rate of change displacement of the coating free surface as the compressive pulse is reflected. By means of a sophisticated digitizer equipment, it is possible to obtain a time resolution of about 0.5 ns for recording displacement fringes. This information is then related to the stress pulse history at the interface. A direct recording of the stress pulse makes this technique useful for interface systems involving ductile components.

References

R.J. Arsenault and R.M. Fisher (1983). *Scripta Met.*, **17**, 67.

R.E. Baier, E.G. Sharfin, and W.A. Zisman (1968). *Science*, **162**, 1360.

L.J. Broutman (1969). In *Interfaces in Composites*, ASTM STP No. 452, American Society of Testing & Materials, Philadelphia.

J.W. Cahn and R.E. Hanneman (1964). *Surf. Sci.*, **94**, 65.

J.W. Cahn (1979). In *Interfacial Segregation*, ASM, Metals Park, OH, p. 3.

C.C. Chamis (1974). In *Composite Materials*, Vol. 6, Academic Press, New York, p. 32.

K.K. Chawla (1997). *Composites Interfaces*, **4**, 287.

K.K. Chawla and M. Metzger (1972). *J. Mater. Sci.*, **7**, 34.

K.K. Chawla and M. Metzger (1978). In *Advances in Research on Strength and Fracture of Materials*, Vol. 3, Pergamon Press, New York, p. 1039.

K.K. Chawla and Z.R. Xu (1994). In *High Performance Composites: Commonalty of Phenomena*, TMS, Warrendale, PA, p. 207.

K.K. Chawla, Z.R. Xu, J.-S. Ha, E. Lara-Curzio, M.K. Ferber, and S. Russ (1995). In *Advances in Ceramic Matrix Composites II*, Amer. Ceram. Soc., Westerville, OH, p. 779.

K.K. Chawla, Z.R. Xu, A. Hlinak, and Y.-W. Chung (1993). In *Advances in Ceramic-Matrix Composites*, Am. Ceram. Soc., p. 725.

D.C. Cranmer (1991). In *Ceramic and Metal Matrix Composites*, Pergamon Press, New York, p. 157.

M.F. Doerner and W.D. Nix (1986). *J. Mater. Res.*, **1**, 601.

L.T. Drzal, M. Madhukar, and M. Waterbury (1994). *Compos. Sci. Tech.*, **27**, 65–71.

L.T. Drzal, M.J. Rich, and P.F. Lloyd (1983). *J. Adhesion*, **16**, 1–30.

L.T. Drzal, N. Sugiura, and D. Hook (1997). In *Composite Interfaces*, **4**, 337.

J.I. Eldridge (1995). In *Mat. Res. Soc. Symp. Proc.*, Vol. 365, Materials Research Society, p. 283.

M.K. Ferber, A.A. Wereszczak, L. Riester, R.A. Lowden, and K.K. Chawla (1993). *Ceramic Sci. & Eng. Proc.*, Amer. Ceram. Soc., Westerville, Oh.

M.K. Ferber, E. Lara-Curzio, S. Russ, and K.K. Chawla (1995). In *Ceramic Matrix Composites—Advanced High-Temperature Structural Materials*, Materials Research Society, Pittsburgh, PA, p. 277.

M.E. Fine, R. Mitra, and K.K. Chawla (1993). *Scripta Met. et Mater.*, **29**, 221.
V. Gupta, A.S. Argon, J.A. Cornie, and D.M. Parks (1990). *Mater. Sci. Eng.*, **A126**, 105.
V. Gupta, A.S. Argon, J.A. Cornie, D.M. Parks (1992). *J. Mech. Phys. Solids*, **4**, 141.
R.G. Hill, R.P. Nelson, and C.L. Hellerich (1969). In *Proceedings of the Refractory Working Group Meeting*, Seattle, WA, Oct.
C.-H. Hsueh, (1992). *J. Am. Ceram Soc.*, **76**, 3041.
N. Iosipescu (1967). *J. Mater.*, **2**, 537.
J. Janczak, G. Bürki, and L. Rohr (1997). *Key Engineering Materials*, 127, 623.
I. Jangehud, A.M. Serrano, R.K. Eby, and M.A. Meador (1993). In *Proc. 21st Biennial Conf. on Carbon*, Buffalo, NY, June 13–18.
R.E. Johnson (1959). *J. Phys. Chem.*, **63**, 1655.
R.J. Kerans, R.S. Hays, N.J. Pagano, and T.A. Parthasarathy (1989). *Amer. Cer. Soc. Bull.*, **68** 429.
R.J. Kerans and T.A. Parthasarathy (1991). *J. Amer. Cer. Soc.*, **74**, 1585.
E. Lara-Curzio and M.K. Ferber (1994). *J. Mater. Sci.*, **29**, 6158.
P. Lawrence (1972). *J. Mater. Sci.*, **7**, 1.
D.B. Marshall and W.C. Oliver (1987). *J. Amer. Ceram. Soc.*, **70**, 542.
D.B. Marshall (1984). *J. Amer. Ceram. Soc.*, **67**, c259.
D.B. Marshall (1989). *J. Am. Ceram. Soc.*, **67**, 7.
D.B. Marshall, M.C. Shaw, and W.L Morris (1992). *Acta Met. et Mater.* **40**, 443.
L.S. Penn and S.M. Lee (1989). *J. Comp. Tech. & Res.*, **11**, 23.
C. Schoene and E. Scala (1970). *Met. Trans*, **1**, 3466.
R.M. Vennett, S.M. Wolf, and A.P. Levitt (1970). *Met Trans.*, **1**, 1569.
M. Vogelsang, R.J. Arsenault, and R.M. Fisher (1986). *Met. Trans. A*, **17**, 379.
J.L. Walter, H.E. Cline, and E. Koch (1969). *Trans. AIME*, **245**, 2073.
T.P. Weihs and W.D. Nix (1991). *J. Amer. Ceram. Soc.*, **74**, 524.
R.N. Wenzel (1936). *Ind. & Eng. Chem.*, **28**, 987.

Suggested Reading

L.A. Carlsson and R.B. Pipes, *Experimental Characterization of Advanced Composite Materials*, Prentice-Hall, Englewood Cliffs, NJ, 1987.
K.T. Faber (1997). *Annual Review of Materials Science*, **27**, 499.
J.-K. Kim and Y.-W. Mai (1998). Engineered Interfaces in Fiber Reinforced Composites, Elsevier, New York.
E.P. Plueddemann (Ed.) (1974). *Interfaces in Polymer Matrix Composites* (Vol. 6 of the series Composite Materials), Academic Press, New York.
H.D. Wagner and G. Marom (Ed.) (1997). *Composite Interfaces (Special issue— Selected papers from the Sixth International Conference on Composite Interfaces (ICCI-6), Israel)*, VSP, Zeist, The Netherlands.

PART II

CHAPTER 5

Polymer Matrix Composites

Polymer matrix composites (PMCs) have established themselves as engineering structural materials, not just as laboratory curiosities or cheap stuff for making chairs and tables. This came about not only because of the introduction of high-performance fibers such as carbon, boron, and aramid, but also because of some new and improved matrix materials (see Chap. 3). Nevertheless, glass fiber reinforced polymers represent the largest class of PMCs. Carbon fiber reinforced PMCs are perhaps the most important structural composites; accordingly, we discuss them separately in Chapter 8. In this chapter, we discuss polymer composite systems containing glass, aramid, polyethylene, and boron fibers.

5.1 Processing of PMCs

Many techniques, originally developed for making glass fiber reinforced polymer matrix composites can also be used with other fibers. Glass fiber reinforced polymer composites represent the largest class of PMCs. As we saw in Chapter 3, polymeric matrix materials can be conveniently classified as thermosets and thermoplastics. Recall that thermosets harden on curing. Curing or crosslinking occurs in thermosets by appropriate chemical agents and/or application of heat and pressure. Conventionally, thermal energy (heating to 200 °C or above) is provided for this purpose. This process, however, brings in the problems of thermal gradients, residual stresses, and long curing times. Residual stresses can cause serious problems in nonsymmetrical or very thick PMC laminates, where they may be relieved by warping of the laminate, fiber waviness, matrix microcracking, and ply delamination. We mentioned electron beam curing in Chapter 3. Electron beam curing offers an alternative that avoids these problems. It is a nonthermal curing process that requires much shorter time cure cycles. Curing by electron beam occurs by electron-initiated reactions at a selectable cure temperature. We describe different methods of fabrication of polymer matrix composites—first thermoset-based composites and then thermoplastic-based composites.

5.1.1   Processing of Thermoset Matrix Composites

There are many processing methods for composites with thermoset matrix materials including epoxy, unsaturated polyester, and vinyl ester.

Hand Lay-Up and Spray Techniques

Hand lay-up and spray techniques are perhaps the simplest polymer-processing techniques. Fibers can be laid onto a mold by hand and the resin (unsaturated polyester is one of the most common) is sprayed or brushed on. Frequently, resin and fibers (chopped) are sprayed together onto the mold surface. In both cases, the deposited layers are densified with rollers. Figure 5.1 shows schematics of these processes. Accelerators and catalysts are frequently used. Curing may be done at room temperature or at a moderately high temperature in an oven.

Filament Winding

Filament winding (Shibley, 1982; Tarnopol'skii and Bail', 1983) is another very versatile technique in which continuous tow or roving is passed through a resin impregnation bath and wound over a rotating or stationary mandrel. A roving consists of thousands of individual filaments. Figure 5.2a shows a schematic of this process, while Figure 5.2b shows a pressure vessel made by filament winding. The winding of roving can be polar (hoop) or helical. In polar winding, the fiber tows do not cross over, while in the helical they do. The fibers are, of course, laid on the mandrel in a helical fashion in both polar and helical windings; the helix angle depends on the shape of the object to be made. Successive layers are laid on at a constant or varying angle until the desired thickness is attained. Curing of the thermosetting resin is done at an elevated temperature and the mandrel is removed. Very large cylindrical (e.g., pipes) and spherical (e.g., for chemical storage) vessels are built by filament winding. Glass, carbon, and aramid fibers are routinely used with epoxy, polyester, and vinyl ester resins for producing filament wound shapes.

There are two types of filament winding processes: wet winding and pre-preg winding. In wet winding, low-viscosity resin is applied to the filaments during the winding process. Polyesters and epoxies with viscosity less than 2 Pa s (2000 centipoise) are used in wet winding. In prepreg winding, a hot-melt or solvent-dip process is used to preimpregnate the fibers. Rigid amines, novolacs, polyimides, and higher-viscosity epoxies are generally used for this process. In filament winding, the most probable void sites are roving cross-overs and regions between layers with different fiber orientations.

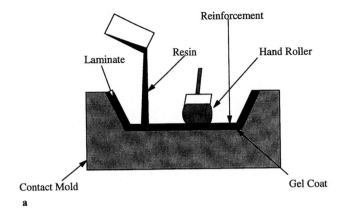

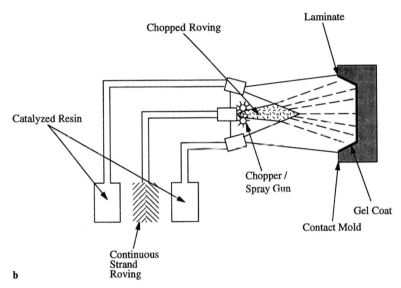

Fig. 5.1. a In hand lay-up, fibers are laid onto a mold by hand, and the resin is sprayed or brushed on. **b** In spray-up, resin and fibers (chopped) are sprayed together onto the mold surface.

Pultrusion

In this process continuous sections of polymer matrix composites with fibers oriented mainly axially are produced. Figure 5.3 shows a schematic of this process. Continuous fiber tows come from various creels. Mat or biaxial fabric may be added to these to provide some transverse strength. These are passed through a resin bath containing a catalyst. After this, the resin-impregnated fibers pass through a series of wipers to remove any excess polymer and then through a collimator before entering the heated die. A thorough wet-out of the rovings is very important. Stripped excess resin is

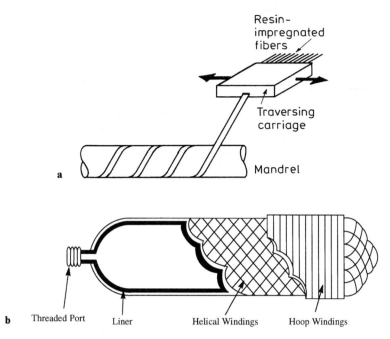

Fig. 5.2. a Schematic of filament winding process. **b** Schematic of a filament wound pressure vessel with a liner; helical and hoop winding are shown.

recirculated to the resin bath. The heated die has the shape of the finished component to be produced. The resin is cured in the die and the composite is pulled out. At the end of the line the part is cut by a flying saw to a fixed length. Typically, the process can produce continuously at a rate of 10 to 200 cm/min. The exact speed depends on the resin type and the cross-sectional thickness of the part being produced. Pultruded profiles as wide as 1.25 m with more than 60% fiber volume fraction can be made routinely.

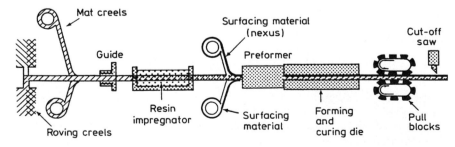

Fig. 5.3. Schematic of the pultrusion process. (Courtesy of Morrison Molded Fiber Glass Co.)

Note that the pultrusion process has a continuous molding cycle. This requires that the fiber distribution be constant and that the cross-sectional shapes not vary, i.e., no bends or tapers are allowed. Main advantages of the process are low labor cost and product consistency. Shapes such as rods, channels, and angle and flat stock are easily produced. Fibrous reinforcements in different forms can be used. Roving, i.e., continuous, fibers are most commonly used. It is easy to saturate such a bundle of fibers with the resin. Continuous strand mat consisting of continuous fiber lengths with random orientation can also be used. They are used to obtain reinforcement action in the transverse direction. Other forms of materials used include chopped strand mat consisting of short (chopped) fibers that can be bonded or stitched to a carrier material, commonly a unidirectional tape, and woven fabrics and braided tapes. Such forms provide reinforcement at $0°$, $90°$, or an arbitrary angle θ to the loading direction. Common resins used in pultrusion are polyester, vinyl ester, and epoxy. Meyer (1985) provides the details regarding this process.

Resin Transfer Molding (RTM)

This is a closed-mold, low-pressure process. A preform made of the desired fiber (carbon, glass, or aramid) is placed inside a mold, and liquid resin such as epoxy or polyester is injected into the mold by means of a pump. Reinforcements can be stitched, but more commonly they are made into a preform that maintains its shape during injection of the polymer matrix. The resin is allowed to cure and form a solid composite. The polymer viscosity should be low (<1 Pa s) enough for the fibers to be wetted easily. Additives to enhance the surface finish, flame retardancy, weather resistance, curing speed, etc. may be added to the resin. Thermoplastics have too high a melting point and too high viscosities (>1 Pa s or 1000 cP) to be processed with RTM. In RTM processing, Darcy's law, which describes the permeability of a porous medium, is of great importance. Darcy's law for single-phase fluid flow says that the volume current density, i.e., volume/(area/time), J of a fluid is given by:

$$J = -\frac{k}{\eta} \nabla P$$

where k is the permeability of the porous medium, η is the fluid viscosity, and P is the pressure that drives the fluid flow. It can be recognized that Darcy's law is an analog of Ohm's law for electrical conduction, i.e., hydraulic permeability is an analog of electrical conductivity. Note that the permeability, k, is a function of the properties of the porous medium, i.e., its microstructure; it does not depend on properties of the fluid.

Among the advantages of RTM, one can cite the following:

- Large, complex shapes can be obtained.
- The level of automation is higher than in other processes.

- Layup is simpler than in manual operations.
- Complex shapes and curvatures can be made easily.
- It take less time to produce.
- With the aid of woven, stitched, or braided preforms, fiber volume fractions as high as 60% can be achieved.
- The process involves a closed mold; therefore styrene emissions can be reduced to a minimum.

In general, RTM produces much fewer emissions compared to hand layup or spray-up techniques.

Mold design is a critical element in the RTM process. Generally, the fibrous preform is preheated and the mold has built-in heating elements to accelerate the process of the resin. Resin flow into the mold and the heat transfer are analyzed numerically to obtain an optimal mold design. The automotive industry has found that the RTM can be a cost-effective, high-volume process for large-scale processing. An important example of the use of RTM in the automobile industry are the composite body parts for the Dodge Viper automobile; they weigh about 90 kg.

Tape-Laying and Fiber Placement Systems

Automation can result in large productivity gains vis à vis manual operations. A completely automated process, with no human being involved, has a great attraction for use in radioactive or clean environments. In the area of fabrication of composites this has led to processing techniques such as automatic tape-laying. Another significant advantage is that rather large structures can be made by this process. The hand lay-up process is limited by the extent of a worker's reach. Such a restriction would be absent in any automatic process. The principle of operation is as follows. The customer's computer aided design (CAD) is the starting point. Using the CAD system, the product or component to be manufactured, for example, a curved part, is developed mathematically onto a flat surface. This is broken down into layers to be fabricated by tape strips laid side by side. Sophisticated software is used to translate strips in each layer via a series of numerical control steps to the final product shape. A spool of fiber tape, preimpregnated with a thermosetting resin and covered with a protective paper on top and protective film underneath, is unwound. The film and paper are peeled off, and the fiber tape, suspended in midair, is cut to the correct shape by cutting blades. The cutting blades can make 6000 cuts per minute. The cut pieces of tape are caught between two new rolls of paper and film and rewound onto a new spool, called a *cassette*. The cassette is then taken to the second machine for lay-up. A laser beam in the tape-laying head is used to accurately lay the tape on the mold.

In fiber placement techniques, individual prepreg tows from spools are fed into the fiber placement head, where they are collimated into a single fiber band and laminated onto the work surface. Each tow is about a 3-mm-wide strand of continuous fibers. A strand, in turn, consists of 12,000 individual filaments impregnated with an epoxy resin. Different tows can be delivered at different speeds, allowing a conformation of a complex structure. For example, when a curved surface is to be laminated, the outer tows of the band pull more strand length than the inner tows. It is possible to cut individual tows and restart the process without stopping the motion of the head. This allows in-process control of the band width, avoiding excess resin build-up and the filling in of any gaps. A compaction roller or shoe consolidates the tape, pressing it onto the work surface. This pressing action serves to remove any trapped air and any minor gaps between individual tows. Figure 5.4 shows a schematic of the fiber placement process.

Autoclave-Based Methods

Autoclave-based methods or bag molding processes (Slobodzinsky, 1982) are used to make large parts. Before we describe this important process, we should digress and describe an important term, viz., prepregs. The term *pre-preg* is a short form of preimpregnated fibers. Prepregs thus represent an intermediate stage in the fabrication of a polymeric composite component. We can define a prepreg as a thin sheet or lamina of unidirectional (or

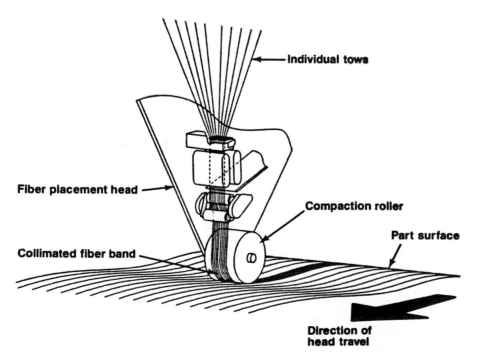

Fig. 5.4. Schematic of the fiber placement process. (Courtesy of Cincinnati Milacron.)

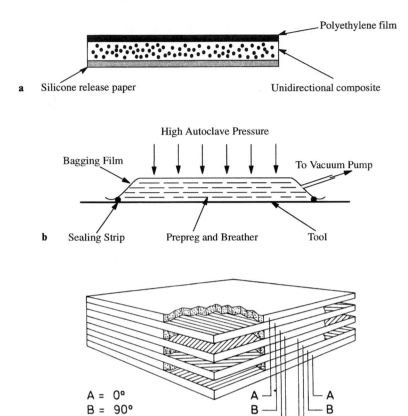

Fig. 5.5. a Schematic of a prepreg. **b** Autoclave process to make a laminated composite. **c** Prepregs of different orientations stacked to form a laminated composite.

sometimes woven) fiber/polymer composite protected on both sides with easily removable separators. Figure 5.5a shows a schematic of a prepreg. An important characteristic of a prepreg made with a thermoset matrix such as epoxy is that the resin is in a partially cured state with a moderately self-adhesive *tack*. This can easily be obtained with epoxies whose cure process can be separated into two stages. In systems having matrix materials that do not go through a two-stage cure, such as polyesters and polyimides, the prepreg tack can be achieved by the addition of liquid rubbers or resins. In prepregs made with a thermoplastic matrix, such a tack is conspicuous by its absence, and they are quite stiff, or *boardy*. Typically, a unidirectional pre-preg can be in the form of a long roll, 300–1500 mm wide, 0.125 mm thick, and 50–250 m long with polymer content by volume of $\sim 35\%$. It is not uncommon to use 50 or more such plies in a component.

Prepregs can be made by a number of techniques. In a process called *solution dip*, the resin ingredients are dissolved in a solvent to 40–50% solids level. The fiber (yarn or fabric) is passed through the solution, and it picks up an amount of solids that depends on the speed of throughput and the solids level in the solution. In another process, *solution spray*, a specified amount of solid resin is sprayed onto the fiber. In both solution dip and solution spray, the impregnated fiber is put through a heat cycle to remove the solvents, and the chemical reaction in the resin proceeds to give the desired tack. In *direct hot-melt*, the resin formulation is put as a high-temperature coat on the fiber. At high temperatures, the viscosity is low enough to obtain a direct coat on the fiber. *Film calendaring* involves casting the resin formulation into a film from either hot-melt or solution. The fiber yarn is sandwiched between two films and calendared so that the film is worked onto the fiber.

Autoclave-based processing of PMCs results in a very high-quality product. An *autoclave* is a closed vessel (round or cylindrical) in which processes (physical and/or chemical) occur under simultaneous application of high temperature and pressure. Heat and pressure are applied to appropriately stacked prepregs. The combined action of heat and pressure consolidates the laminae, removes the entrapped air, and helps cure the polymeric matrix. Autoclave processing of composites thus involves a number of phenomena: chemical reaction (curing of the thermoset resin), resin flow, and heat transfer. A schematic of the autoclave process is shown in Figure 5.5b, while an example of a laminated composite made by this process is shown in Figure 5.5c. Vacuum or pressure bags containing fibers in predetermined orientation in a partially cured matrix (prepreg) are used in an autoclave for densification and curing of resin. Instead of prepregs, chopped fibers mixed with resin can also be used. An autoclave is, generally, a cylindrical oven that can be pressurized. The bags consist of thin and flexible membranes made of rubber that separate the fiber lay-ups from the compressing gases during curing of the resin. Densification and curing are achieved by pressure differentials across the bag walls. In vacuum moldings, the bag contents are evacuated and atmospheric pressure consolidates the composite.

More than one type of fiber may be used to produce so-called *hybrid composites*. A prepreg with fibers parallel to the long dimension is called a 0° lamina or ply. A prepreg that is cut with fibers perpendicular to the long dimension is designated as a 90° lamina, while a prepreg at an intermediate angle θ is designated as a θ-ply. The exact orientation sequence is determined from elasticity theory (see Chap. 11) to give appropriate magnitudes and directions of stresses and to avoid unwanted twisting and/or torsion. This kind of laminate construction, mostly done by hand, is widely used in the aerospace industry. High-fiber volume fractions (60–65%) can be obtained in autoclave-based processing of PMCs.

5.1.2   Thermoplastic Matrix Composites

Thermoplastic matrix composites have several advantages and disadvantages over thermoset matrix composites. We first list these and then we will describe some of the important processes used to form thermoplastic matrix composites.

The advantages of thermoplastic matrix composites include:

- Refrigeration is not necessary with a thermoplastic matrix
- Parts can be made and joined by heating
- Parts can be remolded, and any scrap can be recycled
- Thermoplastics have better toughness and impact resistance than thermosets. This can generally also be translated into thermoplastic matrix composites

The disadvantages include:

- The processing temperatures are generally higher than those with thermosets
- Thermoplastics are stiff and boardy, i.e., they lack the tackiness of the partially cured epoxies

Film Stacking

Laminae of thermoplastic matrix containing fibers with a very low resin content (~ 15 w/o) are used in this process. A low resin content is used because these are very boardy materials. The laminae are stacked alternately with thin films of pure polymer matrix material. This stack of laminae consists of fibers impregnated with insufficient matrix and polymer films of complementary weight to give the desired fiber volume fraction in the end product. These are then consolidated by simultaneous application of heat and pressure.

A good quality laminate must be void-free. This implies that there must be sufficient flow of the thermoplastic matrix between layers as well as within individual tows. Generally, a pressure of 6–12 MPa, a temperature between 275 and 350 °C, and dwell times of up to 30 minutes are appropriate for thermoplastics such as polysulfones and polyetheretherketone (PEEK). Because no time is needed for any curing reaction, the time length of the molding cycle with a thermoplastic matrix is less than that with a thermoset matrix.

An alternative is to use continuous tows of commingled carbon fiber/PEEK from which drapeable prepreg sheets can be made. Heat and pressure are applied simultaneously to these drapeable sheets to produce a composite part.

Hot-forming of laminated sheets of thermoplastics containing fibers is common, with stamping and rolling being frequently used methods.

Diaphragm Forming

This process involves the sandwiching of freely floating thermoplastic pre-preg layers between two diaphragms (Cogswell, 1992). The air between the diaphragms is evacuated and thermoplastic laminate is heated above the melting point of the matrix. Pressure is applied to one side, which deforms the diaphragms and makes them take the shape of the mold. The laminate layers are freely floating and very flexible above the melting point of the matrix, thus they readily conform to the mold shape. After the completion of the forming process, the mold is cooled, the diaphragms are stripped off, and the composite is obtained. One of the advantages of this technique is that components with double curvatures can be formed. The diaphragms are the key to the forming process, and their stiffness is a very critical parameter. Compliant diaphragms do the job for simple components. For very complex shapes requiring high molding pressures, stiff diaphragms are needed. At high pressures, a significant transverse squeezing flow can result, and this can produce undesirable thickness variations in the final composite.

Thermoplastic Tape Laying

Thermoplastic tape laying machines are also available, although they are not as common as the thermosetting tape laying machines. Figure 5.6 shows the schematic of one such machine made by Cincinnati Milacron. A controllable tape head has the tape dispensing and shim dispensing/take-up reels and heating shoes. The hot head dispenses thermoplastic tape from a supply reel. There are three heating and two cooling/compaction shoes. The hot shoes heat the tape to molten state. The cold shoes cool the tape instantly to a solid state.

Commingled fibers

The thermoplastic matrix can be provided in the form of a fiber. The matrix fiber and the reinforcement fiber are commingled to produce a yarn that is

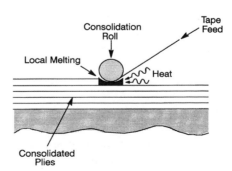

Fig. 5.6. Schematic of a thermoplastic tape laying machine. (Courtesy of Cincinnati Milacron.)

a blend of the thermoplastic matrix and reinforcement yarn. Such a commingled yarn can be woven, knit, or filament wound. The yarn formed into the appropriate shape is then subjected to heat and pressure to melt the thermoplastic matrix component, wet out the reinforcement fibers, and obtain a composite. Vectran (a Hoechst Celanese trademark) is a multifilament yarn of Vectra liquid crystal polymer. It has a low melting point, and thus when commingled with a reinforcement fiber, it can provide the matrix component in the composite.

Injection Molding

As pointed out earlier, thermoplastics soften on heating, and therefore melt flow techniques of forming can be used. Such techniques include injection molding, extrusion, and thermoforming. Thermoforming involves the production of a sheet, which is heated and stamped, followed by vacuum or pressure forming. Generally, discontinuous fibrous (principally glass) reinforcement is used, which results in an increase of melt viscosity. Short fiber-reinforced thermoplastic resin composites can also be produced by a method called *reinforced reaction injection molding* (RRIM) (Lockwood and Alberino, 1981). RRIM is actually an extension of the *reaction injection molding* (RIM) of polymers. In RIM, two liquid components are pumped at high speeds and pressures into a mixing head and then into a mold where the two components react to polymerize rapidly. An important example is a urethane RIM polymer. In RRIM, short fibers (or fillers) are added to one or both of the components. The equipment for RRIM must be able to handle rather abrasive slurries. The fiber lengths that can be handled are generally short, owing to viscosity limitations. Because a certain minimum length of fiber, called the critical length (see Chap. 10), is required for effective fiber reinforcement, more often than not RRIM additives are fillers rather than reinforcements. Most RIM and RRIM applications are in the automotive industry.

5.1.3 Sheet Molding Compound (SMC)

There are some common PMCs that do not contain long, continuous fibers; hence we describe them separately in this section. *Sheet molding compound* (SMC) is the name given to a composite that consists of a polyester resin plus additives. The additives generally consist of fine calcium carbonate particles and short glass fibers. Sometimes calcium carbonate powder is substituted by hollow glass microspheres, which results in a lower density, but makes it more expensive. Figure 5.7 shows a schematic of the SMC processing. Polyester resin can be replaced by vinyl ester to further reduce the weight, but again with a cost penalty. SMC is used in making some auto body parts, such as bumper beams, radiator support panels, and many others. It has been used in the Corvette sports car for many decades. Poly-

Fig. 5.7. Schematic of SMC processing.

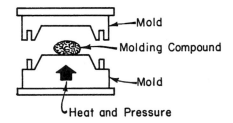

propylene resin can be reinforced with calcium carbonate particles, mica flakes, or glass fibers. Such composites, though structurally not as important as, say, carbon fiber/epoxy composites, do show improved mechanical properties vis à vis unreinforced resin. Characteristics such as strength, stiffness, and service temperature are improved. These materials are used in automotive parts, appliances, electrical components, and so forth.

5.2 Interface in PMCs

We provide below a summary of some important features of the interface region in PMCs with glass, aramid, and polyethylene fibers. For a discussion of the carbon fiber/polymer interface, please see Chapter 8.

5.2.1 Glass Fiber/Polymer

The reader will recall from the description given in Chapters 2 and 3 that inorganic, silica-based glasses are analogous to many organic polymers in that they are amorphous. Recall also that pure, crystalline silica melts at 1800 °C and can be worked in the range of 1600–1800 °C. However, by adding some metal oxides, one can break the Si–O bonds and obtain a series of amorphous glasses with low glass transition temperatures so that they can be processed at much lower temperatures than pure silica. In general, the atomic or molecular arrangement in any material is different at the surface than in the interior. In particular, in the case of silica-based glasses containing a variety of oxides, a complex hydroxyl layer is formed rather easily. Nonhygroscopic oxides absorb water as hydroxyl groups while hygroscopic oxides become hydrated. The activity of a glass surface is thus a function of the hydroxyl content and the cations just below the surface. That is, the surface activity of E-glass will be different from that of fused silica. Invariably, glass fibers are surface treated by applying a *size* on the freshly drawn glass fibers to protect them from the environment, for handling ease, and to avoid introducing surface defects. Common sizes are starch gum, hydrogenated vegetable oil, gelatin, polyvinyl alcohol (PVA), and a variety of nonionic emulsifiers. The size is generally incompatible with the matrix resin and is therefore removed before putting the glass fibers in a resin matrix by heat cleaning at $\sim 350\,^\circ$C for 15 to 20 hours in air, followed by washing with detergent or solvent and drying. After cleaning, organometallic or organosilane coupling agents are applied; an aqueous solution of silane is commonly used for this purpose. The organosilane compounds have the chemical formula

$$R–Si–X_3$$

where R is a resin-compatible group and X represents groups capable of interacting with hydroxylated silanols on the glass surface. Typically, a silane coupling agent will have the following general chemical structure:

$$X_3Si(CH_2)_n Y$$

where n can have a value between 0 and 3, X is a hydrolyzable group on silicon, and Y is an organofunctional group that is resin-compatible. Silane coupling agents are generally applied to glass from an aqueous solution. Hydrolyzable groups are essential for generating intermediate silanols. Examples of coupling agents commonly used are organometallic or organosilane complexes. It is thought by some researchers (Knox, 1982) that the coupling agents create a chemical bridge between the glass surface and the resin matrix. Other researchers (Kardos, 1985) do not subscribe to this view.

The *chemical bridge theory* goes as follows. Silane molecules, as mentioned earlier, are multifunctional groups with a general chemical formula of $R–SiX_3$, where X stands for hydrolyzable groups bonded to Si. For example, X can be an ethoxy group—OC_2H_5—and R is a resin-compatible group. They are hydrolyzed in aqueous size solutions to give trihydroxy silanols (Fig. 5.8a). These trihydroxy silanols get attached to hydroxyl groups at the glass surface by means of hydrogen bonding (Fig. 5.8b). During the drying of sized glass fibers, water is removed and a condensation reaction occurs between silanol and the glass surface and between adjacent silanol molecules on the glass surface, leading to a polysiloxane layer bonded to the glass surface (Fig. 5.8c). Now we can see that the silane coating is anchored at one end, through the R group, to the uncured epoxy or polyester matrix, and at the other end to the glass fiber through the hydrolyzed silanol groups. On curing, the functional groups R either react with the resin or join the resin molecular network (Fig. 5.8d).

Appealing though this chemical bridge model of silane coupling is, there are certain shortcomings. The interface model shown in Fig. 5.8d will result in such a strong bond that it will fail at the very low strains encountered during the curing of the resin and resulting from differential thermal contraction. Also, under conditions of industrial application of silanes from aqueous solution, a covalent reaction to the glass fiber surface does not occur unless a primary or secondary amine is present (Kardos, 1985; Kaas and Kardos, 1971).

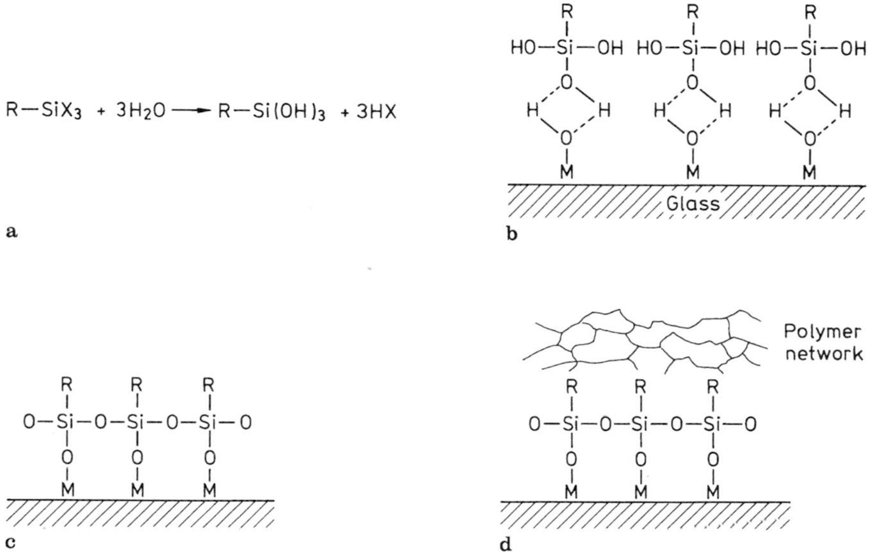

$$R-SiX_3 + 3H_2O \longrightarrow R-Si(OH)_3 + 3HX$$

a

b

c

d

Fig. 5.8. Function of a silane coupling agent: **a** hydrolysis of silane to silanol; **b** hydrogen bonding between hydroxy groups of silanol and a glass surface; **c** polysiloxane bonded to a glass surface; and **d** resin-compatible functional groups R after reacting with a polymer matrix. [From Hull (1981), reprinted with permission.]

Reversible Bond Model

During the processing of PMCs, most polymeric matrix materials will shrink on curing, while the reinforcement fiber remains unaffected. This can lead to large stresses at the fiber/polymer interface. Stresses can also result on cooling due to differences in coefficients of thermal expansion of glass (about $5 \cdot 10^{-6} \text{ K}^{-1}$) and rigid polymer ($\sim 50$–$100 \cdot 10^{-6} \text{ K}^{-1}$). The chemical bridge model provides a rigid bond at the glass/polymer interface that will not be able to withstand the strains involved due to the curing and shrinkage. A clean glass surface under ordinary atmospheric conditions can readily pick up a molecular layer of water. Water can reach the glass/polymer interface by diffusion through the polymer, by penetrating through cracks or by capillary migration along the fibers. Note that hydrophobic mineral fibers such as carbon or silicon carbide are less sensitive to water than glass fiber because there is little tendency for water molecules to cluster at the interface. Thus, the silane coupling agents at the glass/resin interface also have the important function of allowing the composite to accommodate internal stresses.

Plueddeman (1974) pointed out that a silane coupling agent provides a reversible hydrolytic bond between a polymer matrix and an inorganic fiber. Hydrated silanol bonds to the oxides on the glass surface, i.e., $-MOH$, where

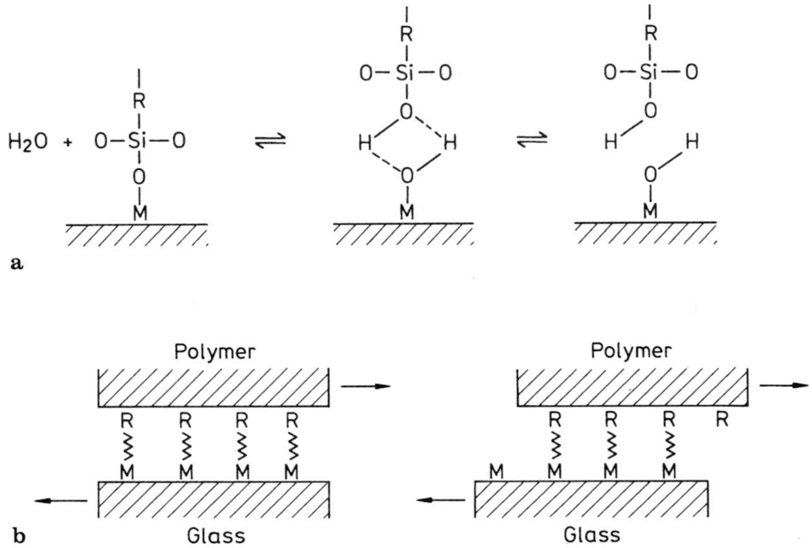

Fig. 5.9. a Plueddemann's reversible bond associated with hydrolysis. **b** Shear displacement at a glass/polymer interface without permanent bond rupture. [From Hull (1981), reprinted with permission.]

M stands for Si, Al, Fe, and so on, with the elimination of water. The dynamic equilibrium mechanism of bonding requires water at a hydrophilic interface to allow relaxation of thermal stresses generated during cooling. Plueddemann's model is shown in Fig. 5.9. In the pressence of water at the interface (it can diffuse in from the resin), the covalent M–O bond hydrolyzes as shown in Fig. 5.9a. If a shear parallel to the interface occurs, the polymer and glass fiber can glide past each other without a permanent bond rupture (Fig. 5.9b). Ishida and Koenig (1978) used infrared spectroscopy to obtain experimental evidence for this reversible bond mechanism. The interface is not a static sandwich of polymer-water-glass. Instead, a dynamic equilibrium prevails that involves making and breaking bonds, which allows relaxation of internal stresses at a molecular scale. Water is therefore necessary to bond rigid polymers to inorganic surfaces such as glass.

5.2.2 Aramid Fiber/Polymer Interface

Most polymers show rather poor adhesion to aramid fibers. This is evidenced by the generally poor interlaminar shear strength and transverse tensile strength values obtained with aramid reinforced PMCs. Typically, aramid/epoxy interfacial strengths are about half of the interfacial strengths of glass/epoxy or carbon/epoxy composites. A highly oriented chain microstructure and skin/core heterogeneity are responsible for this low, poor interfacial strength. This may not be a disadvantage in aramid/polymer composites used to make impact-resistant items such as helmets or body armor, where ease of delamination may be an advantage. However, in high-strength and high-stiffness composites, poor interfacial adhesion can be a disadvantage. Various fiber surface treatments have been tried to alleviate this problem. Among these are the following:

- Bromine water treatment. This also results in a reduced fiber strength.
- Silane coupling agents developed for glass fibers.
- Isocyanate-linked polymer.
- Treatment with reactive chemicals such as acetic acid anhydride, methacrylol chloride, sulfuric acid, among others. Such treatments, however, result in a decrease in the tensile strength of aramid fiber and that of aramid/polyester composites attributable to etching of the aramid fiber.
- Acid (HCl, H_2SO_4) or base (NaOH) hydrolysis of Kevlar aramid yields reactive terminal amino groups to which diepoxide molecules could attach.
- Plasma treatment in vacuum, ammonia, or argon. Plasma treatment in ammonia increases the amine concentration on the fiber surface, which is thought to lead to covalent bonding at the interface. An increase in nitrogen content in the form of amine groups is accompanied by a decrease in oxygen content with increasing exposure time.

5.2.3   Polyethylene Fiber/Polymer Interface

Ultra-high-molecular-weight polyethylene (UHMWPE) fiber is another chemically very inert fiber, and therefore it does not adhere very well to polymeric matrix materials. High-modulus polyethylene fibers such as Spectra or Dyneema are hard to bond with any polymeric matrix. Some kind of surface treatment must be given to the polyethylene fiber to bond with resins such as epoxy, PMMA. By far the most successful surface treatment involves a cold gas (such as air, ammonia, or argon) plasma (Kaplan et al., 1988). A plasma consists of gas molecules in an excited state, i.e., highly reactive, dissociated molecules. When the polyethylene, or any other fiber, is treated with a plasma, surface modification occurs by removal of any surface contaminants and highly oriented surface layers, addition of polar and functional groups on the surface, and introduction of surface roughness, all these factors contribute to an enhanced fiber/matrix interfacial strength (Biro et al., 1992; Brown et al., 1992; Hild and Schwartz, 1992a, 1992b; Kaplan et al., 1988; Li et al., 1992). An exposure of just a few minutes to the plasma is sufficient to do the job.

In a UHMWPE fiber/epoxy composite, a plasma treatment of fiber resulted in an increase in fiber surface roughness and an increased bonding area and interfacial shear strength (Kaplan et al., 1988; Biro et al., 1992). In another work (Brown et al., 1992), it was shown that chemical etching of

UHMWPE fibers resulted in more than a six-fold increase in interfacial shear strength of UHMWPE/epoxy composites.

Chemical treatment with chromic acid and plasma etching in the presence of oxygen are two treatments that are commonly used to modify the surface characteristics of polyethelene fiber with a view to improve their adhesion to polymeric matrix materials.

5.3   Structure and Properties of PMCs

Continuous fiber reinforced polymer composites show anisotropic properties. The properties of a composite will depend on the matrix type, fiber type, amount or volume fraction of fiber (or that of the matrix), fabrication process, and, of course, the fiber orientation.

5.3.1   Structural Defects in PMCs

The final stage in any PMC fabrication is called *debulking*, which serves to reduce the number of voids. Nevertheless, there are some common structural defects in PMCs. Following is a list of these:

- Resin-rich (fiber-poor) regions.
- Voids (e.g., at roving crossovers in filament winding and between layers having different fiber orientations, in general). This is a very serious problem; a low void content is necessary for improved interlaminar shear strength.
- Microcracks (these may form due to curing stresses or moisture absorption during processing).
- Debonded and delaminated regions.
- Variations in fiber alignment.

5.3.2   Mechanical Properties

Some microstructures of polymer matrices reinforced by continuous and discontinuous fibers are shown in Fig. 5.10. A transverse section of continuous glass fiber in an unsaturated polyester matrix is shown in Fig. 5.10a. The layer structure of an injection-molded composite consisting of short glass fibers in a semicrystalline polyethylene terephthalate (PET) is shown in Fig. 5.10b (Friedrich, 1985). Essentially, it is a three-layer structure of different fiber orientations. More fibers are parallel to the mold fill direction (MFD) in the two surface layers, S, than in the transverse direction in the central layer, C. Note the heterogeneity in the microstructure, which results in a characteristically anisotropic behavior.

Laminates of polymer matrix composites made by the stacking of appropriately oriented plies also result in composites with highly anisotropic

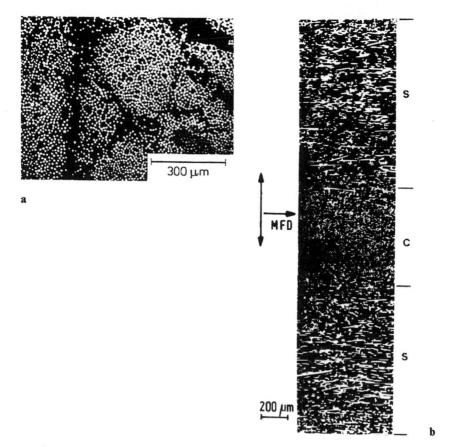

Fig. 5.10. a Continuous glass fibers in a polyester matrix. **b** Discontinuous glass fibers in a semicrystalline polyethylene terephthalate thermoplastic matrix. [From Friedrich (1985), used with permission.]

characteristics. In particular, the properties of continuous fiber reinforced polymers are quite a bit higher in the longitudinal direction than in other directions. It turns out that one generally finds the longitudinal properties of composites being quoted in the literature for comparative purposes. The reader is warned that one must bear in mind this discrepancy when comparing such data of highly anisotropic materials with the data of isotropic materials such as common polycrystalline metals. Besides, composites containing aramid and polyethylene fibers will not have such attractive properties in compression in the longitudinal direction. A summary of some important characteristics of PMCs is presented in Table 5.1 (Hancox, 1983).

Most thermoset matrix composites show elastic behavior right up to fracture, i.e., there is no yield point or plasticity. The strain-to-failure values

Table 5.1. Representative properties of some PMCs[a]

Materials	Density (g cm⁻³)	Tensile Modulus Longitudinal (GPa)	Transverse (GPa)	Shear Modulus (GPa)	Tensile Strength Longitudinal (MPa)	Transverse (MPa)	Compressive Strength Longitudinal (MPa)	Flexure Modulus (GPa)	Flexure Strength (MPa)	ILSS[b] (MPa)	Longitudinal Coefficient of Thermal Expansion (10⁻⁶ K⁻¹)
Unidirectional E glass 60 v/o	2	40	10	4.5	780	28	480	35	840	40	4.5
Bidirectional E glass cloth 35 v/o	1.7	16.5	16.5	3	280	280	100	15	220	60	11
Chopped strand mat E glass 20 v/o	1.4	7	7	2.8	100	100	120	7	140	69	30
Boron 60 v/o	2.1	215	24.2	6.9	1400	63	1760	—	—	84	4.5
Kevlar 29 60 v/o	1.38	50	5	3	1350	—	238	51.7	535	44	—
Kevlar 49 60 v/o	1.38	76	5.6	2.8	1380	30	276	70	621	60	−2.3

[a] The values are only indicative and are based on epoxy matrix at room temperatures.
[b] ILSS, interlaminar shear strength.
Source: Adapted with permission from Hancox (1983).

are rather low, typically, less than 0.5%. Consequently, the work done during fracture is also small. This has some very practical implications for the design engineer, viz., he or she cannot bank on any local yielding to take care of stress concentrations.

In general, the continuous fiber reinforced composites will be stiff and strong along the fiber axis, but off-angle these properties fall rather sharply. The amount of fiber in a composite depends on the fiber alignment. Typically, in a unidirectional PMC, the fiber volume fraction can be 65%. In composites having fibers aligned bidirectionally, this value can fall to 50%, while in a composite containing in-plane random distribution of fibers, the volume fraction will rarely be more than 30%. As a general rule-of-thumb, we can take the Young's modulus of the composite in the longitudinal direction to be given by the following rule-of-mixtures relationship

$$E_{cl} = E_f V_f + E_m V_m$$

where E and V denote the Young's modulus and the volume fraction of a component and the subscripts c, f, m, and l indicate composite, fiber, matrix, and the longitudinal direction, respectively. From the data provided in Chapter 2 on the mechanical properties of fibers, the reader can easily verify that glass fiber/polymer will only give a modest increase in modulus while aramid/polymer will provide a significant increase in stiffness. However, aramid/polymer composites will show a higher creep rate than glass/polymer. Aramid fiber has superior impact characteristics, therefore, aramid fiber–based polymer composites will show better ballistic resistance against impact resistance in general. Similar observations can be made regarding strength characteristics of the polymer matrix composites. As should be clear to the reader, these properties are highly dependent on the fiber properties.

Damping Characteristics

High damping or the ability to reduce vibrations can be very important in many applications; for example, in mechanical equipment that is subject to variable speeds, resonance problems lead to unacceptable levels of noise. Also, in sporting goods such as tennis rackets, fishing rods, and golf clubs, it is desirable to have high damping. In general, aramid fiber/polymer composites provide good damping characteristics. This, of course, stems from the superior damping of aramid fiber.

Environmental Effects in Polymer Matrix Composites

Environmental moisture can penetrate organic materials by a diffusional process. Typically, moisture works as a plasticizer for a polymer, i.e., properties such as stiffness, strength, and glass transition temperature decrease with the ingress of moisture in a polymer. It is now well recognized that the problem of moisture absorption in polymer matrix composites is a very important one. The maximum moisture content under saturated condition,

M_m, as a function of relative humidity, is given by the following relationship:

$$M_m = A(\% \text{ relative humidity})^B$$

The moisture content in a composite, M, can be written as

$$M = M_i + G(M_m - M_i)$$

where M_i is the initial moisture content ($= 0$ for a completely dry material) and G is a dimensionless, time-dependent parameter related to the diffusion coefficient. Generally, one assumes that Fickian diffusion prevails, i.e., Fick's law of diffusion is applicable. Under conditions of Fickian diffusion, water diffuses into the laminate from the two surfaces that are in equilibrium with their surroundings, and the parameter G is given by (Chen and Springer, 1976)

$$G \simeq 1 - (8/\pi^2) \exp(-\pi^2 Dt/S^2)$$

where D is the diffusion coefficient (m^2 s^{-1}) in the direction normal to the laminate surface; $t =$ time (s); $S = h$ (the laminate thickness), if exposed on both sides, and $S = 2h$, if exposed on one side.

A dried specimen, usually in the form of a thin sheet, is placed in a humid environment at a constant temperature, and the mass gain is measured as a function of time. If we plot the mass gained (say, as fractional moisture absorbed) as a function of the square root of time, at a given temperature, we obtain a curve with the following characteristics. After an initial linear portion (i.e., $M \propto \sqrt{t}$), the curve assumes a plateau form in an asymptotic manner as shown in Fig. 5.11. The slope of the linear portion can be used to determine the diffusion coefficient as follows:

$$Slope = \frac{M_2 - M_1}{\sqrt{t_2} - \sqrt{t_1}} = \frac{4M_m\sqrt{D}}{h\sqrt{\pi}}$$

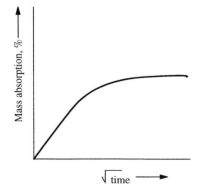

Fig. 5.11. Schematic curve of fractional moisture absorbed as a function of the square root of time, at a given temperature.

or

$$D = \left[\frac{M_2 - M_1}{\sqrt{t_2} - \sqrt{t_1}}\right]^2 \cdot \pi \left[\frac{h}{4M_m}\right]^2$$

The epoxy resins commonly employed as matrices in the composites meant for use in the aerospace industry are fairly impervious to the range of fluids commonly encountered, for example, jet fuel, hydraulic fluids, and lubricants (Anderson, 1984). There are, however, two fundamental effects that must be taken into account when designing components made of PMCs, namely, temperature and humidity. The combined effect of these two, that is, hygrothermal effects, can result in a considerable degradation in the mechanical characteristics of the PMCs. This is especially so in high-performance composites such as those used in the aerospace industry where dimensional tolerances are rather severe.

Degradation owing to ultraviolet radiation is another important environmental effect. Ultraviolet radiation breaks the covalent bonds in organic polymers. Sometimes a prolonged exposure of epoxy laminates to ultraviolet radiation results in a slight increase in strength, attributed to postcuring of the resin, followed by a gradual loss of strength as a result of laminate surface degradation (Bergmann, 1984).

Fracture

Fracture in PMCs, as in other composites, is associated with the characteristics of the three entities: fibers, matrix, and interface. Specifically, fiber/matrix debonding, fiber pullout, fiber fracture, crazing, and fracture of the matrix are the energy-absorbing phenomena that can contribute to the failure process of the composite. Of course, the debonding and pullout processes depend on the type of interface. At low temperatures, the fracture of a PMC involves a brittle failure of the polymeric matrix accompanied by pullout of the fibers transverse to the crack plane. Figure 5.12 shows this kind of fracture at $-80\,^{\circ}$C in the case of a short glass fiber/PET composite. Note the brittle fracture in the matrix. At room temperature, the same polymeric matrix (PET) deformed locally in a plastic manner, showing crazing (Friedrich, 1985). Generally, stiffness and strength of a PMC increase with the amount of stiff and strong fibers introduced in a polymer matrix. The same cannot be said unequivocally for the fracture toughness. The toughness of the matrix and several microstructural factors related to the fibers and the fiber/matrix interface have a strong influence on the fracture toughness of the composite. Friedrich (1985) describes the fracture toughness of short fiber–reinforced thermoplastic matrix composite in an empirical manner by a relationship of the form

$$K_{cc} = MK_{cm}$$

where K_{cm} is the fracture toughness of the matrix and M is a microstructural

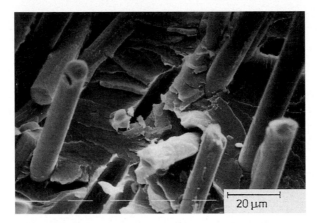

Fig. 5.12. Brittle matrix fracture and fiber pullout in a short glass fiber–reinforced polyethylene terephthalate (PET) composite fractured at −80 °C. [From Friedrich (1985), used with permission.]

efficiency factor. M can be larger than 1 and depends on fiber amount, fiber orientation, and the fiber orientation distribution over the fracture plane, as well as the deformation behavior of the matrix and the relative effectiveness of all the energy-absorbing mechanisms.

5.4 Applications

Glass fiber reinforced polymers are used in a wide variety of industries: from sporting goods to civil construction to aerospace. Tanks and vessels (pressure and nonpressure) in the chemical process industry, as well as process and effluent pipelines, are routinely made of glass fiber reinforced polyester resin. Figure 5.13 shows a wide variety of fiberglass/resin matrix structural shapes made by the pultrusion technique. S-2 glass fibers and Kevlar aramid fibers are used in the storage bins and floorings of civilian aircraft. Other aircraft applications include doors, fairings, and radomes. Kevlar is also used in light load-bearing components in helicopters and small planes. In most applications involving glass fiber reinforced polymers, aramid fibers can be substituted for glass without much difficulty. Racing yachts and private boats are examples of aramid fiber making inroads into the glass fiber fields where performance is more important than cost. Drumsticks made with a pultruded core containing Kevlar aramid fibers and a thermoplastic injection-molded cover are shown in Fig. 5.14. These drumsticks last longer than the wooden ones, are lightweight, will not warp, and are more consistent than wooden sticks. Military applications vary from ordinary helmets to rocket engine cases. One has to guard against using aramid reinforcement in situations involving compressive, shear, or transverse tensile loading paths.

Fig. 5.13. A large variety of fiberglass/resin structural shapes made by pultrusion are available. (Courtesy of Morrison Molded Fiber Glass Co.)

In such situations, one should resort to hybrid composites (using more than one fiber species with aramid in the direction of longitudinal tension). In general, long-term and fatigue characteristics of aramid/epoxy composites are better than those of glass/epoxy composites. The fatigue resistance of aramid PMCs is inferior to that of carbon PMCs. Use of PMCs in military and commercial helicopters has been most impressive. Figure 5.15 shows the use of PMCs, including carbon fiber reinforced polymers, in the UH-60A Black Hawk military helicopter. Note the extensive use of Kevlar and glass fiber reinforced epoxy composites. The main driving force in such applications is that of weight reduction.

Boron/epoxy composites are used in sporting goods (e.g., golf clubs and tennis rackets), in the horizontal stabilizers and tail sections of military aircraft, in the foreflap of the Boeing 707, and so on. The high cost of boron fibers will keep them restricted to specialty items in the aerospace field.

Boeing 757 and 767 were the first large commercial aircraft to make widespread use of structural components made of advanced composites. About 95% of the visible interior parts in 757 and 767 cabins are made from nonconventional materials.

Fig. 5.14. Drumsticks made with a pultruded Kevlar core and a thermoplastic injection-molded cover. (Courtesy of Morrison Molded Fiber Glass Co.)

One of the main reasons for this was the steadily dropping price of carbon fibers. For example, in the mid-1970s the price of carbon fiber was more than $400/kg while in the mid-1980s, it had dropped to less than $50/kg. Similarly, there has been a growing use of composites in army aircraft and helicopters. Weight and cost savings are the driving forces for such applications. Consider the Sikorsky H-69 helicopter. In this helicopter, the fuselage, in

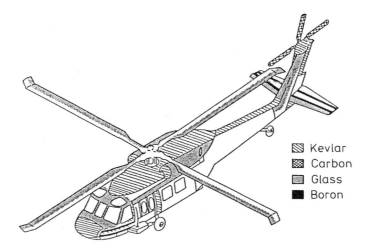

Fig. 5.15. Use of PMCs in a helicopter.

conventional metal construction, is very labor-intensive to make. It involves 856 detail parts, 75 assemblies, and 13,600 fasteners. In comparison, a composite (carbon, aramid, and glass fiber/epoxy) fuselage has only 104 detail parts, 10 assemblies, and 1700 fasteners. It is also 17% lighter and 38% cheaper than the metal fuselage.

Use of lighter composites in aircraft structures results in energy savings. For a given aerodynamic configuration of an aircraft, there is a direct correlation between airplane weight and fuel consumption. Boeing's Materials Technology organization has put out some estimates of the resultant fuel savings, which are worth considering. For a typical route structure of a Boeing 757 or 767 plane, it burns 120 to 150 liters of fuel per kg of weight per year per aircraft. Assuming a weight savings of 1500 kg by use of new lighter materials and a 20-year economic life for each plane, about 4 million liters of fuel will be saved per airplane.

Pressure Vessels

A very important application of PMCs is in the use of natural gas for transportation. Use of natural gas as a fuel results in lower emissions (NO_x, CO_2) vis à vis gasoline fuel. It may also be a cheaper fuel in certain locations. The use of compressed natural gas as a vehicle fuel requires on-board storage of gas at high presure (~ 200 bar). Earlier, steel cylinders were used as pressure vessels. These metallic gas cylinders are quite heavy and thus result in a reduced payload. Much lighter, filament-wound PMC cylinders were developed to replace the steel cylinders. Examples of these include steel or aluminum cylinders hoop-wrapped with glass fiber/polyester and hoop- and polar-wound glass or carbon fiber–reinforced polymer cylinders with a thermoplastic liner. Figure 5.16(a) shows glass fiber– (white) and carbon fiber– (black) reinforced filament–wound cylinders on metallic liners while Figure 5.16(b) shows a vehicle with on-board compressed natural gas cylinders.

Ballistic Protection

Spectra Shield is a product of AlliedSignal that is made by means of woven fabric of Spectra polyethylene fiber. Figure 5.17 shows a schematic of cross-plied ($0°/90°$) Spectra fibers in a resin matrix. Spectra Shield is used to make helmets, hard armor for vehicles, and soft body armor. The helmet manufacture involves a special version of Spectra Shield, a special shell design, and a three-way adjustable liner of shock-absorbing foam padding. These helmets were used by the U.N. peacekeeping troops from France in 1993 and were introduced to police forces in the United States and Europe. Body armor is a high-performance system for protection against rifle bullets. It consists of five Spectra Shield lies that slide into the pockets in the body armor.

Aramid, carbon, and S-glass fibers are used in downhill and cross-country

a   b

Fig. 5.16. a Glass fiber– (white) and carbon fiber– (black) reinforced filament–wound cylinders on metallic liners. **b** A vehicle with on-board compressed natural gas cylinders. (Courtesy of SCI.)

skis, boots, poles, gloves, and as threads to stitch many items. The main advantages that the use of composites brings to the sporting goods industry include safety, lighter weight, and higher strength than conventional materials. Ski poles made of polymer composites are lighter and stiffer than aluminum poles. Frequently, carbon fibers are laid over a small sleeve of aramid. Aramid and carbon fibers and fabrics are used to make the skate shell, which fits around the ankle.

Composites are also used in rifle stocks for biathlons of cross-country skiing and target shooting. Both weight and strength are important factors in the use of the rifles; athletes may have to carry the rifles over distances of up to 20 kilometers.

Polymer composites are also finding applications in civilian infrastructure such as bridges and highway overpasses, and for strengthening damaged structural beams to levels greater than the original values. Although glass fiber–reinforced PMCs are used for such applications, most of them involve carbon fiber-reinforced PMCs. Hence, we describe this new avenue of applications in Chapter 8.

5.5   Recycling of PMCs

When the useful life of a PMC component, be that aircraft-part or a fishing rod, is finished, we need to be concerned about the recycling or reclamation of the components. Service or use is not the only thing that causes a loss in the restorable value of a material. Paint removal of a polymer or composite

Fig. 5.17. Schematic of cross-plied (0°/90°) Spectra fibers in a resin matrix: **a** Soft armor; **b** hard armor.

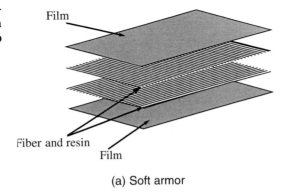

(a) Soft armor

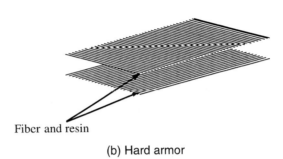

(b) Hard armor

can frequently be a cause of poor recycling of a polymeric material. In thermosetting materials, the crosslinking phenomenon makes recycling very difficult. Continuous fiber reinforced composites pose a great difficulty in recycling; especially the recovery of continuous fibers. Tertiary recycling, i.e. recovery of the monomer, is one possible way. Sheet molding compound (SMC) is particulate/short fiber polymer matrix composite. Such a material can be ground into a fine powder and reused as a filler. Such a technique would not work with continuous fiber reinforced thermoset matrix PMCs. As we know, thermosets are the most common matrix materials in PMCs, and they are not amenable to recycling in a manner similar to metals and thermoplastics. Tertiary or chemical recycling involving conversion of polymer fractions into a gaseous mixture of low molecular weight hydrocarbons with a low temperature (200 °C) catalytic process has been tried (Manson, 1994; Allred et al., 1995). Such chemical recycling breaks polymeric waste into reusable hydrocarbon fractions for use as monomers, chemicals, or fuel. Separate polymer as low molecular weight chemicals and reclaim the fibers and fillers that can be reused. Allred et al. (1995) claim that the although the reclaimed carbon fibers are not continuous, their structural and mechanical characteristics are not altered by the chemical recycling. Analysis of the

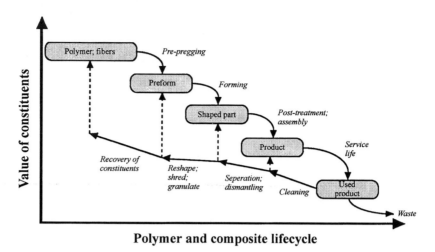

Polymer and composite lifecycle

Fig. 5.18. Change in value of the constituents with increasing lifecycle. [After Manson (1994).]

reclaimed carbon fibers by SEM, XPS, and single fiber tensile tests showed them to have surface texture, chemistry and strength close to the virgin fibers. The epoxy was converted to hydrocarbon fractions. The reader should recognize that these recycling procedures result in a material that is suitable for use in applications less demanding because of a considerably reduced intrinsic value of the material. Figure 5.18 shows schematically the change in value of the constituents with increasing lifecycle (Manson, 1994).

References

R.E. Allred, G.C. Newmeister, T.J. Doak, R.C. Cochran, and A.B. Coons (1997). In *Composites '97 Manufacturing and Tooling*, Paper #EM97-110 Society of Manufacturing Engineering. Dearborn MI 48121.

B.W. Anderson (1984). In *Advances in Fracture Research*, ICF4, New Delhi, Pergamon Press, Oxford, **1**, 607.

H.W. Bergmann (1984). In *Advances in Fracture Research*, ICF4, New Delhi, Pergamon Press, Oxford, p. 569.

D.A. Biro, G. Pleizier, and Y. Deslandes (1992). *J. Mater. Sci. Lett.*, **11**, 698.

J.R. Brown, P.J.C. Chappell, and Z. Mathys (1992). *J. Mater. Sci.*, **27**, 3167.

C.H. Chen and G.S. Springer (1976). *J. Composite Mater.*, **10**, 2.

F.N. Cogswell (1992). *Thermoplastic Aromatic Polymer Composites*, Butterworth-Heinemann, Oxford, p. 136.

K. Friedrich (1985). *Composites Sci. Tech.*, **22**, 43.

N.L. Hancox (1983). In *Fabrication of Composite Materials*, North-Holland, Amsterdam, p. 1.

D.N. Hild and P. Schwartz (1992a). *J. Adhes. Sci. Technol.*, **6**, 879.

D.N. Hild and P. Schwartz (1992b). *J. Adhes. Sci. Technol.*, **6**, 897.

H. Ishida and J.L Koenig (1978). *J. Colloid Interface Sci.*, **64**, 555.

R.L. Kaas and J.L. Kardos (1971). *Polym. Eng. Sci.*, **11**, 11.

S.L. Kaplan, P.W. Rose, H.X. Nguyen, and H.W. Chang (1988). *SAMPE Quarterly*, **19**, 55.

J.L. Kardos (1985). In *Molecular Characterization of Composite Interfaces*, Plenum Press, New York, p. 1.

C.E. Knox (1982). In *Handbook of Composite Materials*, Van Nostrand Reinhold, New York, p. 136.

Z.F. Li, A.N. Netravali, and W. Sachse (1992). *J. Mater. Sci.*, **27**, 4625.

R.J. Lockwood and L.M. Alberino (1981). In *Advances in Urethane Science and Technology*, Technomic Press, Westport, CT.

J.A. Manson (1994). In *High Performance Composite: Commonalty of Phenomena*, K.K. Chawla, P.K. Liaw, and S.G. Fishman (eds.) The Minerals, Metals & Materials Society, Warrendale, PA., 1994, p. 1.

R.W. Meyer (1985). In *Handbook of Pultrusion Technology*, Chapman & Hall, New York.

E.P. Plueddemann (1974). In *Interfaces in Polymer Matrix Composites*, Academic Press, New York, p. 174.

A.M. Shibley (1982). In *Handbook of Composite Materials*, Van Nostrand Reinhold, New York, p. 448.

A. Slobodzinsky (1982). In *Handbook-of Composite Materials*, Van Nostrand Reinhold, New York, p. 368.

Y.M. Tarnopol'skii and A.I. Bail' (1983). In *Fabrication of Composites*, North-Holland, Amsterdam, p. 45.

Suggested Reading

S.G. Advani (Ed.) (1994). *Flow and Rheology in Polymer Composites Manufacturing*, Elsevier, Amsterdam.

P. Ehrburger and J.B. Donnet (1980). "Interface in Composite Materials," *Philos. Trans. R. Soc. London*, **A294**, p. 495.

A. Kelly and S.T. Mileiko (Eds.). (1983). *Fabrication of Composites*, North-Holland, Amsterdam.

E.P. Plueddemann (Ed.) (1974). *Interfaces in Polymer Matrix Composites*, Academic Press, New York.

CHAPTER 6

Metal Matrix Composites

Metal matrix composites consist of a metal or an alloy as the continuous matrix and a reinforcement that can be particle, short fiber or whisker, or continuous fiber. In this chapter, we first describe important techniques to process metal matrix composites, then we describe the interface region and its characteristics, properties of different metal matrix composites, and finally, we summarize different applications of metal matrix composites.

6.1   Types of Metal Matrix Composites

There are three kinds of metal matrix composites (MMCs):

- particle reinforced MMCs
- Short fiber or whisker reinforced MMCs
- continuous fiber or sheet reinforced MMCs

Table 6.1 provides examples of some important reinforcements used in metal matrix composites and their aspect (length/diameter) ratios and diameters. Particle or discontinuously reinforced MMCs have become very important because they are inexpensive vis à vis continuous fiber reinforced composites and they have relatively isotropic properties compared to fiber reinforced composites. Figures 6.1a and b show typical microstructures of continuous alumina fiber/magnesium alloy and silicon carbide particle/aluminum alloy composites, respectively. Use of nanometer-sized fullerenes as a reinforcement has been tried. In addition to being very small, fullerenes (of which C_{60} is the most common) are light and hollow. The important question in this regard is whether they remain stable during processing and service. For the interested reader, we cite work by Barrera et al. (1994), who used powder metallurgy, rf sputtering of a composite (multifullerene) target, and thin film codeposition methods to make fullerene/metal composites. Copper fullerene composites were processed by sputtering copper during fullerene sublimation in an argon atmosphere. Fullerenes were found to withstand the processing.

Table 6.1. Typical reinforcements used in metal matrix composites

Type	Aspect Ratio	Diameter, μm	Examples
Particle	$\sim 1-4$	$1-25$	SiC, Al_2O_3, BN, B_4C
Short fiber or whisker	$\sim 10-1000$	$0.1-25$	SiC, Al_2O_3, $Al_2O_3+SiO_2$, C
Continuous fiber	>1000	$3-150$	SiC, Al_2O_3, C, B, W

However, in the case of aluminum matrix, brittle Al_4C_3 was observed to form at high temperatures.

6.2   Important Metallic Matrices

A variety of metals and their alloys can used as matrix materials. We describe briefly the important characteristics of some of the more common ones.

Aluminum Alloys

Aluminum alloys, because of their low density and excellent strength, toughness, and resistance to corrosion, find important applications in the aerospace field. Of special mention in this regard are the Al-Cu-Mg and Al-Zn-Mg-Cu alloys, very important precipitation-hardenable alloys. Aluminum-lithium alloys form one of the most important precipitation-hardenable aluminum alloys. Lithium, when added to aluminum as a primary alloying element, has the unique characteristic of increasing the elastic modulus and decreasing the density of the alloy. Understandably, the aerospace industry has been the major target of this development. Al-Li alloys are precipitation hardenable, much like the Al-Cu-Mg and Al-Zn-Mg-Cu alloys mentioned earlier. The precipitation hardening sequence in Al-Li alloys is, however, much more complex than that observed in conventional precipitation-hardenable aluminum alloys. Generally, these alloys contain, besides lithium, some copper, zirconium, and magnesium. Vasudevan and Doherty (1989) provide an account of the various aluminum alloys.

Titanium Alloys

Titanium is one of the important aerospace materials. It has a density of 4.5 g cm^{-3} and a Young's modulus of 115 GPa. For titanium alloys, the density can vary between 4.3 and 5.1 g cm^{-3}, while the modulus can have a range of 80 to 130 GPa. High strength/weight and modulus/weight ratios are important. Titanium has a relatively high melting point (1672 °C) and retains strength to high temperatures with good oxidation and corrosion resistance. All these factors make it an ideal material for aerospace applica-

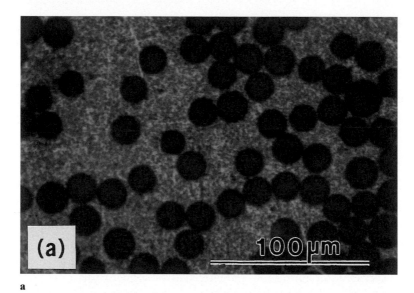

a

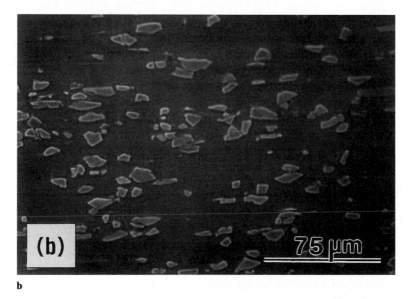

b

Fig. 6.1. a Typical microstructure of continuous alumina fiber/magnesium alloy composite. **b** Typical microstructure of silicon carbide particle/aluminum alloy composite.

tions. Titanium alloys are used in jet engines (turbine and compressor blades), fuselage parts, etc. It is, however, an expensive material.

At supersonic speeds, the skin of an airplane heats up so much that aluminum alloys are no good; titanium alloys must be used at such high temperatures. At speeds greater than Mach 2, the temperatures will be even higher than what titanium alloys can withstand. Titanium aluminides are one of the candidate materials in this case.

Titanium has two polymorphs: alpha (α) titanium has an hcp structure and is stable below 885 °C and beta (β) titanium has a bcc structure and is stable above 885 °C. Aluminum raises the $\alpha \rightarrow \beta$ transformation temperature, i.e., aluminum is an alpha stabilizer. Most other alloying elements (Fe, Mn, Cr, Mo, V, No, Ta) lower the $\alpha \rightarrow \beta$ transformation temperature, i.e., they stabilize the β phase. Thus, three general alloy types can be produced, viz., α, $\alpha + \beta$, and β titanium alloys. The Ti-6%Al-4% V, called the *workhorse* Ti alloy of the aerospace industry, belongs to the $\alpha + \beta$ group. Most titanium alloys are not used in a quenched and tempered condition. Generally, hot working in the $\alpha + \beta$ region is carried out to break the structure and distribute the α phase in an extremely fine form.

Titanium has a great affinity for oxygen, nitrogen, and hydrogen. Parts per million of such interstitials in titanium can change mechanical properties drastically; particularly embrittlement can set in. That is why welding of titanium by any technique requires protection from the atmosphere. Electron beam techniques, in a vacuum, are frequently used.

Magnesium Alloys

Magnesium and its alloys form another group of very light materials. Magnesium is one of the lightest metals; its density is 1.74 g cm^{-3}. Magnesium alloys, especially castings, are used in aircraft gearbox housings, chain saw housings, electronic equipment, etc. Magnesium, being a hexagonal close-packed metal, is difficult to cold work.

Copper

Copper has a face centered cubic structure. Its use as an electrical conductor is quite ubiquitous. It has good thermal conductivity. It can be cast and worked easily. One of the major applications of copper in a composite as a matrix material is in niobium-based superconductors.

Intermetallic Compounds

Intermetallic compounds can be ordered or disordered. Ordered intermetallic alloys possess structures characterized by long-range ordering, i.e., different atoms occupy specific positions in the lattice. Because of their ordered structure, dislocations in intermetallics are much more restricted than in disordered alloys. This results in retention (and, in some cases, even an increase) of strength at elevated temperatures, a very desirable feature. For example, nickel aluminide shows a marked increase in strength up to 800 °C. An undesirable feature of intermetallics is their extremely low ductility. Attempts at ductility enhancement in intermetallics have involved a number of metallurgical techniques. Rapid solidification is one method. Another technique that has met success is the addition of boron to Ni_3Al. With extremely small amounts of boron (0.06 wt. %), the ductility increases from about 2% to about 50%. Long-range order also has significant effects on diffusion-controlled phenomena such as recovery, recrystallization, and grain growth. The activation energy for these processes is increased, and these processes are slowed down. Thus, ordered intermetallic compounds tend to exhibit high creep resistance. Enhancing toughness by making composites with intermetallic matrix materials is a potential possibility.

Chief among the disordered intermetallics is molybdenum disilicide ($MoSi_2$) (Vasudevan and Petrovic, 1992). It has a high melting point and shows good stability at temperatures greater than 1200 °C in oxidizing atmosphere. It is commonly used as a heating element in furnaces. The high oxidation resistance comes from a protective SiO_2 film that it tends to form at high temperatures.

6.3   Processing

Many processes for fabricating metal matrix composites are available. For the most part, these processes involve processing in the liquid and solid state. Some processes may involve a variety of deposition techniques or an in situ process of incorporating a reinforcement phase. We provide a summary of these fabrication processes.

6.3.1   Liquid-State Processes

Metals with melting temperatures that are not too high, such as aluminum, can be incorporated easily as a matrix by liquid route. A description of some important liquid-state processes is given here.

Casting, or *liquid infiltration*, involves infiltration of a fiber bundle by liquid metal (Divecha et al., 1981, 1986; Rohatgi et al., 1986). It is not easy to make MMCs by simple liquid-phase infiltration, mainly because of difficulties with wetting of ceramic reinforcement by the molten metal. When the infiltration of a fiber preform occurs readily, reactions between the fiber and the molten metal can significantly degrade fiber properties. Fiber coatings applied prior to infiltration, which improve wetting and control reactions, have been developed and can result in some improvements. In this case, however, the disadvantage is that the fiber coatings must not be exposed to air prior to infiltration because surface oxidation will alter the positive effects of coating (Katzman, 1987). One liquid infiltration process involving par-

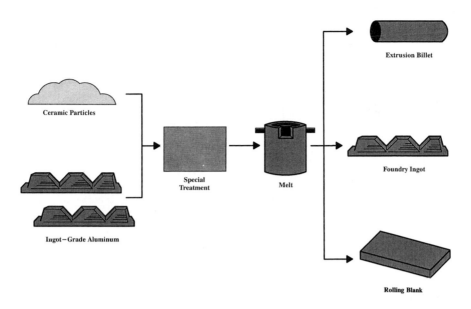

Fig. 6.2. Schematic of the Duralcan process.

ticulate reinforcement, called the *Duralcan* process, has become quite successful. Figure 6.2 shows a schematic of this process. Ceramic particles and ingot-grade aluminum are mixed and melted. The ceramic particles are given a proprietary treatment. The melt is stirred just above the liquidus temperature—generally between 600 and 700 °C. The melt is then converted into one of the following four forms: extrusion blank, foundry ingot, rolling bloom, or rolling ingot. The Duralcan process of making particulate composites by liquid metal casting involves the use of 8–12 μm particles. Too small particles, e.g., 2–3 μm, will result in a very large interface region and thus a very viscous melt. In foundry-grade MMCs, high Si aluminum alloys (e.g., A356) are used while in wrought MMC, Al-Mg-type alloys (e.g., 6061) are used. Alumina particles are typically used in foundry alloys, while silicon carbide particles are used in the wrought aluminum alloys.

For making continuous fiber reinforced MMCs, tows of fibers are passed through a liquid metal bath, where the individual fibers are wet by the molten metal, wiped off excess metal, and a composite wire is produced. Figure 6.3 shows a micrograph of one such wire made of SiC fibers in an aluminum matrix. Note the multifiber cross sections in the broken composite wire. A bundle of such wires can be consolidated by extrusion to make a composite. Another pressureless liquid-metal infiltration process of making MMCs is Lanxide's Primex™ process, which can be used with certain reactive metal alloys such Al-Mg to infiltrate ceramic preforms. For an Al-Mg alloy, the process takes place between 750 and 1000 °C in a nitrogen-rich

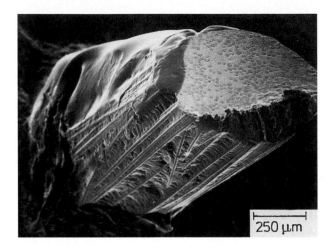

Fig. 6.3. Transverse section of a SiC/Al wire preform.

atmosphere (Aghajanian et al., 1989). Typical infiltration rates are less than 25 cm/h.

Squeeze casting, or *pressure infiltration*, involves forcing the liquid metal into a fibrous preform. Figure 6.4 shows two processes of making a fibrous preform. In one process, a suction is applied to a well-agitated mixture of whisker, binder, and water. This is followed by demolding and drying of the fiber preform (Fig. 6.4a). In the press forming process, an aqueous slurry of fibers is well agitated and poured into a mold, pressure is applied to squeeze the water out, and the preform is dried (Fig. 6.4b). A schematic of the squeeze casting process is shown in Figure 6.5a. Pressure is applied until the solidification is complete. By forcing the molten metal through small pores of fibrous preform, this method obviates the requirement of good wettability of the reinforcement by the molten metal. Figure 6.5b shows the microstructure of Saffil alumina fiber/alumina matrix composite made by squeeze casting. Composites fabricated with this method have minimal reaction between the reinforcement and molten metal because of short dwell time at high temperature and are free from common casting defects such as porosity and shrinkage cavities. Squeeze casting is really an old process, also called *liquid metal forging* in earlier versions. It was developed to obtain pore-free, fine-grained aluminum alloy components with superior properties than conventional permanent mold casting. In particular, the process has been used in the case of aluminum alloys that are difficult to cast by conventional methods, for example, silicon-free alloys used in diesel engine pistons where high-temperature strength is required. Inserts of nickel-containing cast iron, called Ni-resist, in the upper groove area of pistons have also been produced by the squeeze casting technique to provide wear resistance. Use of ceramic fiber reinforced metal matrix composites at locations of high wear

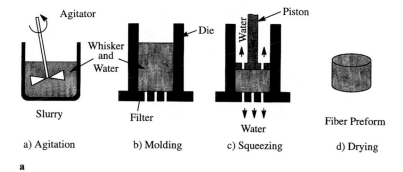

a) Agitation   b) Molding   c) Squeezing   d) Drying

a

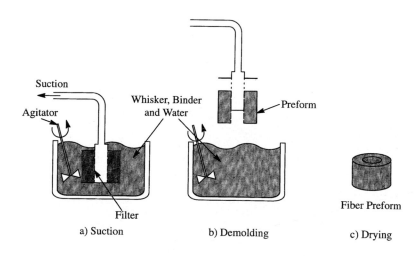

a) Suction   b) Demolding   c) Drying

Suction Forming of Preform

b

Fig. 6.4. a Press forming of a preform. **b** Suction forming of a preform.

and high thermal stress has resulted in a product much superior to the Ni-resist cast iron inserts.

The squeeze casting technique is shown in Figure 6.5a. A porous fiber preform (generally of discontinuous Saffil-type Al_2O_3 fibers) is inserted into the die. Molten metal (aluminum) is poured into the preheated die located on the bed of a hydraulic press. The applied pressure (70–100 MPa) makes the molten aluminum penetrate the fiber preform and bond the fibers. Infiltration of a fibrous preform by means of a pressurized inert gas is another variant of liquid metal infiltration technique. The process is conducted in the controlled environment of a pressure vessel and rather high fiber volume fractions; complex-shaped structures are obtainable (Mortensen et al., 1988;

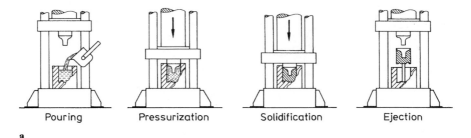

Pouring   Pressurization   Solidification   Ejection

a

b

Fig. 6.5. a Squeeze casting technique of composition fabrication. **b** The microstructure of Saffil alumina fiber/alumina matrix composite made by squeeze casting. (Courtesy of G. Eggeler.)

Cook and Warner, 1991). Although commonly, aluminum matrix composites are made by this technique, alumina fiber reinforced intermetallic matrix composites (e.g., TiAl, Ni_3Al, and Fe_3Al matrix materials) have been prepared by pressure casting (Nourbakhsh et al., 1990). The technique involves melting the matrix alloy in a crucible in a vacuum while the fibrous preform is heated separately. The molten matrix material (at about $100\,°C$ above the T_m) is poured onto the fibers, and argon gas is introduced simultaneously. Argon gas pressure forces the melt to infiltrate the preform. The melt generally contains additives to aid wetting the fibers.

Modeling of Infiltration

When a metal is heated above its melting point (or in the case of an alloy above the liquidus temperature), we get a low-viscosity liquid that has nice fluid flow characteristics so that it can be used to infiltrate a fibrous preform (Michaud, 1993; Cornie et al., 1986). Such an infiltration process can be described by phenomena such as capillarity and permeability. The phenomenon of permeability of a porous medium by a fluid is very important in a variety of fields. Specifically, in the case of MMCs, it is useful to be able to understand the permeability of a porous fibrous or particulate preform by the molten metal. Permeability of a porous medium, k, is commonly described by Darcy's law:

$$J = -\frac{k}{\eta} \nabla P$$

where J is volume current density (i.e., volume/area · time) of the fluid, η is the fluid viscosity, and P is the pressure that drives the fluid flow. It can be recognized that Darcy's law is an analog of Ohm's law for electrical conduction, i.e., hydraulic permeability is an analog of electrical conductivity. Note that the permeability, k, is a function of the properties of the porous medium, i.e., its microstructure; it does not depend on properties of the fluid.

Spray-forming of particulate MMCs involves the use of spray techniques that have been used for some time to produce monolithic alloys (Srivatsan and Lavernia, 1992). A spray gun is used to atomize a molten aluminum alloy matrix. Ceramic particles, such as silicon carbide, are injected into this stream. Usually, the ceramic particles are preheated to dry them. Figure 6.6 shows a schematic of this process. An optimum particle size is required for an efficient transfer. Whiskers, for example, are too fine to be transferred. The preform produced in this way is generally quite porous. The co-sprayed metal matrix composite is subjected to scalping, consolidation, and secondary finishing processes, thus, making it a wrought material. The process is totally computer-controlled and quite fast. It also should be noted that the process is essentially a liquid metallurgy process. One avoids the formation of deleterious reaction products because the time of flight is extremely short. Silicon carbide particles of an aspect ratio (length/diameter) between 3 and 4 and volume fractions up to 20% have been incorporated into aluminum alloys. A great advantage of the process is the flexibility that it affords in making different types of composites. For example, one can make in situ laminates using two sprayers or one can have selective reinforcement. This process, however, is quite expensive, mainly because of the high cost of the capital equipment.

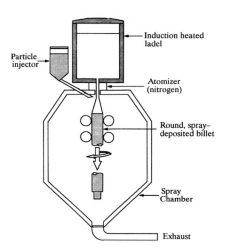

Fig. 6.6. Schematic of the spray forming process.

6.3.2   Solid State Processes

Many solid state techniques are available (Ghosh, 1993). We describe some of the important ones.

Diffusion bonding is a common solid state welding technique used to join similar or dissimilar metals. Interdiffusion of atoms from clean metal surfaces in contact at an elevated temperature leads to welding. There are many variants of the basic diffusion bonding process; however, all of them involve a step of simultaneous application of pressure and high temperature. Matrix alloy foil and fiber arrays, composite wire, or monolayer laminae are stacked in a predetermined order. Figure 6.7a shows a schematic of one such diffusion bonding process, also called the *foil-fiber-foil process*. Figure 6.7b shows the microstructure of SiC fiber/titanium matrix composite made by diffusion bonding. The starting materials in this case were made by sputter coating the SiC fibers with titanium. Filament winding was used to obtain panels, about 250 μm thick. Four such panels were stacked and hot pressed at 900 °C, under a pressure of 105 MPa for 3 hours. Vacuum hot pressing is a most important step in the diffusion bonding processes for metal matrix composites. The major advantages of this technique are: the ability to process a wide variety of matrix metals and control of fiber orientation and volume fraction. Among the disadvantages are: processing times of several hours, high processing temperatures and pressures that are expensive, and objects of limited size can be produced. Hot isostatic pressing (HIP), instead of uniaxial pressing, can also be used. In HIP, gas pressure against a can consolidates the composite piece contained inside the can. With HIP, it is relatively easy to apply high pressures at elevated temperatures over variable geometries.

Deformation processing of metal/metal composites involves mechanical processing (swaging, extrusion, drawing, or rolling) of a ductile two-phase

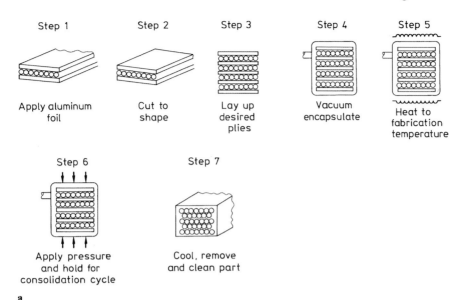

Step 1 — Apply aluminum foil

Step 2 — Cut to shape

Step 3 — Lay up desired plies

Step 4 — Vacuum encapsulate

Step 5 — Heat to fabrication temperature

Step 6 — Apply pressure and hold for consolidation cycle

Step 7 — Cool, remove and clean part

a

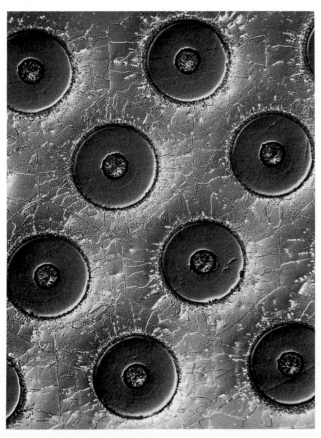

b

Fig. 6.7. a Schematic of diffusion bonding process. **b** The microstructure of SiC fiber/titanium matrix composite made by diffusion bonding. Each fiber is 142 μm in diameter. (Courtesy of J. Baughman.)

material. The two phases co-deform, causing the minor phase to elongate and become fibrous in nature within the matrix. These materials are sometimes referred to as *in situ composites*. The properties of a deformation-processed composite depend largely on the characteristics of the starting material, which is usually a billet of two-phase alloy that has been prepared by casting or powder metallurgy methods. Roll bonding is a common technique used to produce a laminated composite consisting of different metals in the sheet form (Chawla and Collares, 1978; Sherby et al., 1985). Such composites are called sheet laminated metal matrix composites. Other examples of deformation-processed metal matrix composites are the niobium-based conventional filamentary superconductors and the silver-based high-T_c superconductors (see Chapter 9).

Deposition techniques for metal matrix composite fabrication involve coating individual fibers in a tow with the matrix material needed to form the composite followed by diffusion bonding to form a consolidated composite plate or structural shape. The main disadvantage of using deposition techniques is that they are time consuming. However, there are several advantages:

- The degree of interfacial bonding is easily controllable; interfacial diffusion barriers and compliant coatings can be formed on the fiber prior to matrix deposition or graded interfaces can be formed.
- Thin, monolayer tapes can be produced by filament winding; these are easier to handle and mold into structural shapes than other precursor forms—unidirectional or angle-plied composites can be easily fabricated in this way.

Several deposition techniques are available: immersion plating, electroplating, spray deposition, chemical vapor deposition (CVD), and physical vapor deposition (PVD) (Partridge and Ward-Close, 1993). Dipping or immersion plating is similar to infiltration casting except that fiber tows are continuously passed through baths of molten metal, slurry, sol, or organometallic precursors. Electroplating produces a coating from a solution containing the ion of the desired material in the presence of an electric current. Fibers are wound on a mandrel, which serves as the cathode, and placed into the plating bath with an anode of the desired matrix material. The advantage of this method is that the temperatures involved are moderate and no damage is done to the fibers. Problems with electroplating involve void formation between fibers and between fiber layers, adhesion of the deposit to the fibers may be poor, and there are limited numbers of alloy matrices available for this processing. A spray deposition operation, typically, consists of winding fibers onto a foil-coated drum and spraying molten metal onto them to form a monotape. The source of molten metal may be powder or wire feedstock, which is melted in a flame, arc, or plasma torch. The advantages of spray deposition are easy control of fiber alignment and rapid solidification of the molten matrix. In a CVD process, a vaporized component decomposes or reacts with another vaporized chemical on the substrate to form a coating on that substrate. The processing is generally carried out at elevated temperatures.

6.3.3 In Situ Processes

In in situ techniques, one forms the reinforcement phase in situ. The composite material is produced in one step from an appropriate starting alloy, thus avoiding the difficulties inherent in combining the separate components as in a typical composite processing. Controlled unidirectional solidification of a eutectic alloy is a classic example of in situ processing. Unidirectional solidification of a eutectic alloy can result in one phase being distributed in the form of fibers or ribbon in the other. One can control the fineness of distribution of the reinforcement phase by simply controlling the solidification rate. The solidification rate in practice, however, is limited to a range of 1–5 cm/h because of the need to maintain a stable growth front. The stable growth front requires a high temperature gradient. Figure 6.8 shows scanning electron micrographs of transverse sections of in situ composites obtained at different solidification rates (Walter, 1982). The nickel alloy matrix has been etched away to reveal the TaC fibers. At low solidification rates, the TaC fibers are square in cross section, while at higher solidification rates, blades of TaC form. The number of fibers per square centimeter also increased with increasing solidification rate. Table 6.2 gives some important systems that have been investigated. A precast and homogenized rod of a eutectic composition is melted, in a vacuum or inert gas atmosphere. The rod is contained in a graphite crucible, which in turn is contained in a quartz tube. Heating is generally done by induction. The coil is moved up the quartz tube at a fixed rate. Thermal gradients can be increased by chilling the crucible just below the induction coil. Electron beam heating is also used, especially when reactive metals such as titanium are involved. The reader is referred to McLean (1983).

The XD^{TM} process of Lockheed Martin is another in situ process that uses an exothermic reaction between two components to produce a third component. Sometimes such processing techniques are referred to as the *self-propagating high-temperature synthesis* (SHS) process. Specifically, the XD^{TM} process produces ceramic particle reinforced metallic alloy. Generally, a master alloy containing rather high volume fraction of reinforcement is produced by the reaction synthesis. This is mixed and remelted with the base alloy to produce a desirable amount of particle reinforcement. Typical reinforcements are SiC, TiB$_2$, and so on in an aluminum, nickel, or intermetallic matrix (Christodolou et al., 1988).

6.4 Interfaces in Metal Matrix Composites

As we have repeatedly pointed out, the interface region in any composite is very important in determining the ultimate properties of the composite. In

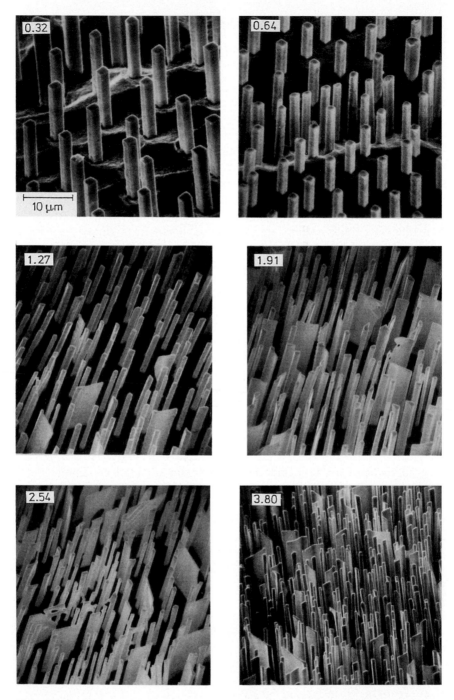

Fig. 6.8. Transverse sections of in situ composites obtained at different solidification rates indicated in left-hand top corners (cm/h). The nickel alloy matrix has been etched away to reveal the TaC fibers. [From Walter (1982), used with permission.]

Table 6.2. Some important in situ composite systems

System	Carbide (vol. %)	T_E[a] (°C)
Co-NbC	12	1365
Co-TiC	16	1360
Co-TaC	10	1402
Ni-HfC	15–28	1260
Ni-NbC	11	1330
Ni-TiC	7.5	1307

[a] T_E is the eutectic temperature.

this section, we give examples of interface microstructure in different metal matrix composite systems and discuss the implications of various interface characteristics on the resultant properties of the composite. Recall that we can use contact angle, θ as a measure of wettability; a small θ indicates good wetting. Most often, the ceramic reinforcement is rejected by the molten metal because of nonwettability or high contact angle. Sometimes the contact angle of a liquid drop on a solid substrate can be decreased by increasing the surface energy of the solid (γ_{SV}) or by decreasing the energy of the interface between the liquid and the solid (γ_{SL}). Thus, under certain circumstances, the wettability of a solid ceramic by a molten metal can be improved by making a small alloy addition to the matrix composition. An example of this is the addition of lithium to aluminum to improve the wettability in the alumina fiber/aluminum composite (Champion et al., 1978; Chawla, 1989). However, in addition to wettability, there are other important factors such as chemical, mechanical, thermal, and structural ones that affect the nature of bonding between reinforcement and matrix. As it happens, these factors frequently overlap, and it may not always be possible to isolate these effects.

6.4.1 Major Discontinuities at Interfaces in MMCs

As we said earlier, at interface a variety of discontinuities can occur. The important parameters that can show discontinuities in MMCs at a ceramic reinforcement/metal matrix interface are as follows:

- *Bonding*: A ceramic reinforcement will have an ionic or a mixed ionic/covalent bonding while the metal matrix will have a metallic bonding.
- *Crystallographic*: The crystal structure and the lattice parameter of the matrix and the reinforcement will be different.
- *Moduli*: In general, the elastic moduli the matrix and the reinforcement will be different.
- *Chemical potential*: The matrix and the reinforcement will not be in thermodynamic equilibrium at the interface, i.e., there will be a driving force for a chemical reaction.

- *Coefficient of Thermal Expansion (CTE)*: The matrix and the reinforcement will, in general, have different CTEs.

6.4.2   Interfacial Bonding in Metal Matrix Composites

Here we provide a summary of salient features of the interfacial region in some of the most important metal matrix composites.

Crystallographic Nature

In crystallographic terms, ceramic/metal interfaces in composites are, generally, incoherent, high-energy interfaces. Accordingly, they can act as very efficient vacancy sinks, provide rapid diffusion paths, segregation sites, sites of heterogeneous precipitation, and sites for precipitate free zones. Among the possible exceptions to this are the eutectic composites (Cline et al., 1971) and the XD^{TM}-type particulate composites (Mitra et al., 1993). In eutectic composites, X-ray and electron diffraction studies show preferred crystallographic growth directions, preferred orientation relationships between the phases, and low-index habit planes. Boundaries between in situ components are usually semicoherent, and the lattice mismatch across the interfaces can be accommodated by interface dislocations. Figure 6.9 shows a network of dislocations between the NiAl matrix and a chromium rod in a unidirectionally solidified NiAl-Cr eutectic (Cline et al., 1971).

Mechanical Bonding

Some bonding must exist between the ceramic reinforcement and the metal matrix for load transfer from matrix to fiber to occur. Two main categories of bonding are mechanical and chemical. Mechanical keying between two

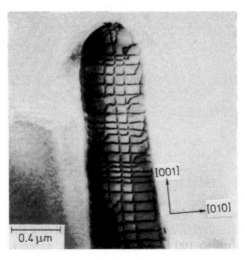

Fig. 6.9. A network of dislocations between the NiAl matrix and a chromium rod in a unidirectionally solidified NiAl-Cr eutectic. (Reprinted with permission from *Acta Met.*, **19**, H.E. Cline et al., © 1971, Pergamon Press Ltd.)

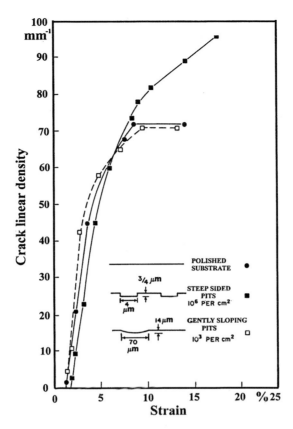

Fig. 6.10. The linear density of cracks inalumina as a function of strain in an alumina/aluminum composite for different degrees of interface roughness.

surfaces can lead to bonding. Hill et al. (1969) confirmed this experimentally for tungsten filaments in an aluminum matrix while Chawla and Metzger (1978) observed mechanical gripping effects at Al_2O_3/Al interfaces. The results of Chawla and Metzger (1978) are shown in Figure 6.10 in the form of linear density of cracks in alumina as a function of strain in an alumina/ aluminum composite for different degrees of interface roughness. The main message of this figure is that the crack density continues to increase to larger strain values in the case of a rough interface (deeply etched pits) vis à vis a smooth or not very rough interface, i.e., the rougher the interface, the stronger the mechanical bonding.

We can make an estimate of the radial stress, σ_r, at the fiber/matrix interface due to roughness induced gripping by using the following expression (Kerans and Parthasarathy, 1991):

$$\sigma_r = \frac{-E_m E_f}{E_f(1 + v_m) + E_m(1 - v_f)} \left[\frac{A}{r}\right]$$

where E is the Young's modulus, v is the Poisson's ratio, A is the amplitude of roughness, r is the radius of the fiber, and the subscripts m and f indicate matrix and fiber, respectively. For a given composite, the compressive radial stress increases with the roughness amplitude and decreases with the fiber radius. An important example of such an MMC, i.e., nonreacting components with purely a mechanical bond at the interface, is the filamentary superconducting composite consisting of niobium-titanium alloy filaments in a copper matrix.

Chemical Bonding

Ceramic/metal interfaces are generally formed at high temperatures. Diffusion and chemical reaction kinetics are faster at elevated temperatures. One needs to have knowledge of the chemical reaction products and, if possible, their properties. Molten iron, nickel, titanium, low alloy steels, austenitic and ferritic stainless steels, and nickel-based superalloys react with silicon-containing ceramics to form eutectics, with the reaction products being mainly metal silicides and carbide. It is thus imperative to understand the thermodynamics and kinetics of reactions in order to control processing and obtain optimum properties. We provide some examples.

Most metal matrix composite systems are nonequilibrium systems in the thermodynamic sense; that is, there exists a chemical potential gradient across the fiber/matrix interface. This means that given favorable kinetic conditions (which in practice means a high enough temperature or long enough time), diffusion and/or chemical reactions will occur between the components (see Fig. 4.4 in Chapter 4). Two common morphologies of reaction products at an interface in common metal matrix composites are: (a) a reaction layer that covers the ceramic reinforcement more or less uniformly and (b) a discrete precipitation, particle or needle shaped, around the reinforcement.

Type (a) reaction is controlled by diffusion of elements in the reaction layer. Typically, such a diffusional growth of the reaction layer scales as $x^2 \simeq Dt$, where x is the layer thickness, t is the time, and D is the diffusion coefficient. Examples of systems showing such interfacial reactions include B/Al and SiC/Ti. Fujiwara et al. (1995) observed such a parabolic growth of the reaction zone in SCS-6 silicon carbide fiber/Ti-4.5Al-3V-2Fe-2 Mo (wt. %) composites at three different temperatures. Figure 6.11 shows the reaction zone thickness squared (x^2) as a function of reaction time for three different boron fiber/titanium matrix composites. Note that the B_4C coating is more effective for the B/Ti system.

Silicon carbide fiber reinforced titanium matrix composites are attractive for some aerospace applications. Titanium and its alloys are very reactive in the liquid state; therefore, only solid state processing techniques such as diffusion bonding are used to make these composites (Partridge and Ward-Close, 1993). In particular, a titanium alloy matrix containing SCS-6-type

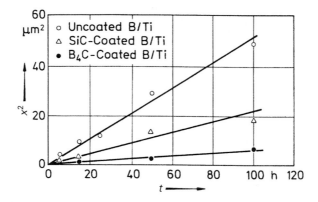

Fig. 6.11. Reaction zone thickness squared (x^2) as a function of reaction time for three different boron fibers in a titanium matrix. A B_4C carbide coating on the boron is most effective for the B/Ti system. [From Naslain et al. (1976), used with permission.]

silicon carbide fiber can have a very complex interfacial chemistry and microstructure. A schematic of the interface region in these composites is shown in Figure 6.12 (Gabryel and McLeod, 1991).

Type (b) reaction is controlled by nucleation process, and discrete precipitation occurs at the reinforcement/matrix interface. Examples include alumina/magnesium, carbon/aluminum, and alumina-zirconia/aluminum. Figure 6.13 shows an example, a dark field TEM micrograph, of the reaction

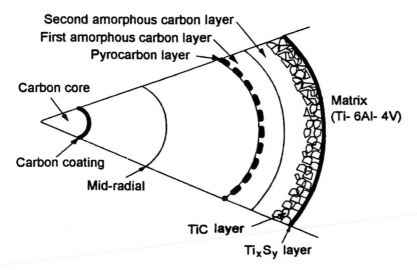

Fig. 6.12. Schematic of the interface region in silicon carbide fiber reinforced titanium matrix composites. [Adapted from Gabryel and McLeod (1991).]

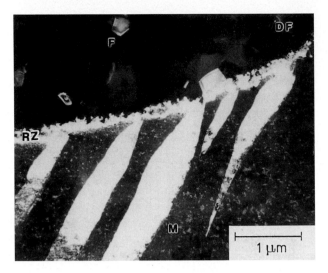

Fig. 6.13. A controlled amount of interfacial reaction at the Al_2O_3 fiber/Mg alloy matrix interface. *M*, *F*, and *RZ* denote matrix, fiber, and reaction zone. Dark field (DF) transmission electron micrograph.

zone between alumina fiber and magnesium matrix. Also to be seen in this figure are the deformation twins in the matrix, the result of thermal stresses on cooling during liquid metal infiltration. Aluminum oxide can react with magnesium present in an aluminum alloy (Jones et al., 1993):

$$3\,Mg + Al_2O_3 \leftrightarrow 3\,MgO + 2\,Al$$
$$3\,Mg + 4\,Al_2O_3 \leftrightarrow 3\,MgOAl_2O_3 + 2\,Al$$

At high levels of Mg and low temperatures, MgO is expected to form, while the spinel forms at low levels of magnesium (Pfeifer et al., 1990).

A good example of obtaining a processing window by exploiting the kinetics of interfacial reaction between the fiber and matrix can be seen in the work of Isaacs et al. (1991). These authors examined the interface structure in an aluminum matrix reinforced with an (alumina + zirconia) fiber. The composite made by pressure infiltration of the fibrous preform by liquid aluminum at 973 K (700 °C), with a dwell time of 13 minutes, showed faceted $ZrAl_3$ platelets growing from the fiber into a matrix. However, they could suppress the kinetics of interfacial reaction by minimizing the high-temperature exposure. No interfacial reaction product was observed when they processed the composite with the initial fibrous preform temperature below the melting point of aluminum and the solidification time less than 1 minute.

Carbon fiber reacts with molten aluminum to form aluminum carbide, which is a very brittle compound and highly susceptible to corrosion in humid environments. Thus, it becomes imperative to use a barrier coating on carbon fibers before bringing them in contact with the molten aluminum. The carbon fibers are coated with a co-deposition of Ti + B (presumably giving TiB$_2$). The starting materials for the coating process are: TiCl$_4$ (g), BCl$_3$ (g), and Zn (v); the possible reaction products are: TiB$_2$, TiCl$_2$, TiCl$_3$, and ZnCl$_3$. Any residual chloride in the coating is highly undesirable from a corrosion-resistance point of view.

In the case of carbon fibers in aluminum, poor wettability is a major problem. The wettability seems to improve somewhat with temperature increasing above 1000 °C (Manning and Gurganus, 1969; Rhee, 1970), but it turns out that above 500 °C a deleterious reaction occurs between the carbon fibers and the aluminum matrix, whereby Al$_4$C$_3$, a very brittle intermetallic compound, is known to form (Baker and Shipman, 1972). Fiber surface coatings have also been tried. Such coatings do allow wetting by low-melting-point metals, but these coatings are also unstable in molten metals. Fiber degradation leading to reduced composite strength generally results. In the case of carbon fiber/aluminum composites, co-deposition of titanium and boron onto carbon fibers before incorporating them into an aluminum matrix came to be established as a commercial method for carbon fiber surface treatment. Figure 6.14 shows a schematic of this process (Meyerer et al., 1978). Multiple yarn creels result in increased capacity over single yarn creels. The polyvinyl alcohol (PVA) size on carbon fibers (meant for polymeric matrices; see Chap. 9) is removed in a furnace, designated as a PVA furnace in Fig. 6.14. The first CVD furnace in Fig. 6.14 is a precoat fur-

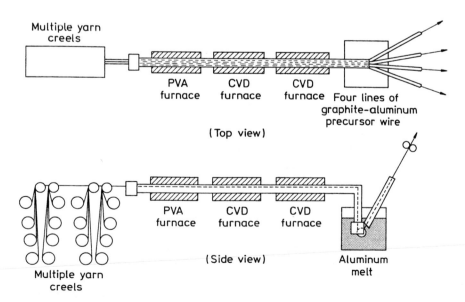

Fig. 6.14. Schematic of the Ti/B co-deposition process. [From Meyerer et al. (1978), used with permission.]

nace for cleaning and activating the carbon yarn surface. The Ti/B coating is deposited in the second CVD furnace, followed by drawing through a molten aluminum bath.

Warren and Andersson (1984) have reviewed the thermodynamics of chemical equilibria between SiC and some common metals. They divide these systems into *reactive* and *stable* types. In reactive systems, SiC reacts with the metal to form silicides and/or carbides and carbon. No two-phase field exists in the ternary phase diagram showing SiC and the metal in equilibrium. For example, with nickel and titanium, the following reactions are possible:

$$SiC + Ni \longrightarrow Ni_xSi_y + C$$

$$SiC + Ti \longrightarrow Ti_xSi_y + TiC \quad \text{(low SiC fractions)}$$

Such reactions are thermodynamically possible between SiC and nickel or titanium, but in practice, reaction kinetics determine the usefulness of the composite. In stable systems, SiC and a metallic matrix alloy can coexist thermodynamically, that is, a two-phase field exists. Examples are SiC fibers in alloys of aluminum, gold, silver, copper, magnesium, lead, tin, and zinc. This does not imply that SiC will not be attacked. In fact, a fraction of SiC fibers in contact with molten aluminum can dissolve and react to give Al_4C_3:

$$SiC + Al \longrightarrow SiC + (Si) + Al_4C_3$$

This reaction can happen because the section SiC-Al lies in a three-phase field. Such reactions can be avoided by prior alloying of aluminum with silicon.

The interface product(s) formed because of a reaction will generally have characteristics different from those of either component. It should be pointed out, however, that at times, some controlled amount of reaction at the interface, such as that shown in Fig. 6.13, may be desirable for obtaining strong bonding between the fiber and the matrix, but, too thick an interaction zone will adversely affect the composite properties.

Silicon carbide particle reinforced aluminum composites have been investigated extensively. An important processing technique for these MMCs involves liquid metal infiltration of a particulate preform. In a silicon-free aluminum alloy matrix, silicon carbide and molten aluminum can react as follows:

$$4\,Al(l) + 3\,SiC(s) \leftrightarrow Al_4C_3(s) + 3\,Si(s)$$

The forward reaction will add silicon to the matrix. As the silicon level increases in the molten matrix, the melting point of the alloy decreases with time. The reaction can be made to go leftward by using high silicon alloys. This, of course, restricts the choice of Al alloys for liquid route processing. Table 6.3 gives a summary of interfacial reactions in some important MMCs.

Table 6.3. Interfacial reaction products in some important MMCs

Reinforcement	Matrix	Reaction Product(s)
SiC	Ti alloy	TiC, Ti_5Si_3
	Al alloy	Al_4C_3
Al_2O_3	Mg alloy	MgO, $MgAl_2O_4$ (spinel)
C	Al alloy	Al_4C_3
B	Al alloy	AlB_2
$Al_2O_3 + ZrO_z$	Al alloy	$ZrAl_3$
W	Cu	None
C	Cu	None
Al_2O_3	Al	None

In general, ceramic reinforcements (fibers, whiskers, or particles) have a coefficient of thermal expansion greater than that of most metallic matrices. This means that when the composite is subjected to a temperature change, thermal stresses will be generated in both of the components. This observation is true for all composites—polymer, metal, and ceramic matrix composites. What is unique of metal matrix composites is the ability of a metal matrix to undergo plastic deformation in response to the thermal stresses generated and thus alleviate them. Chawla and Metzger (1972), working with a single crystal copper matrix containing large diameter tungsten fibers, showed the importance of thermal stresses in MMCs. Specifically, they employed a dislocation etch-pitting technique to delineate dislocations in single-crystal copper matrix and showed that near the fiber the dislocation density was much higher in the matrix than the dislocation density far away from the fiber. The situation in the as-cast composite can be depicted as shown schematically in Figure 6.15, where a primary plane section of the composite is shown having a hard zone (high dislocation density) around each fiber and a soft zone (low dislocation density) away from the fiber (Chawla, 1975). The enhanced dislocation density in the copper matrix near the fiber arises because of the plastic deformation in response to the thermal stresses generated by the thermal mismatch between the fiber and the matrix. The intensity of the gradient in dislocation density will depend on the inter-fiber spacing. The dislocation density gradient will decrease with a decrease in the interfiber spacing. The existence of a plastically deformed zone containing high dislocation density in the metallic matrix in the vicinity of the reinforcement has since been confirmed by transmission electron microscopy by a number of researchers, both in fibrous and particulate metal matrix composites (Arsenault and Fisher, 1983; Rack, 1987; Christman and Suresh, 1988). Such high dislocation density in the matrix can alter the precipitation behavior, and, consequently, the aging behavior in MMCs in those composites that have a precipitation hardenable alloy matrix.

We mentioned the roughness induced radial compression stress at the

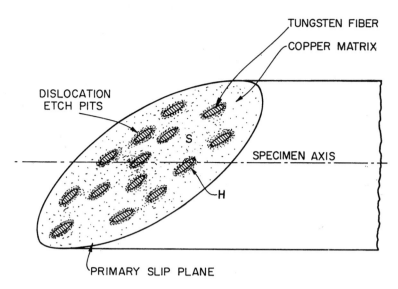

Fig. 6.15. A primary plane section of the composite is shown as having a hard zone around each fiber and a soft zone away from the fiber.

fiber/matrix interface in Section 4.2. As discussed earlier, the thermal mismatch between the components can lead to thermal residual stresses. Specifically, a radial stress component will be introduced which can be positive or negative. To obtain the net radial stress at the fiber/matrix interface, one should add the stresses algebraically from the two sources, viz., the contribution due to thermal mismatch between the reinforcement and the matrix and the stress arising due to the roughness induced gripping, as mentioned previously. A combined expression for the radial stress from these two sources can be written as (Kerans and Parthasarathy, 1991)

$$\sigma_r = \frac{-qE_mE_f}{E_f(1+v_m) + E_m(1-v_f)} \left[\Delta\alpha\Delta T + \frac{A}{r} \right]$$

where q is an adjustable parameter (equal to 1 for an infinite matrix), E is the Young's modulus, v is the Poisson's ratio, ΔT is the temperature change, $\Delta\alpha$ is the thermal mismatch between the fiber and the matrix $= (\alpha_m - \alpha_f)$, A is the amplitude of roughness, r is the radius of the fiber, and the subscripts m and f refer to matrix and fiber, respectively. Thus, in the absence of chemical bonding, one can control the degree of interfacial bonding by controlling the degree of interfacial roughness, the thermal mismatch between the matrix and the reinforcement, and the amplitude of temperature change.

Metal matrix composites with reinforcement in the form of a short fiber, whisker, or particle (also called *discontinuously reinforced MMCs*) have become important because of their low cost. In particular, aluminum matrix composites have been commercialized. Such MMCs can be subjected to conventional metalworking processes (e.g., rolling, forging, extrusion, machining, and swaging). Figure 6.16a shows a perspective montage (SEM) of a rolled plate of SiC (20 v/o)/2124 Al. The rolling direction is perpendicular to the original extrusion direction. The very small particles in Fig. 6.16a are the precipitates in the matrix after aging, while the large irregular particles are probably (Fe, Mn) Al$_6$. The distribution of SiC whiskers in the aluminum matrix, as seen in a transmission electron microscope, is shown in Fig. 6.16b, while a closeup of the whisker/matrix interface region is shown in Fig. 6.16c. Note the waviness of the interface. Fu et al. (1986) examined the interface chemistry and crystallography of C/Al and SiC$_w$/Al MMCs in the TEM. They observed an oxide at some of the SiC/Al interfaces. Figures 6.17a and

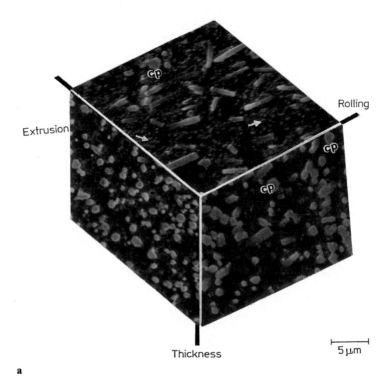

a

Fig. 6.16. a A perspective montage of a SiC/Al composite. The smallest particles (indicated by arrows on the top face) are precipitates from the matrix as a result of aging. Silicon carbide whiskers are the intermediate particles, both long and narrow and equiaxial ones having a diameter similar to the width of the long, narrow ones. The large irregular particles are (Fe, Mn) Al$_6$-type constituent particles (some of these are marked cp). The liquid-phase hot-pressed billet was first extruded in the extrusion direction, forged along the thickness direction, and finally rolled along the rolling direction. (Courtesy of D.R. Williams and M.E. Fine.) **b** Distribution of SiC$_w$ in an aluminum matrix (TEM). **c** A higher magnification of the whisker/matrix interface (TEM). (Courtesy of J.G. Greggi and P.K. Liaw.)

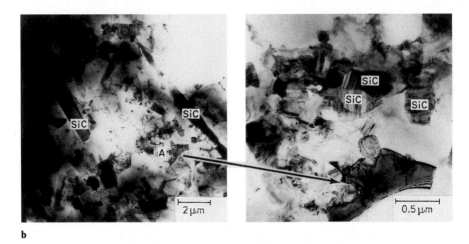

b

c

Fig. 6.16 (*continued*)

b show the TEM bright and dark field micrographs of the interface region of 20 v/o SiC$_w$/Al. The crystalline γ-Al$_2$O$_3$ phase is about 30 nm. It was not uniformly present, however, at every interface. In C/Al composites, both fine-grained γ-Al$_2$O$_3$ and coarse-grained Al$_4$C$_3$ were found at the interfaces. On heat treating, some of the Al$_4$C$_3$ grew into and along the porous sites at the carbon fiber surface.

6.5 Properties

We describe some of the important mechanical and physical properties of metal matrix composites in this section.

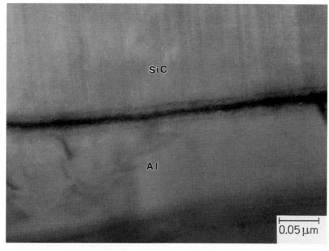

a

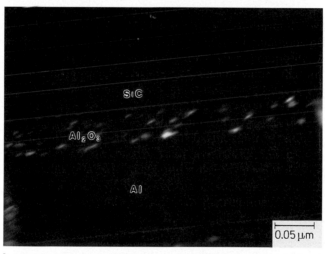

b

Fig. 6.17. Interface in SiC$_w$/Al composite: **a** bright field TEM, **b** dark field TEM showing the presence of Al$_2$O$_3$ at the interface. [From Fu et al. (1986), ©ASTM, reprinted with permission.]

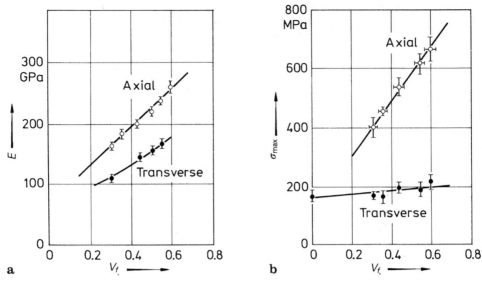

Fig. 6.18. Properties of Al_2O_3/Al-Li composites as a function of fiber fraction (V_f): **a** axial and transverse Young's modulus versus fiber volume fraction, **b** axial and transverse ultimate tensile strength versus fiber volume fraction. [From Champion et al. (1978), used with permission.]

Modulus

Unidirectionally reinforced continuous fiber reinforced metal matrix composites show a linear increase in the longitudinal Young's modulus of the composite as a function of the fiber volume fraction. Figure 6.18 shows an example of modulus and strength increase as a function of fiber volume fraction for alumina fiber reinforced aluminum-lithium alloy matrix (Champion et al., 1978). The increase in the longitudinal Young's modulus is in agreement with the rule-of-mixtures value while the modulus increase in a direction transverse to the fibers is very low. Particle reinforcement also results in an increase in the modulus of the composite, the increase, however, is much less than that predicted by the rule of mixtures. This is understandable inasmuch as the rule of mixtures is valid only for continuous fiber reinforcement. Figure 6.19 shows the increase in Young's modulus of an MMC as a function of reinforcement volume fraction for different forms of reinforcement viz., continuous fiber, whisker, or particle. Note the loss of reinforcement efficiency as one goes from continuous fiber to particle. Metal matrix particulate composites, such as SiC particle reinforced aluminum can offer a 50 to 100% increase in modulus over that of unreinforced aluminum, i.e., a modulus equivalent to that of titanium but density about 33% less. Also, unlike the fiber reinforced composites, the stiffness enhancement in particulate composites is reasonably isotropic.

Fig. 6.19. Increase in Young's modulus of an MMC as a function of reinforcement volume fraction for continuous fiber, whisker, or particle reinforcement.

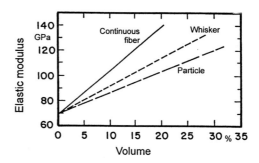

Strength

Prediction of strength of an MMC is more complicated than the prediction of modulus. Consider an aligned fiber reinforced metal matrix composite under a load P_c in the direction of the fibers. This load is distributed between the fiber and the matrix. Thus,

$$P_c = P_m V_m + P_f V_f$$

where P_m and P_f are the loads on the matrix and the fiber, respectively. This equation can be converted to the following rule-of-mixtures relationship under conditions of isostrain (i.e., the strain in the fiber, matrix, and composite is the same):

$$\sigma_c = \sigma_f V_f + \sigma_m V_m$$

where σ is the stress, V is the volume fraction, and the subscripts c, f, and m denote the composite, fiber, and matrix, respectively. This equation, commonly referred to as the *Rule of Mixtures*, says that the strength of the composite is a volume-weighted average of the strengths of the fiber and the matrix. On the face of it, there is nothing wrong with this statement. The problem is that one needs the in situ values of σ_f and σ_m. If the fiber remains essentially elastic up to the point of fracture, then σ_f, the fiber strength in the composite, is the same as that determined in an isolated test of the fiber. This is particularly ture of ceramic fibers. The same, however, cannot be said for the metallic matrix. The matrix strength in the composite (the in situ strength) will not be the same as that determined from a test of an unreinforced matrix sample in isolation. This is because the metal matrix can suffer several microstructural alterations during processing and, consequently, changes in its mechanical properties. In view of the fact that, in general, ceramic reinforcements have a coefficient of thermal expansion greater than that of most metallic matrices, thermal stresses will be generated in the two components, the fiber, and the matrix. A series of events can take place in response to the thermal stresses (Chawla and Metzger, 1972; Chawla, 1973a, 1973b, 1974):

Table 6.4. Representative properties of some MMCs

Composite and direction	Fiber Volume Fraction (%)	Density (g cm^{-3})	σ_{max} (Mpa)	E (GPa)
B/Al				
0°	50	2.65	1500	210
90°	50	2.65	140	150
SiC/Al				
0°	50	2.84	250	310
90°	50	2.84	105	—
SiC/Ti-6 Al-4V				
0°	35	3.86	1750	300
90°	35	3.86	410	—
Al$_2$O$_3$/Al-Li				
0°	60	3.45	690	262
90°	60	3.45	172–207	152
C/Mg alloy (Thornel 50)	38	1.8	510	—
C/Al	30	2.45	690	160

1. Plastic deformation of the ductile metal matrix (slip, twinning, cavitation, grain boundary sliding and/or migration).
2. Cracking and failure of the brittle fiber.
3. An adverse reaction at the interface.
4. Failure of the fiber/matrix interface.

We will discuss the subject of thermal stresses in Chapter 10. A discussion of the effect of thermal stresses on the strength properties of composites will be presented in Chapter 12. A summary of typical properties of long or continuous fiber reinforced MMCs is given in Table 6.4. Next we discuss the subject of strengthening in particulate MMCs.

There is some controversy in explaining the strengthening mechanisms operating in particle reinforced composites. Arsenault and Shi (1986) and Shi and Arsenault (1991, 1993) analyzed various contributing factors in particulate MMCs, in the absence of a shear lag strengthening mechanism, which would be operational in the case of long fibrous reinforcement. They enumerated the following strengthening mechanisms in particulate MMCs:

• Orowan strengthening, which is given by Gb/l, where G is the shear modulus of the matrix, b is the Burgers vector of the matrix, and l is the particle spacing.
• Grain and substructure strengthening, following a Hall-Petch-type relationship, i.e., strength varying as $d^{-1/2}$, where d is the grain or subgrain size in the matrix. The Hall-Petch slope k for Al is in the range of 0.1 to 0.15, i.e., it is very low, while for steels, k is very high. Thus, a grain diameter $d < 10\,\mu m$ will give significant strengthening, while $d < 1\,\mu m$ can give very good strengthening.

- Quench strengthening with thermal strain in the matrix given by $e_m = \Delta\alpha\Delta T$. Theoretically, the dislocation density resulting from the coefficient of thermal expansion mismatch is

$$\rho_{\text{CTE}} = (AeV_p)/b(1 - V_p)d$$

where A is a geometric constant, e is the thermal misfit strain, A is the particle area, b is the Burgers vector, and V_p is the particle volume fraction.

The corresponding stress is given by

$$\sigma_q = \alpha Gb(\rho_{\text{CTE}})^{1/2}$$

where α is a constant. This contribution to strength can be significant.
- Work hardening of the matrix. Particles affect the matrix work hardening rate.

Thus, for total strength of the composite from all these contributions we can write

$$\sigma_c = \sigma_o + \sigma_{gb} + \sigma_q + \sigma_{WH} + \sigma_m$$

where σ_m is the matrix strength.

In powder metallurgy processed composites, fine oxide from the SiC particle's surface can enter the matrix. For a 20% V_p 10-μm diameter, the estimated σ_o is about 10 MPa. The contribution of grain boundary strengthening is given by

$$\sigma_{gb} = kd^{-1/2}$$

Grain boundary strengthening can be high in spray cast and powder metallurgy processed composites.

An analysis (Nardone and Prewo, 1986) that takes into account tensile loading at the particle ends gives the following expression for the yield strength of a particulate composite

$$\sigma_{yc} = \sigma_{ym}[1 + (L + t)/4L]V_p + \sigma_{ym}(1 - V_p)$$

where σ_{ym} is the yield stress of the unreinforced matrix, V_p is the particle volume fraction, L is the length of the particle perpendicular to the applied load and t is the length of the particle parallel to the loading direction (see also Chapter 10).

What is important to realize is that the matrix in fiber composites is not merely a kind of glue or cement to hold the fibers together [Chawla, 1985]. The characteristics of the matrix, as modified by the introduction of fibers, must be evaluated and exploited to obtain an optimum set of properties of the composite. The nature of the fibers, fiber diameter, and distribution, as well as conventional solidification parameters, influence the final matrix microstructure. Porosity is one of the major defects in cast MMCs, owing to the shrinkage of the metallic matrix during solidification. At high fiber volume fractions, the flow of interdendritic liquid becomes difficult, and large-scale movement of semisolid metal may not be possible. More importantly, the microstructure of the metallic matrix in a fiber composite can differ significantly from that of the unreinforced metal. Mortensen et al. (1986) have shown that the presence of fibers influences the solidification of the matrix alloy. Figure 6.20a shows a cross section of an SCS-2 silicon carbide fiber/Al–4.5% Cu matrix. Note the normal dendritic cast structure in the unreinforced region, whereas in the reinforced region the dendritic morphology is controlled by the fiber distribution. Figure 6.20b shows the same system at a higher magnification. The second phase (θ) appears preferentially at the fiber/matrix interface or in the narrow interfiber spaces Kohyama et al. [1985] observed that shrinkage cavities present in the matrix in SiC/Al composites were the predominant crack-initiation sites. They also observed a wavy interface structure between SiC and the aluminum matrix, indicating some mechanical bonding in addition to any chemical and physical bonding. In the system SiC/Mg (AZ19C), they observed a magnesium-rich interfacial layer that acted as the fracture-initiating site.

Superior high-temperature properties of MMCs have been demonstrated in a number of systems. For example, silicon carbide whiskers (SiC_w) improve the high-temperature properties of aluminum considerably. Figure 6.21 compares the elastic modulus, yield stress, and ultimate tensile strength of SiC_w (21% V_f)/2024 Al composites to those of the unreinforced aluminum alloy (Phillips, 1978). Figure 6.22 shows the fracture surface after tensile testing of a Nicalon fiber/aluminum composite at two temperatures. Note the more or less planar fracture with no fiber pullout at room temperature (Fig. 6.22a). At 500 °C, a loss of adhesion between the fiber and the matrix occurs with the accompanying fiber/matrix separation and fiber pullout. The fiber pullout has left a hole in the center; see Figure 6.22b.

High-modulus carbon fiber/aluminum composites combine a very high stiffness with a very low thermal expansion due mainly to the almost zero longitudinal expansion coefficient of carbon fibers. Carbon/aluminum composites, however, are susceptible to galvanic corrosion between carbon and aluminum. Carbon is cathodic in nature, while aluminum is anodic. Thus, galvanic corrosion can be a serious problem in joining aluminum to a carbon fiber composite. A common solution is to have an insulating layer of glass between the aluminum and the carbon composite. Any welding or joining involving localized heating could also be a potential source of problems because aluminum carbide may form as a result of overheating, which will be detrimental to the mechanical and corrosion properties.

We mentioned earlier the class of MMCs called in situ composites. An important example of such a system is a composite made by directional solidification of a eutectic alloy. The strength, σ, of such an in situ metal matrix composite is given by a relationship similar to the Hall-Petch relationship used for grain boundary strengthening of metals (Meyers and Chawla, 1984):

$$\sigma = \sigma_o + k\lambda^{-1/2}$$

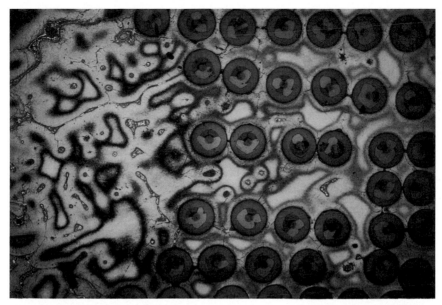

a

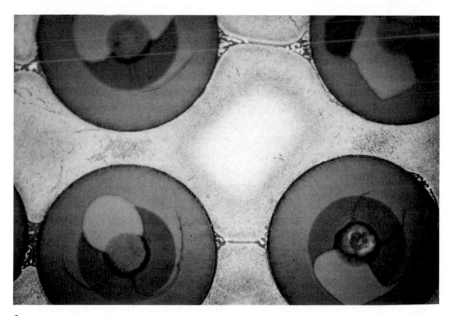

b

Fig. 6.20. Transverse section of an SCS-2 SiC fiber in an Al–4.5% Cu matrix. **a** Note the difference in the dendritic structure of the unreinforced and the fiber-rich regions of the matrix. **b** the second phase appears preferentially at the fiber/matrix interface or in the narrow interfiber region. The fiber diameter is 142 μm in both a and b. (Reprinted with permission from *Journal of Metals*, Vol. 37, No. 6, pp. 45, 47, and 48, a publication of the Metallurgical Society, Warrendale, Pennsylvania.)

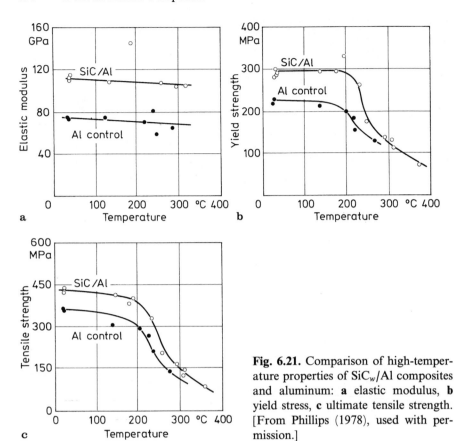

Fig. 6.21. Comparison of high-temperature properties of SiC$_w$/Al composites and aluminum: **a** elastic modulus, **b** yield stress, **c** ultimate tensile strength. [From Phillips (1978), used with permission.]

where σ_o is a friction stress term, k is a material constant, and λ is the spacing between rods or lamellae. It turns out that one can vary the interfiber spacing, λ, rather easily by controlling the solidification rate, R, because

$$\lambda^2 R = \text{constant}$$

van Suchtelen (1972) has classified eutectic or in situ composites into two broad categories from an electronic property viewpoint:

1. *Combination-type properties.* This can be further subdivided into (a) sum type and (b) product type. In *sum type*, properties of the constituent phases contribute proportionately to their amount. Examples are heat conduction, density, and elastic modulus. In the *product type*, the physical output of one phase serves as input for the other phase: for example, conversion of a magnetic signal into an electrical signal in a eutectic composite with one phase magnetostrictive and the other piezoelectric.

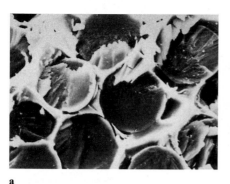 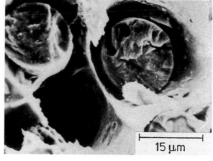

15 µm

a   b

Fig. 6.22. Tensile fracture in Nicalon fiber/aluminum: **a** at room temperature showing a planar fracture, **b** at 500 °C showing fiber/matrix separation and fiber pullout leaving a hole. (Courtesy of K. Okamura.)

2. *Morphology-dependent properties.* In this case, the properties depend on the periodicity and anisotropy of the microstructure, the shape and size of the phases, and the amount of interface area between the phases. A good example is InSb-NiSb, which is a quasibinary system with a eutectic at 1.8% NiSb. Unidirectional solidification of a eutectic melt at a growth rate of 2 cm h^{-1} results in an aligned composite consisting of an InSb semiconducting matrix containing long hexagonal fibers of the NiSb phase. The magnetoresistance of the InSb-NiSb composite becomes extremely large if the directions of the metallic fibers, the electric current, and the magnetic field are mutually perpendicular. This characteristic has been exploited in making contactless control devices; a practical example is described in Section 6.6.

Toughness

Toughness can be regarded as a measure of energy absorbed in the process of fracture, or more specifically as the resistance to crack propagation, K_{Ic}. The toughness of MMCs depends on the following factors:

- the matrix alloy composition and microstructure
- the reinforcement type, size, and orientation
- processing, insofar as it affects microstructural variables (e.g., distribution of reinforcement, porosity, segregation, and so on)

For a given V_f, the larger the diameter of the fiber, the tougher the composite. This is because the larger the fiber diameter, for a given fiber volume fraction, the larger the amount of tough, metallic matrix in the interfiber region that can undergo plastic deformation and thus contribute to the toughness.

Unidirectional fiber reinforcement can lead to easy crack initiation and easy propagation vis à vis the unreinforced alloy matrix. Braiding of fibers can make the crack propagation toughness increase tremendously due to extensive matrix deformation, fiber bundle debonding, and pullout (Majidi and Chou, 1987). The general range of K_{Ic} values for particle-reinforced aluminum-type MMCs is between 15 and 30 MPa $m^{1/2}$, while short fiber- or whisker-reinforced MMCs have $\simeq 5$–10 MPa $m^{1/2}$.

Among the explanations for low toughness values of the composites are the following variables: the type of intermetallic particles, inhomogeneous internal stress, and particle or whisker distribution. Improvements in fracture toughness of SiC_w/Al composites and K_{Ic} values levels equivalent to those of 7075 Al(T6) have been obtain by using a cleaner matrix powder, better mixing, and increased mechanical working during fabrication (McDanels, 1985). McDanels' work also reinforces the idea that the metallic matrix is not merely a medium to hold the fibers together. Figure 6.23 shows that the type of aluminum matrix used in 20 v/o SiC_w/Al composites was the most important factor affecting yield strength and ultimate tensile strength of these composites. Higher-strength aluminum alloys showed higher strengths but lower ductilities. Composites with 6061 Al matrix showed good strength and higher fracture strain. The 5083 Al, in Fig. 6.23, is a nonheat-treatable alloy.

Thermal Characteristics

In general, ceramic reinforcements (fibers, whiskers, or particles) have a coefficient of thermal expansion greater than that of most metallic matrices.

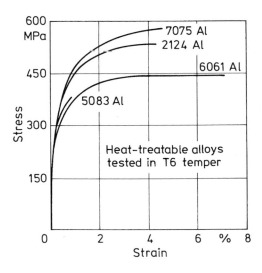

Fig. 6.23. Effect of aluminum matrix type on stress-strain behavior of 20 v/o SiC_w/Al composites. [From McDaniels (1985), used with permission.]

This means that when the composite is subjected to a temperature change, thermal stresses will be generated in both the components.

Chawla and Metzger (1972), working with a single crystal copper matrix containing large-diameter tungsten fibers, showed the importance of thermal stresses in MMCs. Specifically, they employed a dislocation etch-pitting technique to delineate dislocations in a single crystal copper matrix and showed that near the fiber the dislocation density was much higher in the matrix than the dislocation density far away from the fiber. The situation in the as-cast composite can be depicted as was shown schematically in Figure 6.15, where a primary plane section of the composite is shown having a hard zone (high dislocation density) around each fiber and a soft zone (low dislocation density) away from the fiber. The enhanced dislocation density in the copper matrix near the fiber arises because of the plastic deformation in response to the thermal stresses generated by the thermal mismatch between the fiber and the matrix. The intensity of the gradient in dislocation density will depend on the interfiber spacing. The dislocation density gradient will decrease with a decrease in the interfiber spacing. The existence of a plastically deformed zone containing high dislocation density in the metallic matrix near the reinforcement has been confirmed by transmission electron microscopy by a number of researchers, both in fibrous and particulate metal matrix composites.

Thermal mismatch is indeed something that is difficult to avoid in any composite. By the same token, however, one can control the overall thermal expansion characteristics of a composite by controlling the proportion of reinforcement and matrix and the distribution of the reinforcement in the matrix. Many researchers have proposed models to predict the coefficients of thermal expansion of composites, determined experimentally these coefficients, and analyzed the general thermal expansion characteristics of metal matrix composites; we describe some of these in Chapter 10. It would be appropriate to point out here some special aspects of thermal characteristics of MMCs, especially particulate MMCs. This is because of their applications in electronic packaging, where their superior thermal characteristics play an important role; this makes it very important to understand the effects of thermal stresses and thermal cycling in MMCs. Prediction of CTE is complicated because of the structure of the composite (particles, whiskers, or fibers), interface, and the matrix plastic deformation due to internal thermal stresses. Expansion characteristics of isotropic composites are a function of V_p or V_f, size, and morphology of the reinforcement. Phenomena such as hysteretic deformation during thermal cycling, microcracking, and fiber intrusion on the surface lead to surface roughness. Rezai-Aria et al. (1993) observed surface roughness when they subjected Saffill fiber reinforced aluminum to thermal cycling. It is important to realize that the CTE of a material is not an absolute physical constant, but that it varies with temperature. This is especially so with MMCs, where the matrix can undergo plastic deformation as a result of thermal stresses (Vaidya and Chawla, 1994). In addition, there are many microstructural factors that come into play. For example, a micromechanical analysis of the thermal expansion of two-dimensional SiC_p/Al composites showed the importance of reinforcement continuity (Shen et al., 1994). Balch et al. (1996) investigated the role of reinforcement continuity on the thermal expansion of SiC/Al and the effect of the presence of voids on CTE. Their analysis showed that a high matrix yield strength, a high interfacial strength, and a convex reinforcement shape would minimize variations in CTE of a SiC_p/Al-type composite during any thermal excursion. In foam reinforced composites, a weak interface could be exploited to produce void and thus minimize the average CTE of the composite, but hysteretic thermal expansion would be observed.

Aging

Frequently the metal matrix alloy used in an MMC has precipitation hardening characteristics, i.e., such an alloy can be hardened by a suitable heat treatment, called *aging treatment*. It has been shown by many researchers (Dunand and Mortensen, 1991a, 1991b; Dutta et al., 1988; Dutta and Bourell, 1989, 1990; Esmaeili et al., 1991; Nieh and Karlak, 1984; Parrini and Schaller, 1994) that the microstructure of the metallic matrix is modified by the presence of ceramic reinforcement and consequently the standard aging treatment for, say, an unreinforced aluminum alloy will not be valid (see Suresh and Chawla (1993) for a review of this topic). The particle- or whisker-type reinforcements, such as SiC, B_4C, Al_2O_3, are unaffected by the aging process. These reinforcements, however, can affect the precipitation behavior of the matrix quite significantly. In particular, as we pointed out earlier, a higher dislocation density in the matrix metal or alloy than that in the unreinforced metal or alloy is produced. The higher dislocation density in the matrix has its origin in the thermal mismatch ($\Delta\alpha$) between the reinforcement and the metallic matrix. This thermal mismatch can be quite large, for example, in the case of SiC and aluminum, it has a high value of $21 \times 10^{-6}/K$. The precipitate-hardenable aluminum alloy matrix has a more important role to play in these composites than a nonhardenable matrix material. A considerable strength increment results due to the age-hardening treatments. One would expect that the high dislocation density will also affect the precipitation kinetics in a precipitation-hardenable matrix such as 2XXX series aluminum alloy. Indeed, faster precipitation kinetics have been observed in the matrix than in the bulk unreinforced alloy. This has very important practical implications. One may not use the standard heat treatments commonly available in various handbooks, say, for monolithic aluminum alloys for the same alloy used as a matrix in an MMC.

Fatigue and Creep

Fatigue is the phenomenon of mechanical property degradation leading to failure of a material or a component under cyclic loading. Many high-

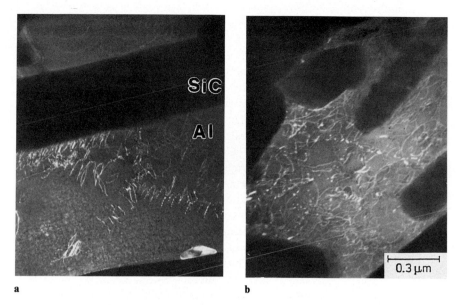

Fig. 6.24. Dislocation distribution in the aluminum matrix of a SiC$_w$/Al composite: **a** inhomogeneous dislocation distribution before testing, **b** uniform dislocation distribution after fatigue testing. [From Williams and Fine (1985b), used with permission.]

volume applications of composite materials involve cycling loading situations, e.g., automobile components. For a general discussion of fatigue behavior of composites, see Chapter 12. We mention here but one example of the importance of matrix microstructure on the fatigue behavior of MMCs.

Williams and Fine (1985a) investigated the fatigue behavior of SiC whisker reinforced 2124 Al alloy composites. They found that unbonded SiC$_p$ and non-SiC intermetallics were the fatigue crack-initiation sites. The unbonded SiC particles occur when clusters of SiC are present. Thus, not unexpectedly, reducing the clustering of SiC and the number and size of the intermetallics resulted in increased fatigue life. Cyclic loading results in a uniform distribution of dislocations in the metal matrix. Figure 6.24 shows the dislocation distribution in an aluminum matrix before and after fatigue testing of a SiC$_w$/Al composite (1985b). Note the uniform dislocation distribution after fatigue testing.

The phenomenon of creep refers to time-dependent deformation. In practice, at least for most metals and ceramics, the creep behavior becomes important at high temperatures and, thus, sets a limit on the maximum application temperature. In general, this limit increases with the melting point of a material. We describe the fatigue and creep behavior of MMCs in chapter 12.

6.6 Applications

It is convenient to divide the applications of metal matrix composites into aerospace and nonaerospace categories. In the category of aerospace applications, low density and other desirable features such as a tailored thermal expansion and conductivity and high stiffness and strength are the main drivers. Performance rather than cost is the important item. Inasmuch as continuous fiber reinforced MMCs deliver superior performance than particle reinforced composites, the former are frequently used in aerospace applications. In the nonaerospace applications, cost and performance are important, i.e., an optimum combination of these items is required. It is thus understandable that particle reinforced MMCs are increasingly finding structural and nonstructural applications in nonaerospace applications. By far, the largest commercial application of MMCs is in the form of filamentary superconducting composites. We devote a whole chapter to such composites (Chapter 9) because of their commercial importance. We provide a brief description of various other applications of MMCs.

Reduction in the weight of a component is a major driving force for any application in the aerospace field. For example, in the Hubble telescope, pitch-based continuous carbon fiber reinforced aluminum was used for waveguide booms because this composite is very light, has a high elastic modulus, E, and has a low coefficient of thermal expansion, α. Other aerospace applications of MMCs involve replacement of light but toxic beryllium. For example, in the US Trident missile, beryllium has been replaced by SiC_p/Al composite.

One of the important applications of MMCs in the automotive area is in diesel piston crowns (Donomoto et al., 1983). This application involves incorporation of short fibers of alumina or alumina + silica in the crown of the piston. The conventional diesel engine piston has an Al-Si casting alloy with crown made of a nickel cast iron. The replacement of the nickel cast iron by aluminum matrix composite resulted in a lighter, more abrasion-resistant, and cheaper product. Figure 6.25 shows a picture of such an application; the arrow indicates the piston crown that is made of short fiber reinforced aluminum composite. Another application in the automotive sector involves the use of carbon fiber and alumina fibers in an aluminum alloy matrix for use as cylinder liners in the Prelude model of Honda Motor Co. Figure 6.26 shows the microstructure of the cylinder liner showing the unreinforced and fiber-reinforced portions.

An important potential commercial application of particle reinforced aluminum composite is to make automotive driveshafts. In the design of a driveshaft, one needs to consider the speed at which it becomes dynamically unstable. It turns out that in terms of geometric parameters, a shorter shaft length and larger diameter will give a higher critical speed, N_c, while in terms

Fig. 6.25. A cutout of a squeeze cast piston with MMC inserts indicated by dotted lines. (Courtesy of Toyota Motor Co.)

of material parameters, the higher the specific stiffness (E/ρ), the higher N_c will be. One can make changes in the driveshaft geometry (increase the length or reduce the diameter) while maintaining a constant critical speed. A decrease in the driveshaft diameter can be important because of under-chassis space limitations.

Particulate metal matrix composites, especially with light metal matrix composites such as aluminum and magnesium, also find applications in automotive and sporting goods. In this regard, the price per kilogram becomes the driving force for application. An excellent example involves the use of Duralcan particulate MMCs to make mountain bicycles. A company called Specialized Bicycle in the United States sells these bicycles with frames made from extruded tubes of 6061 aluminum containing about 10% alumina particles. The major advantage is the gain in stiffness.

An interesting example of a sheet laminate composite is a nonvibration sheet steel, made by Kawasaki Steel under the tradename Nonvibra™. A thin film of resin is sandwiched between two metal sheets. A chromate coating is applied on the inner surface of the skin metal. The chromate coating is conducive to cementing the bond between the resin and the outer skin metal. Such a laminated composite muffles noise over a broad range of frequencies, and it can be used in the temperature range of 0–100 °C. Examples of applications include oil pans, locker covers, dashboard panels, floor panels, electrical machinery and appliances, and office equipment.

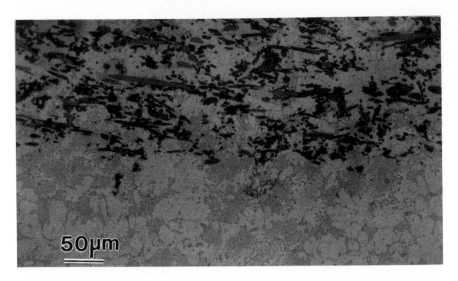

Fig. 6.26. The microstructure of the cylinder liner showing the unreinforced Al alloy (bottom) and fiber (alumina + carbon) reinforced (top) portions.

Electronic-Grade MMCs

Metal matrix composites can be tailored to have optimal thermal and physical properties to meet the requirements of electronic packaging systems (e.g., cores, substrates, carriers, and housings). Continuous boron fiber reinforced aluminum composites made by diffusion bonding have been used as heat sinks in chip carrier multilayer boards.

Unidirectionally aligned, pitch-based carbon fibers in an aluminum matrix can have very high thermal conductivity along the fiber direction. The conductivity transverse to the fibers is about two-thirds that of aluminum. Such a C/Al composite can find applications in heat transfer applications where weight reduction is an important consideration, for example, in high-density, high-speed integrated-circuit packages for computers and in base plates for electronic equipment. Can the reader think of a potential application in the disk drive of a computer?

Example

Discuss the use of carbon fiber/copper matrix as a high-thermal-conductivity metal matrix composite for applications that require high thermal conductivity and high strength.

Answer

Pitch-based carbon fibers can have very high thermal conductivity, higher than even copper. Putting such fibers in a copper matrix will result in a high-strength, high-thermal-conductivity metal matrix composite. The problem with carbon fibers and metallic matrixes is that the surface energy considerations preclude wetting of the carbon fibers by the molten metals. The surface energy of carbon fiber is about 100 mJ/m^2, while most molten metals have surface energy of about 1000 mJ/m^2. Recall from Chapter 2 that wetting occurs when the surface energy of the substrate (fiber) is higher than that of the molten metal. Thus, one does not expect the wetting of carbon fibers by molten copper. Two possible routes around this difficulty are

- Fiber surface modification
- Matrix modification

It would appear that modification of carbon fiber surface by making it rough could result in a reasonable mechanical bond between carbon fiber and copper matrix.

Recycling of Particle Reinforced Metal Matrix Composites

The impressive gains made by particle reinforced metal matrix composites (MMCs) can be attributed to the fact that particulate MMCs are more cost-effective than continuous- or short-fiber reinforced MMCs. Continuous fiber reinforcement of metals is expensive for two main reasons: fiber and processing costs. Also, one must recognize that with continuous fiber reinforced composites, secondary fabrication is very difficult. The ultimate objective is always to obtain a part or component that performs a function in a cost-effective manner. One item that stands out in this regard is the problem of recycling of composites made of alumina or silicon carbide particles in an aluminum matrix. There are two aspects of this (Lloyd, 1994), viz.,

- *Recycling* for reuse as a composite.
- *Reclamation* to obtain the individual components of the composite, i.e., aluminum and ceramic particles, separately.

The problem of recycling aluminum-based MMCs is particularly important for the aluminum industry, which has a notable reputation for recycling soda pop cans made of aluminum. Two types of alloys, 3000 and 5000 series, are commonly used. Both can be remelted to give the body composition alloy. Composite recycling is not that simple or straightforward. One of the problems with aluminum-based composites is that a mass of unreinforced metal is indistinguishable from a mass of reinforced material. Various methods of separating the metal and ceramic particles, such as crushing, shredding, and gravity separation, can be useful. Addition of virgin metal to the reinforced material to obtain a composite with a desired particle volume fraction will generally be required.

Cast MMCs can be remelted and reused as a composite. Although some of the issues, such as particle/molten metal reaction, are the same in remelting, it should be appreciated that scrap sorting and melt cleanliness and degassing can be rather tricky problems in MMCs. Special fluxing and degassing techniques have been developed at Duralcan. It would appear that the powder-processed MMCs will also involve remelting. The reclamation route also involves remelting. According to Lloyd (1994), by using combined argon and salt fluxing, silicon carbide particles can be removed from the melt and 85 to 90% of the aluminum can be reclaimed.

References

M.K. Aghajanian, J.T. Burke, D.R. White, and A.S. Nagelberg (1989). *SAMPE Quart.*, **34**, 817–823.

R.J. Arsenault and R.M. Fisher (1983). *Scripta Met.*, **17**, 67.

R.J. Arsenault and N. Shi (1986). *Mater. Sci. Eng.*, **81**, 175.

D.K. Balch, T.J. Fitzgerald, V.J. Michaud, A. Mortensen, Y.-L. Shen, and S. Suresh (1996). *Metall. and Mater. Trans. A.*, **27A**, 3700.

A.A. Baker and C. Shipman (1972). *Fibre Sci. & Tech.*, **5**, 282.

E.V. Barrera, J. Sims, D.L. Callahan, V.J. Provenzano, J. Milliken, and R.L. Holtz (1994). *J. Mater. Res.*, **9**, 2662.

A.R. Champion, W.H. Krueger, H.S. Hartman, A.K. Dhingra (1978). In *Proc.: Intl. Conf. Composite Materials (ICCM/2)*, TMS-AIME, New York, p. 883.

K.K. Chawla (1973a). *Metallography*, **6**, 155.

K.K. Chawla (1973b). *Philos. Mag.*, **28**, 401.

K.K. Chawla (1974). In *Grain Boundaries in Eng. Mater.*, Claitor's Publishing, Baton Rouge, LA, 435.

K.K. Chawla (1975). *Fibre Sci. & Tech.* **8**, 49.

K.K. Chawla (Dec., 1985). *J. of Metals,* **37**, 25.

K.K. Chawla (1989). In *Precious and Rare Metal Technologies*, Elsevier, Amsterdam, p. 639.

K.K. Chawla (1991). In *Metal Matrix Composites; Properties.* Academic Press, San Diego.

K.K. Chawla and C.E. Collares (1978). In *Proceedings of the 1978 International Conference on Composite Materials (ICCM/2)*, TMS-AIME, New York, p. 1237.

K.K. Chawla, A.H. Esmaeili, A.K. Datye, and A.K. Vasudevan (1991). *Scripta Met. et Mater.*, **25**, 1315–1319.

K.K. Chawla and L.B. Godefroid (1984). In *Proc.: 6th Int. Conf. on Fracture*, Pergamon Press, Oxford, p. 2873.

K.K. Chawla and P.K. Liaw (1979). *J. Mater. Sci.* **14**, 2143.

K.K. Chawla and M. Metzger (1972). *J. Materials Sci.*, 7, 34.

K.K. Chawla and M. Metzger (1978). In *Fracture 1977: Proc. of the 4th Int. Conf. Fracture*, Pergamon Press, vol. 3, p. 1039.

T. Christman and S. Suresh (1988). *Mater. Sci. Eng.* **A102**, 211.

L. Christodolou, P.A. Parrish, and C.R. Crowe (1988). *Mat. Res. Soc. Symp. Proc.*, **120**, 29–34.

H.E. Cline, J.L Walter, E.F Koch, L.M. Osika (1971). *Acta. Met.*, **19**, 405.

A.J. Cook and P.S. Warner (1991). *Mater. Sci. Eng.* **A144**, 189.

J.A. Cornie, Y.-M. Chiang, D.R. Uhlmann, A.S. Mortensen, and J.M. Collins (1986). *Ceram. Bull.*, **65**, 293.

A.P. Divecha, S.G. Fishman, and S.D. Karmarkar (September 1981). *J. Metals*, **9**, 12.

T. Donomoto, N. Miura, K. Funatani, and N. Miyake (1983). *SAE Tech. Paper* No. 83052, Detroit, MI.

D.C. Dunand and A. Mortensen (1991a). *Acta Metall. Mater.*, **39**, 127.

D.C. Dunand and A. Mortensen (1991b). *Acta Metall. Mater.*, **39**, 1405.

I. Dutta, D.L. Bourell, and D. Latimer (1988). *J. Comp. Mater.*, **22**, 829.

I. Dutta and D.L. Bourell (1988). *Mater. Sci. Eng.*, **A112**, 67.

I. Dutta and D.L. Bourell (1990). *Acta Metall.*, **38**, 2041.

A.H. Esmaeili, K.K. Chawla, A.K. Datye, and A.K. Vasudevan (1991). In *Proc. of the Intl. Conf. on Comp. Mater., ICCM/VII*, Honolulu, HI, p. 17F1.

S.G. Fishman (1986). *ASTM Standardization News*, 46.

L.-J. Fu, M. Schmerling, and H.L. Marcus (1986). In *Composite Materials: Fatigue and Fracture*, ASTM STP 907, ASTM, Philadelphia p. 51.

C. Fujiwara, M. Yoshida, M. Matsuhama, and S. Ohama (1995). In *Proc. International Conf. on Composite Materials (ICCM-10)*, p. II–687.

C.M. Gabryel and A.D. McLeod (1991). *Met Trans.*, **23A**, 1279.

A.K. Ghosh (1993). In *Metal Matrix Composites*, Butterworth-Heinemann, Boston, MA, p. 119.

R.G. Hill, R.P. Nelson, and C.L. Hellerich (1969). In *Proc. of the 16th Refractory Working Group Meeting*, Seattle, WA.

W.H. Hunt (1994). In *Processing and Fabrication of Adv. Materials*, TMS, Warrendale, PA, p. 663.

W.H. Hunt, T.M. Osman, and J.J. Lewandowski (Mar. 1991). *J. Miner. Metal Mater. Soc.*, 30.

J.A. Isaacs, F. Taricco, V.J. Michaud, and A. Mortensen (1991). *Met. Trans.*, **22A**, 2855.

C. Jones, C.J. Kiely, and S.S. Wang (1993). *J. Mater. Res.*, **4**, 327.

H.A. Katzman (1987). *J. Mater. Sci.*, **22**, 144.

A. Kelly and K.N. Street (1972). *Proc. R. Soc. London*, **328A**, 283.

A. Kelly and W.R. Tyson (1966). *J. Mech. Phys. Solids*, **14**, 177.

R.J. Kerans and T.A. Parthasarathy (1991). *J. Am. Ceram. Soc.*, **74**, 1585.

A. Kohyama, N. Igata, Y. Imai, H. Teranishi, and T. Ishikawa (1985). In *Proc.: Fifth International Conference on Composite Materials (ICCM/V)*, TMS-AIME, Warrendale, PA, p. 609.

D.J. Lloyd (1994). *International Materials Review*, **39**, 1.

A.P. Majidi and T.W. Chou (1987). *Proc. ICCM VI*, **2**, 422.

M. Manoharan, L. Ellis, and J.J. Lewandowski (1990). *Scripta Met. et Mater.*, **24**, 1515.

C. Manning and T. Gurganus (1969). *J. Am. Ceram. Soc.*, 52, 115.

D.L. McDanels (1985). *Met. Trans.*, **16A**, 1105.

M. McLean (1983). *Directionally Solidified Materials for High Temperature Service*, The Metals Soc., London.

W. Meyerer, D. Kizer, S. Paprocki, and H. Paul (1978). In *Proc.: 1978 Intl Conf. Composite Materials (ICCM/2)*, TMS-AIME, New York, p. 141.

M.A. Meyers and K.K. Chawla (1984). *Mechanical Metallurgy*, Prentice-Hall, Englewood Cliffs, NJ, 494.

V.J. Michaud (1993). In *Metal Matrix Composites*, Butterworth-Heinemann, Boston, MA, p. 3.

R. Mitra, W.A. Chiou, M.E. Fine, and J.R. Weertman (1993). *J. Mater. Res.*, **8**, 2300.

A. Mortensen, J.A. Cornie, and M.C. Flemings (1988). *J. Metals*, **40**, 12.

A. Mortensen, M.N. Gungor, J.A. Cornie, and M.C. Flemings (Mar., 1986). *J. of Metals*, **38**, 30.

V.C. Nardone and K.M. Prewo (1986). Scipta Met., **20**, 43.

R. Naslain, J. Thebault, and R. Pailler (1976). In *Proceedings of the 1975 International Conference on Composite Materials*, vol. 1, TMS-AIME, New York, p. 116.

T.G. Nieh and R.F. Karlak (1984). *Scripta Met.*, **17**, 67.

S. Nourbakhsh, F.L. Liang, and H. Margolin (1990). *Met. Trans. A*, **21A**, 213.

L. Parrini and R. Schaler (1994). *J. of Alloys and Compounds*, **211**, 402.

P.G. Partridge and C.M. Ward-Close (1993). *Int. Mater. Rev.*, **38**, 1.

M. Pfeifer, J.M. Rigsbee, and K.K. Chawla (1990). *J. Mater. Sci.*, **25**, 1563.

W.L. Phillips, (1978). In *Proc.: 1978 Intl Conf. Composite Materials (ICCM/2)*, TMS-AIME, New York, p. 567.

S. Pickard and D.B. Miracle (1995). *Mater. Sci. & Eng. A*, **A203**, 59.

H.J. Rack (1987). In *Sixth International Conference on Composite Materials*, New York: Elsevier Applied Science, p. 382.

F. Rezai-Aria, T. Liechti, and G. Gagnon (1993). *Scripta Metall. Mater.*, **28**, 587.

S. Rhee, (1970). *J. Am. Ceram. Soc.*, **53**, 386.

P.K. Rohatgi, R. Asthana, and S. Das (1986). *Intl. Met. Rev.*, **31**, 115.

Y.-L. Shen, A Needleman, and S. Suresh (1994). *Metall. Mater. Trans. A.*, **25A**, 839.

O. Sherby, S. Lee. R. Koch, T. Sumi, and J. Wolfenstine (1985). *Mater. Manufact. Processes*, **5**, 363.

N. Shi and R.J. Arsenault (1991). *J. of Comp. Tech. and Res.*, **13**, 211.

N. Shi and R.J. Arsenault (1993). *Met. Trans.*, **24A**, 1879.

J. van Suchtelen (1972). *Philips Res. Rep.*, **27**, 28.

T.S. Srivatsan and E.J. Lavernia (1992). *J. Materials Sci.*, **27**, 5965.

N.S. Stoloff (year). In *Advances in Composite Materials Applied Sci. Pub.*, London, p. 247.

S. Suresh and K.K. Chawla (1993). In *Metal Matrix Composites*, Butterworth-Heinemann, Boston, MA, p. 119.

R.U. Vaidya and K.K. Chawla (1994). *Composites Sci. and Tech.*, **50**, 13.

J. van Suchtelan (1972). Philips Res. Report, **27**, 28.

E. Tempelman, W.L. Dalmijn, and A. Vlot (April, 1996). *JOM*, **48**, 62.

S. Towata and Y. Yamada (1995). *Trans. Japan Inst. of Metals*, **27**, 709.

A.K. Vasudevan and R.D. Doherty (Eds.) (1989). *Aluminum Alloys - Contemporary Research and Applications*, Academic Press, Boston.

J.L. Walter (1982). In *In Situ Composites IV*, Elsevier, New York, p. 85.

R. Warren and C.-H. Andersson (1984). *Composites*, **15**, 101.

D.R. Williams and M.E. Fine (1985a). In *Proc. 5th Int. Conf. Composite Mater.*, *(ICCM/V)*, TMS, Warrendale, PA. p. 275.

D.R. Williams and M.E. Fine (1985b). In *Proc.: Fifth Intl Conf. Composite Materials (ICCM/V)*, Warrendale, PA.: TMS-AIME, p. 369.

Z.R. Xu, K.K. Chawla, R. Mitra, and M.E. Fine (1994). *Scripta Met. et Mater.* **31**, 1277.

Z.R. Xu, A. Neuman, K.K. Chawla, A. Wolfenden, and G.M. Liggett and N. Chawla (1995). *Mater. Sci. & Eng. A.* **A203**, 75.

Suggested Reading

T.W. Clyne and P.J. Withers (1993). *An Introduction to Metal Matrix Composites*, Cambridge University Press, Cambridge.

S. Suresh, A. Needleman, and A. Mortensen (Eds.) (1993). *Metal Matrix Composites*, Butterworth-Heinemann, Boston.

M. Taya and R.J. Arsenault (1990). *Metal Matrix Composites*, Pergamon Press, Oxford.

CHAPTER 7

Ceramic Matrix Composites

Ceramic materials in general have a very attractive package of properties: high strength and high stiffness at very high temperatures, chemical inertness, low density, and so on. This attractive package is marred by one deadly flaw, namely, an utter lack of toughness. They are prone to catastrophic failures in the presence of flaws (surface or internal). They are extremely susceptible to thermal shock and are easily damaged during fabrication and/or service. It is therefore understandable that an overriding consideration in ceramic matrix composites (CMCs) is to toughen the ceramics by incorporating fibers in them and thus exploit the attractive high-temperature strength and environmental resistance of ceramic materials without risking a catastrophic failure. It is worth pointing out at the very outset that there are certain basic differences between CMCs and other composites. The general philosophy in nonceramic matrix composites is to have the fiber bear a greater proportion of the applied load. This load partitioning depends on the ratio of fiber and matrix elastic moduli, E_f/E_m. In nonceramic matrix composites, this ratio can be very high, while in CMCs, it is rather low and can be as low as unity. Another distinctive point regarding CMCs is that because of limited matrix ductility and generally high fabrication temperature, thermal mismatch between components has a very important bearing on CMC performance. The problem of chemical compatibility between components in CMCs has ramifications similar to those in, say, MMCs. We first describe some of the processing techniques for CMCs, followed by a description of some salient characteristics of CMCs regarding interface and mechanical properties and, in particular, the various possible toughness mechanisms, and finally a description of some applications of CMCs.

7.1 Processing of CMCs

Ceramic matrix composites (CMCs) can be processed either by conventional powder processing techniques used for making polycrystalline ceramics or by some new techniques developed specifically for making CMCs. We describe below some of the important processing techniques for CMCs.

7.1.1   Cold Pressing and Sintering

Cold pressing of the matrix powder and fiber followed by sintering is a carry-over from conventional processing of ceramics. Generally, in the sintering step, the matrix shrinks considerably and the resulting composite has many cracks. In addition to this general problem of shrinkage associated with sintering of any ceramic, certain other problems arise when we put high-aspect ratio (length/diameter) reinforcements in a glass or ceramic matrix material and try to sinter. Fibers and whiskers can form a network that may inhibit the sintering process. Depending on the difference in thermal expansion coefficients of the reinforcement and matrix, a hydrostatic *tensile* stress may develop in the matrix on cooling, which will counter the driving force (surface energy minimization) for sintering (Raj and Bordia, 1989; Kellet and Lange, 1989). Thus the densification rate of the matrix will, in general, be retarded in the presence of reinforcements (Bordia and Raj, 1988; De Jonghe et al., 1986; Sacks et al., 1987; Rahaman and De Jonghe, 1987; Prewo, 1986). Whiskers or fibers may also give rise to the phenomenon of bridging, which is a function of the orientation and aspect ratio of the reinforcement.

7.1.2   Hot Pressing

Some form of hot pressing is frequently resorted to in the consolidation stage of CMCs. This is because a simultaneous application of pressure and high temperature can accelerate the rate of densification and a pore-free and fine-grained compact can be obtained. A common variant, called the *slurry infiltration* process, is perhaps the most important technique used to produce continuous fiber reinforced glass and glass-ceramic composites (Phillips, 1983; Cornie et al., 1986; Prewo and Brennan, 1980; Brennan and Prewo, 1982; Sambell et al., 1974). The slurry infiltration process involves two stages:

1. Incorporation of a reinforcing phase into an unconsolidated matrix.
2. Matrix consolidation by hot pressing.

Figure 7.1 shows a schematic of this process. In addition to incorporation of the reinforcing phase, the first stage involves some kind of fiber alignment. A fiber tow or a fiber preform is impregnated with a matrix-containing slurry by passing it through a slurry tank. The impregnated fiber tow or preform sheets are similar to the prepregs used in polymer matrix composites. The slurry consists of the matrix powder, a carrier liquid (water or alcohol), and an organic binder. The organic binder is burned out prior to consolidation. Wetting agents may be added to ease the infiltration of the fiber tow or pre-form. The impregnated tow or prepreg is wound on a drum and dried. This is followed by cutting and stacking of the prepregs and consolidation in a hot press. The process has the advantage that, as in PMCs, the prepregs can be arranged in a variety of stacking of sequences, e.g., unidirectional, cross-

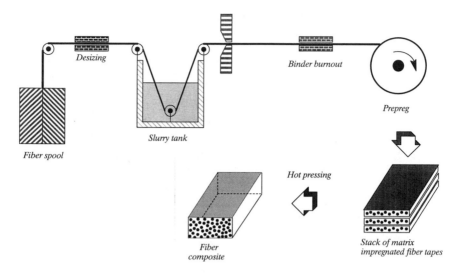

Fig. 7.1. Schematic of the slurry impregnation process.

plied ($0°/90°/0°/90°$ etc.), or angle-plied ($+\theta/-\theta/+\theta/-\theta$ etc.). Figure 7.2a shows an optical micrograph of a transverse section of a unidirectional Nicalon fiber/glass matrix composite. Some porosity can be seen in this picture. Figure 7.2b shows the pressure and temperature schedule used during hot pressing of a typical CMC.

The slurry infiltration process is well suited for glass or glass-ceramic matrix composites, mainly because the processing temperatures for these materials are lower than those used for crystalline matrix materials. Any hot pressing process has certain limitations on producing complex shapes. The fibers should suffer little or no damage during handling. Application of a very high pressure can easily damage fibers. Refractory particles of a crystalline ceramic can damage fibers by mechanical contact. The reinforcement can also suffer damage from reaction with the matrix at very high processing temperatures. The matrix should have as little porosity as possible in the final product as porosity in a structural ceramic material is highly undesirable. To this end, it is important to completely remove the fugitive binder and use a matrix powder particle smaller than the fiber diameter. The hot pressing operational parameters are also important. Precise control within a narrow working temperature range, minimization of the processing time, and utilization of a pressure low enough to avoid fiber damage are important factors in this final consolidation part of the process. Fiber damage and any fiber/matrix interfacial reaction, along with its detrimental effect on the bond strength, are unavoidable attributes of the hot pressing operation.

In summary, the slurry infiltration process generally results in a fairly uniform fiber distribution; low porosity and high strength values can be

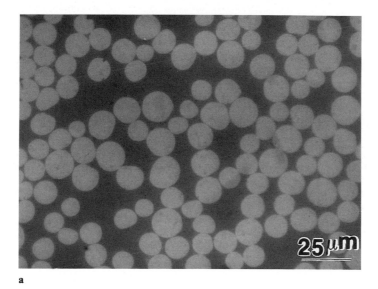

a

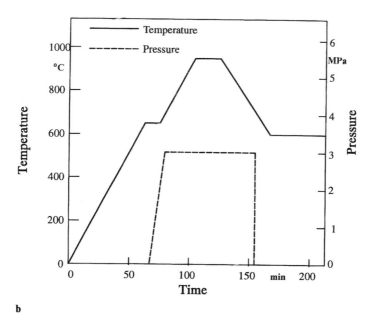

b

Fig. 7.2. a An optical micrograph of a transverse section of a unidirectional Nicalon fiber/glass matrix composite. b Pressure and temperature schedule used during hot pressing of a typical composite.

obtained. The main disadvantage of this process is that one is restricted to relatively low melting or low softening point matrix materials.

Whisker reinforced CMCs are generally made by mixing the whiskers with a ceramic powder slurry, then drying and hot-pressing. Sometimes hot isostatic pressing (HIPing) rather than uniaxial hot pressing is used. Whisker agglomeration in a green body is a major problem. Mechanical stirring and adjustment of pH level of the suspension (matrix powder+whiskers in water) can be of help in this regard. Addition of whiskers to a slurry can result in very high viscosity. Also, whiskers with large aspect ratios (>50) tend to form bundles and clumps (Liu et al., 1991). Obtaining well-separated and deagglomerated whiskers is of great importance for reasonably high-density composites. Use of organic dispersants (Barclay et al., 1987), techniques such as agitation mixing assisted by an ultrasonic probe, and deflocculation by a proper pH control (Yang and Stevens, 1990) can be usefully employed. Most whisker reinforced composites are made at temperatures in the 1500–1900 °C range and pressures in the 20–40 MPa range (Homeny et al., 1987; Shalek et al., 1986).

7.1.3   Reaction Bonding Processes

Reaction bonding processes similar to the ones used for monolithic ceramics can be used to make ceramic matrix composites. A reaction bonding process has the great advantage that problems with matrix shrinkage during densification are avoided. The other advantages are:

- Rather large volume fractions of whiskers or fiber can be used.
- Multidirectional, continuous fiber preforms can be used.
- The reaction bonding temperatures for most systems are generally lower than the sintering temperatures, so that fiber degradation can be avoided.

One great disadvantage of this process is that high porosity is difficult to avoid.

A hybrid process involving a combination of hot pressing and the reaction bonding technique can also be used (Bhatt, 1986; Bhatt, 1990). Silicon cloth is prepared by attrition milling a mixture of silicon powder, a polymer binder, and an organic solvent to obtain a dough of proper consistency. This dough is then rolled to make a silicon cloth of desired thickness. Fiber mats are made by filament winding of silicon carbide with a fugitive binder. The fiber mats and silicon cloth are stacked in an alternate sequence, debinderized, and hot pressed in a molybdenum die in a nitrogen or vacuum environment. The temperature and pressure are adjusted to produce a handleable preform. At this stage, the silicon matrix is converted to silicon nitride by transferring the composite to a nitriding furnace between 1100 and 1400 °C. Typically, the silicon nitride matrix has about 30% porosity, which is not unexpected in reaction bonded silicon nitride.

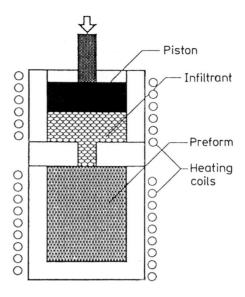

Fig. 7.3. Schematic of the melt infiltration process. [After Cornie et al. (1986), used with permission.]

7.1.4 Infiltration

Infiltration of a preform made of a reinforcement can be done with a matrix material in solid, liquid, or gaseous form.

Liquid Infiltration

This technique is very similar to liquid polymer or liquid metal infiltration (Fig. 7.3). Proper control of the fluidity of the liquid matrix is, of course, the key to this technique. It yields a high-density matrix, i.e., no pores in the matrix. Almost any reinforcement geometry can be used to produce a virtually flaw-free composite. The temperatures involved, however, are much higher than those encountered in polymer or metal processing. Processing at such high temperatures can lead to deleterious chemical reactions between the reinforcement and the matrix. Thermal expansion mismatch between the reinforcement and the matrix, the rather large temperature interval between the processing temperature and room temperature, and the low strain to failure of ceramics can add up to a formidable set of problems in producing a crack-free CMC. Viscosities of ceramic melts are generally very high, which makes the infiltration of preforms rather difficult. Wettability of the reinforcement by the molten ceramic is another item to be considered. Hillig (1988) has discussed the melt infiltration processing of ceramic matrix composites in regard to chemical reactivity, melt viscosity, and wetting of the reinforcement by the melt. A preform made of reinforcement in any form (for example, fiber, whisker, or particle) having a network of pores can be infiltrated by a ceramic melt by using capillary pressure. Application of pressure or processing in vacuum can aid in the infiltration process. Assuming that the preform consists of a bundle of regularly spaced, parallel channels, one can use Poissuelles's equation to obtain the infiltration height, h:

$$h = \sqrt{\frac{\gamma r t \cos \theta}{2\eta}}$$

where r is the radius of the cylindrical channel, t is the time, γ is the surface energy of the infiltrant, θ is the contact angle, and η is the viscosity. The penetration height is proportional to the square root of time, and the time required to penetrate a given height is inversely proportional to the viscosity of the melt. Penetration will also be easier if the contact angle is low (i.e., better wettability), and the surface energy, γ, and the pore radius, r, are large. If the radius, r, of the channel is made too large, the capillarity effect will be lost.

Infiltration of a fibrous preform by a molten intermetallic matrix material under pressure has been successfully done (Nourbakhsh and Margolin, 1989). Alumina fiber reinforced intermetallic matrix composites (e.g., TiAl, Ni$_3$Al, and Fe$_3$Al matrix materials) have been prepared by *pressure casting*, also called *squeeze casting* (Nourbakhsh and Margolin, 1989; Nourbakhsh et al., 1990). The matrix alloy is melted in a crucible in a vacuum while the fibrous preform is heated separately. The molten matrix material (at about 100 °C above the melting temperature, T_m) is poured onto the fibers and argon gas is introduced simultaneously. Argon gas pressure forces the melt to infiltrate the preform. The melt generally contains additives to aid wetting of the fibers.

We may summarize the advantages and disadvantages of different melt infiltration techniques as follows. The advantages are:

- The matrix is formed in a single processing step.
- A homogeneous matrix can be obtained.

The disadvantages of infiltration techniques are:

- High melting points of ceramics mean a greater likelihood of reaction between the melt and the reinforcement.
- Ceramics have higher melt viscosities than metals; therefore, infiltration of preforms is relatively difficult.
- The matrix is likely to crack because of the differential shrinkage between the matrix and the reinforcement on solidification. This can be minimized by choosing components with nearly equal coefficients of thermal expansion.

7.1.5 Directed Oxidation, or the Lanxide™ Process

Yet another version of liquid infiltration is the directed oxidation process, or the Lanxide™ process, developed by Lanxide Corp. (Urquhart, 1991). One of the Lanxide processes is called DIMOX™, which stands for directed

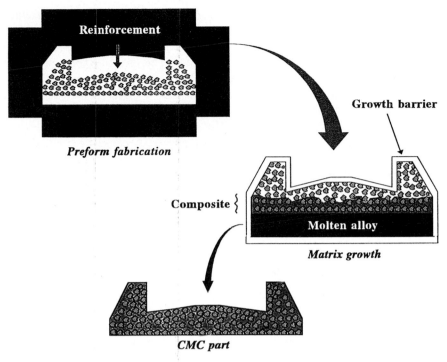

Fig. 7.4. Schematic of the directed metal oxidation process of Lanxide Corp. (Courtesy of Lanxide Corp.)

metal oxidation process. A schematic of the directed metal oxidation process is shown in Figure 7.4. The first step in this process is to make a preform, which in the case of a particulate composite can be a ceramic green body. In the case of a fibrous composite, filament winding or a fabric lay-up may be used to make a preform. A barrier to stop growth of the matrix material is placed on the preform surfaces. In this method, a molten metal is subjected to directed oxidation, i.e., the desired reaction product forms on the surface of the molten metal and grows outward. The metal is supplied continuously at the reaction front by a wicking action through channels in the oxidation product. For example, molten aluminum in air will get oxidized to aluminum oxide. If one wants to form aluminum nitride, then molten aluminum is reacted with nitrogen. The reaction can be represented as follows:

$$Al + air \rightarrow Al_2O_3$$

$$Al + nitrogen \rightarrow AlN$$

The end product in this process is a three-dimensional, interconnected network of a ceramic material plus about 5–30% of unreacted metal. When filler particles are placed next to the molten metal surface, the ceramic network forms around these particles. As we said earlier, a fabric made of a continuous fiber can also be used. The fabric is coated with a proprietary coating to protect the fiber from highly reducing aluminum and to provide a weak interface, which is desirable for enhanced toughness. Some aluminum (6–7 wt. %) remains at the end of the process. This must be removed if the composite is to be used at temperatures above the melting point of aluminum (660 °C). On the other hand, the presence of a residual metal can be exploited to provide some fracture toughness in these composites.

Proper control of the reaction kinetics is of great importance in this process. The process is potentially a low-cost process because near-net shapes are possible. Also, good mechanical properties (such as strength and toughness) have been reported (Urquhart, 1991). Figure 7.5 shows some fiber reinforced ceramic components made by Lanxide Corp. Figure 7.5a shows some fiber reinforced ceramic composites for applications in high-temperature gas turbine engine components, while Fig. 7.5b shows heat exchanger and radiant burner tubes, flame tubes, and other high-temperature furnace parts made of particle reinforced ceramic composites.

The main disadvantages of the Lanxide processes are:

- It is difficult to control the chemistry and produce an all-ceramic matrix by this method. There is always some residual metal, which is not easy to remove completely.
- It is difficult to envision the use of such techniques for large, complex parts, such as those required, say, for aerospace applications.

7.1.6   In Situ Chemical Reaction Techniques

In situ chemical reaction techniques to produce CMCs are extensions of those used to produce monolithic ceramic bodies. We describe below some of the more important techniques, viz., chemical vapor deposition (CVD) and chemical vapor infiltration (CVI) and different types of reaction bonding techniques.

Chemical Vapor Deposition and Chemical Vapor Impregnation

When the chemical vapor deposition (CVD) technique is used to impregnate rather large amounts of matrix material in fibrous preforms, it is called *chemical vapor impregnation* (CVI). Common ceramic matrix materials used are SiC, Si_3N_4, and HfC. The CVI method has been used successfully by several researchers to impregnate fibrous preforms (Fitzer and Hegen, 1979; Fitzer and Schlichting, 1980; Fitzer and Gadow, 1986; Stinton et al., 1986; Burkland et al., 1988). The preforms can consist of yarns, woven fabrics, or three-dimensional shapes.

In very simple terms, in the CVI process a solid material is deposited from gaseous reactants onto a heated substrate. A typical CVD or CVI process would require a reactor with the following parts:

a

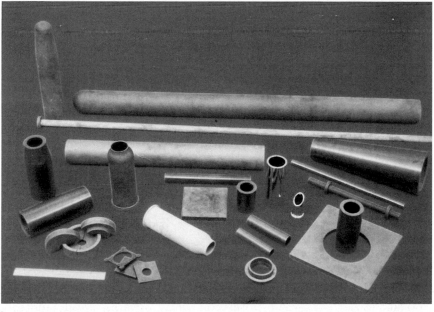

b

Fig. 7.5. Some commercially available fiber-reinforced ceramic components made by Lanxide Corp. **a** Fiber reinforced ceramic composites for applications in high-temperature gas turbine engine components. **b** Heat exchanger and radiant burner tubes, flame tubes, and other high-temperature furnace parts made of particle-reinforced ceramic composites. (Courtesy of Lanxide Corp.)

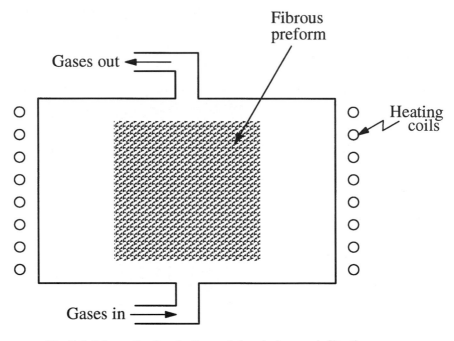

Fig. 7.6. Schematic of an isothermal chemical vapor infiltration process.

1. A vapor feed system.
2. A CVD reactor in which the substrate is heated and gaseous reactants are fed.
3. An effluent system where exhaust gases are handled.

The chemistry for making ceramic fibers by CVD was given in Chapter 2. The basic chemistry to make a bulk ceramic matrix in and around fibers in a preform remains the same. One can synthesize a variety of ceramic matrixes such as oxides, glasses, ceramics, and intermetallics by CVD. Commonly, the process involves an isothermal decomposition of a chemical compound in the vapor form to yield the desired ceramic matrix on and in between the fibers in a preform. Figure 7.6 shows a schematic of such an isothermal process. For example, methyltrichlorosilane (CH_3SiCl_3), the starting material to obtain SiC, is decomposed at between 1200 and 1400 K:

$$CH_3Cl_3Si(g) \rightarrow SiC(s) + 3\ HCl(g)$$

The vapors of SiC deposit as solid phase on and between the fibers in a free-standing preform to form the matrix. An example of the microstructure of a SiC/SiC composite obtained by CVI is shown in Figure 7.7. The CVI process is very slow because it involves diffusion of the reactant species to the fibrous substrate, followed by outflow of the gaseous reactant products. The CVI process of making a ceramic matrix is, indeed, a kind of low-stress and low-

Fig. 7.7. An example of the microstructure of a Nicalon fiber/SiC matrix composite obtained by CVI. (Courtesy of R.H. Jones.)

temperature CVD process, and thus avoids some of the problems associated with high-temperature ceramic processing. However, when the CVI process is carried out isothermally, surface pores tend to close first, restricting the gas flow to the interior of the preform. This necessitates multiple cycles of impregnation, surface machining, and reinfiltration to obtain an adequate density. One can avoid some of these problems to some extent by using a forced gas flow and a temperature gradient. A schematic of one version of this process is shown in Figure 7.8 (Stinton et al., 1986). A graphite holder in contact with a water-cooled metallic gas distributor holds the fibrous preform. The bottom and side surfaces thus stay cool while the top of the fibrous preform is exposed to the hot zone, creating a steep thermal gradient. The reactant gaseous mixture passes unreacted through the fibrous preform because of the low temperature. When these gases reach the hot zone, they decompose and deposit on and between the fibers to form the matrix. As the matrix material gets deposited in the hot portion of the preform, the preform density and thermal conductivity increase and the hot zone moves progressively from the top of the preform toward the bottom. When the composite is formed completely at the top and is no longer permeable, the gases flow radially through the preform, exiting from the vented retaining ring.

This variant of CVI, which combines forced gas flow and temperature gradient, avoids some of the problems mentioned earlier. Under these modified conditions, 70 to 90% dense SiC and Si_3N_4 matrices can be impregnated

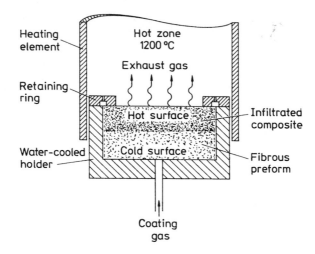

Fig. 7.8. Schematic of a chemical vapor infiltration process with pressure and temperature gradients. [After Stinton et al. (1986).]

in SiC and Si_3N_4 fibrous preforms in less than a day. Under conditions of plain CVI, it would take several weeks to achieve such densities, i.e., one can reduce the processing time from several days to less than 24 hours. One can also avoid using binders in this process with their attendant problems of incomplete removal. The use of a graphite holder simplifies the fabrication of the preform, and the application of a moderate pressure to the preform can result in a higher-than-normal fiber volume fraction in the final product. The final obtainable density in a ceramic body is limited by the fact that closed porosity starts at about 93–94% of theoretical density. It is difficult to impregnate past this point.

Advantages of a CVI technique or any variant thereof include:

- Good mechanical properties at high temperatures.
- Large, complex shapes can be produced in a near-net shape.
- Considerable flexibility in the fibers and matrices that can be used (oxide and nonoxide).

Among the disadvantages, one should mention that the process is slow and expensive.

Reactive Consolidation or Liquid-Phase Sintering

Siliconized silicon carbide is the name give to a composite of SiC grains in a silicon matrix. Molten silicon reacts with carbon fibers to form SiC. The original geometry of the carbon fibers is retained. Carbon fiber in the form of cloth, tow, felt, or matte is used as a precursor. A preform is made of carbon fiber and infiltrated with liquid silicon. Silicon reacts with carbon fibers to form SiC fibers in a Si matrix. Typical composition of the resultant composite is Si (30–50%) + SiC fiber. The silicon matrix limits the use temperature to about 1400 °C. A big advantage of SiC/Si composites is that the constituents are in chemical equilibrium and they have closely matched thermal expansion coefficients.

In another version of this process, a liquid phase forms as a result of an exothermic reaction between elemental powders. A good example is that from the field of intermetallics, e.g., nickel aluminides. The following steps are involved:

1. Mix nickel and aluminum in stoichiometric proportions.
2. Cold isostatic press to 70% theoretical density to obtain a green body.
3. Vacuum encapsulate the green body in a 304 stainless-steel can.
4. Subject the canned material to reactive hot isostatic pressing.

7.1.7 Sol-Gel and Polymer Pyrolysis

Sol-gel and polymer pyrolysis techniques, which have been used to make conventional ceramic materials, can also be used to make ceramic matrix materials in the interstices of a fibrous preform. We described the sol-gel technique in Chapter 2. Very briefly, a solution containing metal compounds, e.g., a metal alkoxide, acetate, or halide, is reacted to form a sol. The sol is converted to a gel, which in turn is subjected to controlled heating to produce the desired end product: a glass, a glass-ceramic, or a ceramic. Characteristically, the gel-to-ceramic conversion temperature is much lower than that required in a conventional melting or sintering process. Some of the advantages of these techniques for making composites are the same as the ones for monolithic ceramics, viz., lower processing temperatures, greater compositional homogeneity in single-phase matrices, potential for producing unique multiphase matrix materials among others. Specifically, in regard to composite material fabrication, the sol-gel technique allows processing via liquids of low viscosity such as the ones derived from alkoxides. Covalent ceramics, for example, can be produced by pyrolysis of polymeric precursors at temperatures as low as 1400 °C and with yields greater than those in CVD processes. Among the disadvantages of sol-gel are high shrinkage and low yield compared to slurry techniques. The fiber network provides a very high surface area to the matrix gel. Consequently, the shrinkage during the drying step, frequently, results in a large density of cracks in the matrix. Generally, repeated impregnations are required to produce a substantially dense matrix.

It is easy to see that many of the polymer handling and processing techniques can be used for sol-gel as well. Impregnation of fibrous preforms in a vacuum and filament winding are two important techniques. In filament winding, fiber tows or rovings are passed through a tank containing the sol, and the impregnated tow is wound on a mandrel to a desired shape and thickness. The sol is converted to gel and the structure is removed from the mandrel. A final heat treatment then converts the gel to a ceramic or glass matrix.

The sol-gel technique can also be used to prepare prepregs by the slurry infiltration method. The sol in the slurry acts as a binder and coats fibers and glass particles. The binder burnout step is thus eliminated because the binder, being of the same composition as the matrix, becomes part of the glass matrix. An advantage of this sol-gel-based slurry method is that consolidation can be done at lower temperatures. Among the problems, one should mention that the coating layer on the fiber is porous, carbon-rich, and nonuniform.

Polymeric precursors can also be used to form a ceramic matrix in a composite. The matrix selection criteria include:

- High char yield
- Low shrinkage
- Good mechanical properties
- Easy fabrication

To make a silicon carbide matrix, polycarbosilane would appear to be a natural choice as a precursor to produce a silicon carbide matrix, in view of its successful use to make Nicalon (SiC) fiber. Polysilane is another possibility. Its char yield is about 60%. One can use fillers to reduce the shrinkage.

Repeated infiltration and in situ thermal decomposition of porous reaction-bonded ceramics, such as silicon carbide and silicon nitrate with silazanes and polycarbosilanes, give Si_3N_4/SiC composites. The organic silicon compound is thermally decomposed in situ. Fitzer and Gadow (1986), for example, followed these steps given to make such composites:

1. Porous SiC or Si_3N_4 fibrous preform with some binder phase is prepared.
2. Fibrous preform is evacuated in an autoclave.
3. Samples are infiltrated with molten precursors, silazanes, or polycarbosilanes, at a high temperature (780 K) and the argon or nitrogen pressure is slowly increased from 2 to 40 MPa. The high temperature results in a transformation of the oligomer silane to polycarbosilane and simultaneous polymerization at high pressures.
4. Infiltrated samples are cooled and treated with solvents.
5. Samples are placed in an autoclave and the organosilicon polymer matrix is thermally decomposed in an inert atmosphere at a high pressure and at a temperature in the 800–1300 K range.
6. Steps 2 through 5 are repeated to attain an adequate density.
7. To produce an optimum matrix crystal structure, the material is annealed in the 1300–1800 K range.

In all these methods involving use of polymeric precursors, one must resort to repeated impregnations to improve the density. Typically, the amount of porosity will reduce from 35% to less than 10% after about five impregnations.

7.1.8 Self-Propagating High-Temperature Synthesis (SHS)

The SHS technique involves synthesis of compounds without an external source of energy. One exploits exothermic reactions to synthesize ceramic compounds, which are difficult to fabricate by conventional techniques. For example, one can mix titanium powder and carbon black, cold press the mixture, and ignite the compact at the top in a cold-walled vessel. A combustion wave will pass through the compact, giving titanium carbide.

Among the salient features of SHS are:

- High combustion temperature (up to 4000 °C).
- Simple, low-cost equipment.
- Good control of chemical composition.
- Different shapes and forms can be obtained.

This technique can be used to produce a variety of refractory materials. The main disadvantage is that SHS products are very porous, because of the fairly large porosity in the original mix of reactants and because of the large volume change that results when the reactants transform to the products. Any adsorbed gases at the elevated temperatures used during this process can also add to the porosity of the final product. Synthesis concomitant with densification can improve the situation to some extent. This involves application of high pressure during the combustion or immediately after the completion of the combustion reaction, when the product temperature is still quite high. Hot pressing, rolling, and shock waves are some of the techniques used to apply the necessary pressure.

Many ceramics, such as borides, carbides, nitrides, silicides, and sialons, and composites, such as $SiC_w + Al_2O_3$, have been synthesized by means of SHS. The SHS process gives a weakly bonded compact. Therefore, the process is generally followed by breaking the compact, milling, and consolidation by some technique such as HIPing. Explosive or dynamic compaction can result in a relatively dense product. A good example of an SHS process to make composites is the proprietary process of Martin Marietta Corp., called the XD™ process, wherein exothermic reactions are used to produce multiphase alloy powders. These are hot pressed at ~ 1450 °C to full density. Reinforcement in the form of particles, whiskers, and platelets can be added to the master alloy to make a composite. A good example is that of TiB_2 particles, about 1 μm in diameter, distributed in intermetallic matrixes such as TiAl, TiAl + Ti$_3$Al, and NiAl.

7.2 Interface in CMCs

In general, for CMCs one must satisfy the following compatibility requirements: thermal expansion compatibility and chemical compatibility. Ceramics have a limited ductility, and in the fabrication of CMCs one uses high temperatures. Thus, thermal mismatch on cooling from high temperatures can cause matrix (or fiber) cracking. Thermal strain in composites is proportional to $\Delta\alpha\Delta T$, where $\Delta\alpha = \alpha_f - \alpha_m$, where α_f and α_m are the linear expansion coefficients of the fiber and matrix, respectively, and ΔT is the temperature interval. There is, of course, another complication, namely, fiber expansion coefficients are also sometimes not equal in the axial and radial directions. Carbon fiber in particular has the following axial and radial coefficients:

$$\alpha_a \approx 0$$

$$\alpha_r \approx 8 \times 10^{-6}\,\mathrm{K}^{-1}$$

If $\Delta\alpha$ is positive, the matrix is compressed on cooling, which is beneficial because it leads to an increase in the tensile stress at which matrix cracking will occur. Conversely, if $\Delta\alpha$ is negative, the matrix experiences tension, which, if ΔT is sufficiently large, can cause matrix cracking. In the radial direction, if $\Delta\alpha$, is positive, the fibers tend to shrink away from the matrix on cooling, which results in a reduced interfacial bond strength. If, however, $\Delta\alpha$ is negative, the fiber matrix bond strength can even be improved.

Matrix cracking resulting from thermal mismatch is a more serious problem in short fiber composites than in continuous fiber composites. The reason is that in a ceramic matrix containing aligned continuous fibers, transverse microcracks appear in the matrix, but fibers continue to hold the various matrix blocks together and the composite can still display a reasonable amount of strength. In a randomly oriented short fiber composite, owing to increased stress at the fiber ends, matrix cracking occurs in all directions and the composite is very weak. We can define a thermal expansion mismatch parameter for axial and radial directions in an aligned fiber composite as (Sambell et al., 1972):

$$\phi_a, \phi_r = \Delta\alpha\Delta T(E_m/\sigma_m)$$

Table 7.1 makes a comparison of damage resulting from thermal expansion mismatch in some carbon fiber reinforced ceramic matrix composites. Note that only glass and glass-ceramic matrices show no damage.

Chemical compatibility between the ceramic matrix and the fiber involves the same thermodynamic and kinetic considerations as with other composite types. Quite frequently, the bond between fiber and ceramic matrix is simple mechanical interlocking. During fabrication (by hot pressing) or during subsequent heat treatments, the fiber/matrix bond could be affected by the high temperatures attained because of any chemical reaction between the fiber and matrix or because of any phase changes in either one of the components. Sambell et al. (1972) studied the zirconia-reinforced magnesia composite system in which there occurs a chemical reaction at 1600 °C. At temperatures less than 1600 °C, the composites showed a weak fiber/matrix interface and fiber pullout occurred during mechanical polishing. Upon heat

Table 7.1 A comparison of damage resulting from the thermal expansion mismatch in some carbon fiber–reinforced systems[a]

Matrix	α_m[b] $(10^{-6}\,{}^\circ C^{-1})$	T_c[c] $({}^\circ C)$	E (GPa)	σ_{mu}[d] (MPa)	ϕ_a[e]	ϕ_r[e]	Damage
MgO	13.6	1200	300	200	25	10	Severe cracking
Al$_2$O$_3$ (80% dense)	8.3	1400	230	300	9	0.3	Severe cracking
Soda-lime glass	8.9	480	60	100	2.6	0.3	Localized cracks
Borosilicate glass	3.5	520	60	100	1.1	−1.4	Uncracked
Glass-ceramic	1.5	1000	100	100	1.5	−6.5	Uncracked

[a] Type I carbon fibers $\alpha_a \approx 0$, $\alpha_r \approx 8 \times 10^{-6}\,K^{-1}$.
[b] α_m is the matrix thermal expansion coefficient.
[c] T_c is the temperature below which little stress relaxation can occur.
[d] σ_{mu} is the matrix strength.
[e] ϕ_a and ϕ_r are the thermal expansion mismatch parameters.
Source: Adapted with permission from Phillips (1983).

treating at 1600 °C, however, because of the interfacial reaction and the resultant improved bonding, no damage was observed upon mechanical polishing. Heat treatment at 1700 °C resulted in the complete destruction of the zirconia fibers and the distribution of zirconia to grain boundaries in magnesia. Thus, as we noted in the case of MMCs (Chap. 6), it is of the utmost importance to be able to control the interfacial bond by means of controlled chemical reactions between components. Chokshi and Porter (1985) also observed a reaction layer on the fiber surface after creep testing in air. Auger electron microscopy analysis showed a mullite layer with large glassy phase regions along grain boundaries. The following interfacial reaction was proposed:

$$2\,SiC + 3\,Al_2O_3 + 4\,O = Al_6Si_2O_{13} + 2\,C$$

and reaction kinetics were modeled by an equation of the form

$$x^2 \simeq Dt$$

where x is the reaction zone thickness, $D = D_o \exp(-Q/kT)$, Q is the activation energy, k is Boltzmann's constant, T is the temperature in kelvin, and t is the time in seconds.

The nature of the bond between fiber and ceramic matrix is thought to be predominantly mechanical. In carbon fiber reinforced glass or glass-ceramics there is little or no chemical bonding (Davidge, 1979; Phillips et al., 1972; Prewo, 1982). Evidence for this is the low transverse strength of these composites and the fact that one does not see matrix material adhering to the fibers on the fracture surface. The bond is thought to be entirely mechanical with the ceramic matrix penetrating the irregularities present on the carbon surface. Shear strength data on carbon fiber in borosilicate glass and lithium-aluminosilicate glass-ceramics (LAS) show that the borosilicate glass composites have double the shear strength of LAS composites. The reason for this is the different radial shrinkage of fibers from the matrix during cooling. Calculated radial contractions of fiber from the matrix are 2.4×10^{-8} m for the LAS ceramic and 0.9×10^{-8} m for the glass-ceramic composite (Phillips, 1983). Shrinkage reduces the mechanical interlocking and thus the fiber/matrix bond.

In ceramic matrix composites, interfacial roughness-induced radial stress will affect the interface debonding, the sliding friction of debonded fibers, and the fiber pullout length. Fiber pullout is one of the important energy dissipating fracture processes in fiber reinforced ceramic or glass matrix composites. An absence of strong chemical bond and a purely mechanical bond at the fiber/matrix interface is highly desirable for the fiber pullout to occur. In regard to the mechanical bonding, a number of researchers have pointed out the importance of interfacial roughness in ceramic matrix composites (CMCs) (Jero, 1990; Jero and Kerans, 1990; Carter et al., 1991; Jero et al., 1991; Kerans and Parthasarathy, 1991; Mackin et al., 1992; Mumm and Faber, 1992; Venkatesh and Chawla, 1992; Chawla et al., 1993a, 1993b; Sorensen, 1993). As shown by Chawla and coworkers (Venkatesh and Chawla, 1992; Chawla et al., 1993a, 1993b), even when the coefficient of thermal expansions of the coating, fiber, and matrix are such that a radial tensile stress exists at the fiber/coating interface after cooling from an elevated processing temperature, fiber pullout may not occur because of a strong mechanical bonding due to a roughness-induced clamping at the fiber/matrix. Thus, a tensile thermal stress in the radial at the interface is desirable factor. A radial tensile stress at the interface will encourage fiber debonding and slippage, which in turn result in high toughness and high work of fracture. The sliding resistance has been expressed as (Hutchinson and Jenson, 1990)

$$\tau = \tau_o - \mu\sigma_n \qquad (\sigma < 0)$$

$$\tau = \tau_o \qquad (\sigma_n \geq 0)$$

where σ_n is the stress normal to the interface, μ is the coefficient of friction, and τ_o is the sliding resistance when σ_2 is positive (tensile). Kerans and Parthasarathy (1991) included interface surface roughness in a detailed treatment of fiber debonding and sliding during both pushout and pullout experiments. They modeled the effect of fiber surface roughness as an increase in the interfacial strain mismatch, leading to an increase in the interfacial normal pressure. The effective normal stress σ_n at the interface due to residual thermal stresses and roughness-induced stresses can be written as

$$\sigma_n = \sigma_t + \sigma_r$$

where σ_t and σ_r are thermal and roughness-induced stresses, respectively.

The thermal stress and roughness-induced stress can be expressed as

$$\sigma_t = \beta(\Delta\alpha\Delta T)$$

$$\sigma_r = \beta(A/r)$$

where $\Delta\alpha$ is the difference in thermal expansion coefficients between matrix and fiber, ΔT is the temperature differential $(T_2 - T_1)$, β is a term containing elastic constants, A is the amplitude of roughness, and r is the fiber radius.

7.3 Properties of CMCs

As mentioned earlier, the relative elastic modulus values of fiber and matrix are very important in CMCs. The ratio E_f/E_m determines the extent of matrix microcracking. Typically, the strain-to-fracture value of a ceramic matrix is very low. Thus, in MMCs and in thermoplastic PMCs, the matrix failure strain (ε_m) is considerably greater than that of fibers. Most unreinforced metals show $\varepsilon_m > 10\%$ while most polymers fail between 3 and 5% strain. Thus, in both MMCs and PMCs fiber failure strain controls the composite failure strain. Typically, fibers such as boron, carbon, and silicon carbide show failure strain values of $\sim 1\%$. Compare this with the failure strains of less than 0.05% for most ceramic matrix materials. The situation in regard to fiber and/or matrix failure is shown in a simplified manner in Figure 7.9. The crack-free original situation is shown in Figure 7.9a. In the case of MMCs and PMCs, fibers fail first at various weak points distributed along their lengths. The composite will fail along a section that has a large number of fiber fractures; see Figure 7.9b. In a strongly bonded CMC, fiber

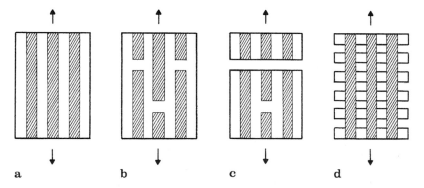

a   b   c   d

Fig. 7.9. a Original crack-free situation. b In the case of MMCs and PMCs, fibers fall first at various weak points distributed along their lengths. The composite will fail along a section that has a large number of fiber fractures. c In a strongly bonded CMC, fiber and matrix would fail simultaneously at matrix failure strain. d In a weakly bonded CMC, however, the matrix will start cracking first and the fibers will be bridging the matrix blocks.

and matrix would fail simultaneously at matrix failure strain and a situation similar to that shown in Figure 7.9c will prevail. In a weakly bonded CMC, however, the matrix will start cracking first and the fibers will be bridging the matrix blocks (Fig. 7.9d). Thus, from a toughness point of view, in general, we do not want too strong a bond in a CMC because it would make a crack run through the specimen. A weak interface, however, would lead to fiber-bridging of matrix microcracks. Consider the simple isostrain model that predicts a rule-of-mixtures type relationship (see Chap. 10) for a unidirectional composite. Let the composite be subjected to a strain e_c. The isostrain condition implies:

$$e_c = e_f = e_m$$

where subscripts c, f, and m denote composite, fiber, and matrix, respectively. This results in a rule-of-mixtures relationship for strength in the longitudinal direction (see Chap. 10), namely,

$$\sigma_c = \sigma_f V_f + \sigma_m V_m$$

where σ is the stress, V is the volume fraction, and the subscripts have the same meaning as given earlier. Note that $V_f + V_m = 1$. Then, assuming elastic behavior for both the matrix and the fiber, we can write

$$\frac{\text{stress carried by fiber}}{\text{stress carried by matrix}} = \frac{\sigma_f V_f}{\sigma_m V_m} = \frac{E_f V_f}{E_m V_m}$$

As the composite is loaded, matrix failure strain being smaller than that of the fiber, it will start showing microcracks at some stress σ_0, as shown in Figure 7.10. We can write

$$\sigma_0 = \sigma_f V_f + \sigma_{mu}(1 - V_f)$$

where σ_{mu} is the matrix stress at its breaking strain.

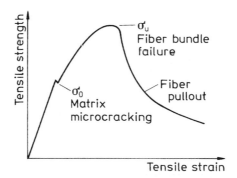

Fig. 7.10. Tensile stress-strain curve of an aligned CMC in the longitudinal direction showing a damage-tolerant behavior. [Evans (1985), used with permission]. As the composite is loaded, matrix failure strain being smaller than that of the fiber, microcracks appear in the matrix at some stress σ_0, followed by fiber bundle failure and pullout.

Table 7.2 Theoretical matrix cracking stresses for a borosilicate glass and magnesia reinforced with 60% of high-modulus carbon fibers ($E = 360$ GPa)

Material	E_m (GPa)	σ_{mu} (MPa)	σ_0 (MPa)
Borosilicate glass	60	100	400
Magnesia	250	120	151

Source: Adapted with permission from Phillips (1983).

Rearranging, we have

$$\sigma_0 = \sigma_{mu}\left[\frac{\sigma_f V_f}{\sigma_{mu}} + (1 - V_f)\right] = \sigma_{mu}\left[1 + V_f\left(\frac{E_f}{E_m} - 1\right)\right]$$

Table 7.2 shows some relevant parameters for borosilicate glass and magnesia matrix materials containing 60% V_f of carbon fibers (Phillips, 1983). We note that in high matrix modulus composites (magnesia in the current case), matrix cracking will be expected to occur at much lower stresses. Thus, it would appear that low-modulus glasses and ceramics offer some advantages over high-modulus ceramic matrices. A CMC with even a microcracked ceramic matrix can retain some reasonable strength ($\sigma_c \simeq \sigma_f V_f$) and there are applications, such as bushings, where such a damage-tolerant characteristic would be very valuable because in the absence of fibers bridging the cracks, the monolithic matrix would simply disintegrate. The disadvantage, of course, is that matrix microcracking provides an easy path for environmental attack of the fibers and the fiber/matrix interface. Let us focus attention on the tensile stress-strain curve of an aligned CMC in the longitudinal direction, as shown schematically in Figure 7.10 (Evans, 1985). This figure shows that CMCs have damage-tolerant characteristics in uniaxial tension. At a stress σ_0, the stress-strain curve shows a dip, indicating the incidence of periodic matrix cracking. Because the fibers have enough strength to support the load in the presence of a damaged matrix (a very desirable feature indeed), the stress-strain curve continues to rise until, at a stress marked σ_u, the fiber bundle fails. At this point, the phenomenon of fiber pullout starts. The extent of this fiber pullout region depends critically on the interfacial frictional resistance. The fiber/matrix interface has a lot to do with the form of the stress-strain curve. If the bonding is too strong, matrix cracking will be accompanied by a small amount of fiber pullout, which is an undesirable characteristic from the toughness viewpoint, as we shall see in Section 7.4. Both σ_0 and σ_u are insensitive to specimen or component size because both strength levels are independent of matrix flaws (Evans, 1985). This is in distinct contrast to the behavior of monolithic ceramic materials, which show a significant size dependence. Increased stiffness and strength were observed in a unidirectionally aligned composite consisting of continuous carbon fibers (50% V_f) in a glass matrix compared to the unreinforced glass matrix (Davidge, 1979). But more importantly, a large increase in the work of fracture occurred. The increased work of fracture is a result of the controlled fracture behavior of the composite, while the unreinforced matrix failed in a catastrophic manner.

Fiber length, or more precisely, the fiber aspect ratio (length/diameter), fiber orientation, relative strengths and moduli of fiber and matrix, thermal expansion mismatch, matrix porosity, and fiber flaws are the important variables that control the performance of CMCs. Sambell et al. (1974) showed that, for ceramic matrix materials containing short, randomly distributed carbon fibers, a weakening effect occurred rather than a strengthening effect. This was attributed to the stress concentration effect at the extremities of randomly distributed short fibers and thermal expansion mismatch. Aligned continuous fibers do lead to a real fiber reinforcement effect. The stress concentration at fiber ends is minimized and higher fiber volume fractions can be obtained. At very high fiber volume fractions, however, it becomes difficult to remove matrix porosity. Figure 7.11 shows a linear increase in strength with fiber volume fraction V_f up to $\sim 55\%$ (Phillips et al., 1972). Beyond 55% V_f, the matrix porosity increased. The Young's modulus also increased

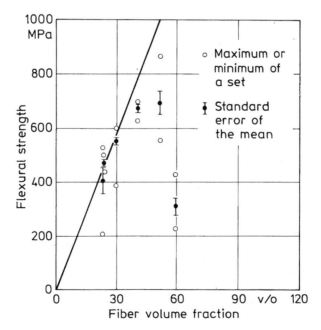

Fig. 7.11. A linear increase in strength with fiber volume fraction V_f up to $\sim 55\%$. [After Phillips et al. (1972), used with permission.]

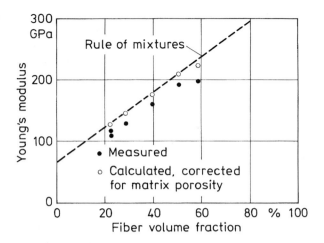

Fig. 7.12. A linear increase in Young's modulus with fiber volume fraction, V_f. At higher V_f, it deviates from linear owing to matrix porosity and possible fiber misalignment. [Phillips et al., (1972).]

linear with V_f (see Fig. 7.12), but at higher V_f it deviated from linear owing to matrix porosity and possible fiber misalignment (Phillips et al., 1972). Silicon carbide SiC whisker reinforced alumina is also more creep-resistant than polycrystalline alumina (Chokshi and Porter, 1985); see Figure 7.13. Note that the composite has a higher-stress exponent than the unreinforced matrix. The higher-stress exponent indicated a change in the operating creep mechanism. The exponent for polycrystalline alumina is about 2, and this is rationalized in terms of some kind of diffusion creep being the controlling mechanism. A stress exponent of about 5, which was observed for the composite, is indicative of a dislocation creep mechanism being in operation. Observation of the specimens deformed in creep in transmission electron microscope showed dislocation activity. Figure 7.14 shows dislocations emanating from the whisker tips, probably resulting from high stress concentrations at these sites.

7.4   Toughness of CMCs

Many concepts have been proposed for augmenting the toughness of ceramic matrix materials (see Chawla, 1993, for a summary). Table 7.3 lists some of these concepts and gives the basic requirements for the models to be valid. Clearly, more than one toughness mechanism may be in operation at a given time. Matrix microcracking, fiber/matrix debonding leading to crack deflection and fiber pullout, and phase transformation toughening

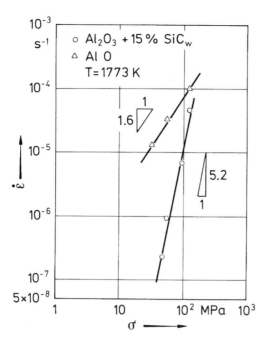

Fig. 7.13. Silicon carbide SiC whisker–reinforced alumina is more creep-resistant than poly-crystalline alumina (Chokshi and Porter, 1985). Note that the composite has a higher stress exponent than the unreinforced matrix.

are all basically energy-dissipating processes that can result in an increase in toughness or work of fracture. Figure 7.15 shows schematically some of these toughness mechanisms or energy-dissipating mechanisms that can be brought to play in CMCs. Impressive work on carbon and SiC fiber rein-forced glass and glass-ceramic composites has been done by Prewo and coworkers (Brennan and Prewo, 1982; Prewo et al., 1986; Fitzer and Hegen,

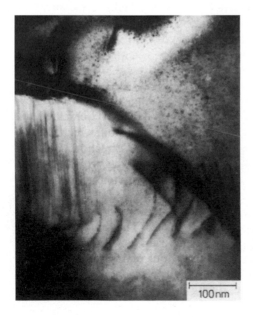

Fig. 7.14. Dislocations emanating from the SiC whisker tips, probably resulting from high stress concen-trations at these sites: TEM. (Courtesy of A.H. Chokshi.)

Table 7.3. Ceramic matrix composite toughening mechanisms

Mechanism	Requirement
1. Compressive prestressing of the matrix	$\alpha_f > \alpha_m$ will result in an axial compressive prestressing of the matrix after fabrication.
2. Crack impeding	Fracture toughness of the second phase (fibers or particles) is greater than that of the matrix locally. Crack is either arrested or bows out (line tension effect).
3. Fiber (or whisker) pullout	Fibers or whiskers having high transverse fracture toughness will cause failure along fiber/matrix interface leading to fiber pullout on further straining.
4. Crack deflection	Weak fiber/matrix interfaces deflect the propagating crack away from the principal direction.
5. Phase transformation toughening	The crack tip stress field in the matrix can cause the second-phase particles (fibers) at the crack tip to undergo a phase transformation causing expansion ($\Delta V > 0$). The volume expansion can squeeze the crack shut.

1979; Prewo, 1982). Their work showed that tough and strong CMCs could be made. Extensive fiber pullout and a controlled fracture behavior of the CMC (i.e., a damage-tolerant fracture behavior) were the marked characteristics. A scanning electron micrograph of a hot-pressed A Ba-Si-Al-O-N glass-ceramic containing Nicalon fibers composite is shown in Figure 7.16. (Herron and Risbud, 1986). The fracture surface of this composite (Fig. 7.17) showed the phenomenon of fiber pullout, indicating a weak fiber/matrix bond. The fibers were bonded to the matrix by an amorphous layer whose characteristics changed with heat treatment; a carbon-rich layer was also observed on the fiber surface (Herron and Risbud, 1986). Remnants of the interfacial amorphous layer adhering to the Nicalon fibers can be seen in Figure 7.17. Another illustration of Nicalon fiber pullout in a Nicalon/ pyrolitic carbon coating/SiC composite is shown in Figure 7.18 (Chawla et al., 1994). Among oxide ceramic matrix materials, alumina and mullite have attracted the most attention. In particular, SiC whisker reinforced alumina composites (20–30% by volume of SiC whiskers in alumina, made by hot pressing) showed impressive gains in toughness and strength (Becher and Wei, 1984; Wei and Becher 1984; Tiegs and Becher, 1986); see Figures 7.19 and 7.20. A typical fine-grained monolithic alumina has a toughness (K_{Ic}) of 4–5 MPa m$^{1/2}$ and a flexural strength between 350 and 450 MPa. Al$_2$O$_3$ containing 20% by volume of SiC whiskers showed a K_{Ic} of 8–8.5 MPa m$^{1/2}$ and a flexural strength of 650 MPa; these levels were maintained up to about 1000 °C.

If the crack growth can be impeded by some means, then a higher stress would be required to make it move. Fibers (metallic or ceramic) can play the role of toughening agents in ceramic matrices. Metallic fiber–reinforced ceramics will clearly be restricted to lower temperatures than ceramic fiber–reinforced ceramic matrices. Glass and glass-ceramic matrices containing carbon fibers (Phillips et al., 1972; Davidge, 1979; Prewo and Brennan,

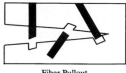

Fiber Pullout

Fig. 7.15. Schematic of some toughness or energy-dissipating mechanisms that can be brought to play in CMCs.

Fiber Bridging

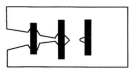

Interface Debonding
+
Crack Deflection

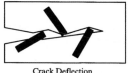

Crack Deflection

Crack Pinning

1980; Prewo, 1982; Brennan and Prewo, 1982; Prewo et al., 1986) have been shown to have fiber pullout as the dominant toughening mechanism. Basically, fiber pullout requires that the strength transferred to the fiber during the ceramic matrix fracture be less than the fiber ultimate strength, σ_{fu}, and that an interfacial shear stress be developed that is greater than the fiber/matrix interfacial strength, τ_i; that is the interface must fail in shear. For a given fiber of radius r, we have the axial tensile stress in the fiber (see Chap. 10)

$$\sigma_f = 2\tau_i \left(\frac{l_c}{r}\right)$$

where l_c is the critical fiber length and $\sigma_f < \sigma_{fu}$. The tensile stress increases from a minimum at both fiber ends and attains a maximum along the central portion of the fiber (see Fig. 10.13). Fibers that bridge the fracture plane and whose ends terminate within $l_c/2$ from the fracture plane will suffer fiber pullout, while those with ends further away will fracture when $\sigma_f = \sigma_{fu}$. A

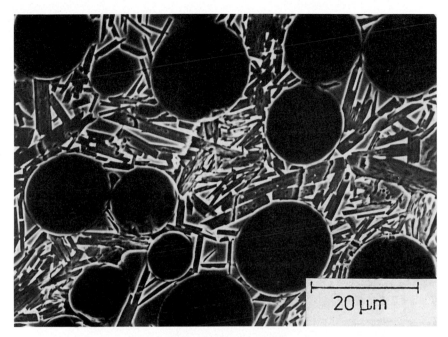

Fig. 7.16. Scanning electron micrograph of a hot-pressed Ba-Si-Al-O-N glass-ceramic matrix/Nicalon fibers composite. [From Herron and Risbud (1986), used with permission.]

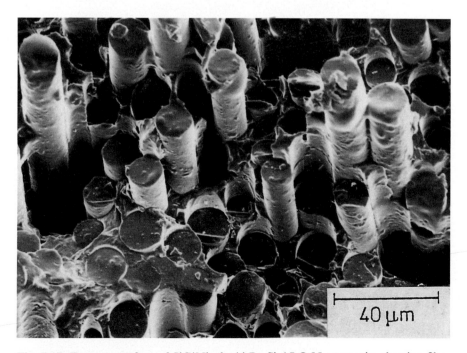

Fig. 7.17. Fracture surface of SiC(Nicalon)/ Ba-Si-Al-O-N composite showing fiber pullout. [Herron and Risbud (1986), used with permission.]

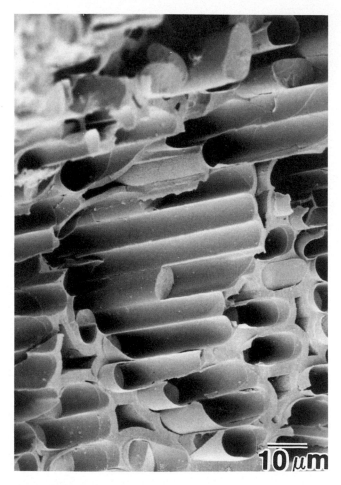

Fig. 7.18. Fiber pullout in a Nicalon/pyrolitic carbon coating/SiC matrix composite. [Courtesy of N. Chawla.]

crack deflection mechanism also requires a weak fiber/matrix bond so that as a matrix crack reaches the interface, it gets deflected along the interface rather than passing straight through the fiber. This is illustrated in Figure 7.21 (Harris, 1980). The original state involving frictional gripping of the fiber by the matrix is shown in Fig. 7.21a. On stressing the composite, a crack initiates in the matrix and starts propagating in the matrix normal to the interface. As it approaches the interface, the crack is momentarily halted by the fiber; see Figure 7.21b. If the fiber/matrix interface is weak, then interfacial shear and lateral contraction of fiber and matrix will result in debonding and crack deflection away from its principal direction (normal to the interface); see Figure 7.21c. A further increment of crack extension in the

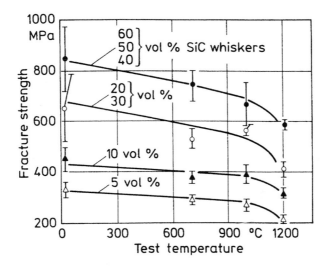

Fig. 7.19. High-temperature strength increases as a function of the SiC whiskers volume fraction in SiC$_w$/Al$_2$O$_3$. [From Tiegs and Becher (1986), used with permission.]

principal direction will occur after some delay. On continuing stressing of the composite, the fiber/matrix interface delamination continues and fiber failure will occur at some weak point along its length; see Figure 7.21d. This is followed by broken fiber ends being pulled out against the frictional resistance of the interface and finally causing a total separation.

DiCarlo (1985) has discussed the requirements for strong and tough CMCs. Avoiding processing-related flaws in the matrix and in the fiber

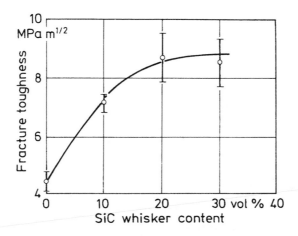

Fig. 7.20. Toughness gains are obtained by incorparating SiC whiskers in alumina. [Tiegs and Becher (1986), used with permission.]

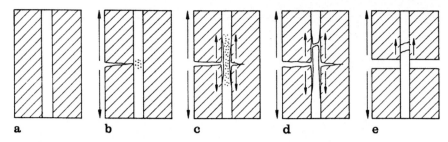

Fig. 7.21. The crack deflection mechanism requires a weak fiber/matrix bond so that as a matrix crack reaches the interface, it gets deflected along the interface rather than passing straight through the fiber. This is illustrated in after Harris (1980). **a** The original state involving frictional gripping of the fiber by the matrix; **b** the crack in the matrix is momentarily halted by the fiber; **c** interfacial shear and lateral contraction of fiber and matrix result in debonding and crack deflection along the interface; **d** further debonding, fiber failure at a weak point, and further extension; **e** broken fiber ends are pulled out against frictional resistance of interface, leading to total separation. [Reprinted from Harris (1981), used with permission.]

would appear to be an elementary and straightforward recommendation. If processing results in large flaws in the matrix, the composite fracture strain will be low. In this respect, fiber bridging of cracks in a CMC will result in a reduced flaw size in the matrix. This, in turn, will help achieve higher applied strains before crack propagation in the matrix than in an unreinforced, monolithic ceramic (Aveston et al., 1971). A weak interfacial bond, as

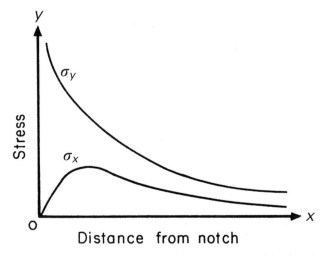

Fig. 7.22. Stress distribution at a crack tip. [After Cook and Gordon (1964).]

pointed out earlier, leads to crack deflection at the fiber/matrix interface and/or fiber pullout. Use of a high-volume fraction of continuous fibers stiffer than the matrix will give increased stiffness, which results in a higher stress level being needed to produce matrix microcracking and a higher composite ultimate tensile stress, as well as a high creep resistance. A high-volume fraction and a small fiber diameter also provide a sufficient number of fibers for crack bridging and postponing crack propagation to higher strain levels. A small-diameter fiber also translates into a small l_c, the critical length for effective load transfer from matrix to fiber (see Chap. 10). Although a weak interface is desirable from a toughness point of view, it provides a short circuit for environmental attack. An ability to maintain a high strength level and high inertness at high temperatures and in aggressive atmospheres is highly desirable.

Crack Deflection Criteria

Cook and Gordon (1964) analyzed the phenomenon of crack deflection or the formation of secondary cracks at a weak interface in terms of the state of stress at the crack tip. Let us consider a fiber/matrix interface, perpendicular to the main advancing crack. Cook and Gordon estimated the strength of the interface necessary to cause a diversion of the crack from its original direction when both fiber and matrix have the same elastic constant. At the tip of any crack, a triaxial state of stress (plane strain) or a biaxial stress (plane stress) is present. Figure 7.22 shows schematically the stress distribution at a crack tip. The main applied stress component, σ_y, has a very high value at the crack tip, and decreases sharply with distance from the crack tip. The stress component acting normal to the interface, σ_x, is zero at the crack tip; it rises to a maximum value at a small distance from the crack tip and then falls off in a manner similar to σ_y. Now, it is easy to visualize that if the interface tensile strength is less than the maximum value of σ_x, then the interface will fail in front of the crack tip. According to the estimates of Cook and Gordon, an interfacial strength of 1/5 or less than that of the main stress component, σ_y, will cause the opening of the interface in front of the crack tip.

More sophisticated analyses of crack interaction with an interface have been proposed (He and Hutchinson, 1989; Evans and Marshall, 1989; Ruhle and Evans, 1988; Gupta, 1991; Gupta et al., 1993). He and Hutchinson's results give the conditions for fiber/matrix debonding in terms of the energy requirements; see Figure 7.23. A plot is made of G_i/G_f vs. α, where G_i is the mixed-mode interfacial fracture energy of the interface, G_f is the mode I fracture energy of the fiber, and α is a measure of elastic anisotropy as defined in the Fig. caption. The plot in Figure 7.23 shows the conditions under which the crack will deflect along the interface or propagate through the interface into the fiber. For all values of G_i/G_f below the dashed line area, interface debonding is predicted. For the special case of zero elastic

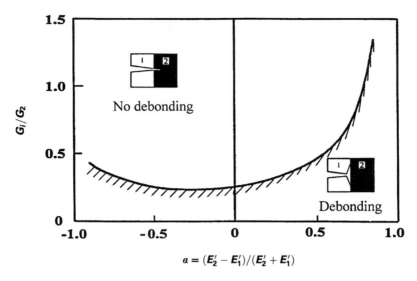

Fig. 7.23. Fiber/matrix debonding criterion in terms of the energy requirements. G_i is the mixed-mode interfacial fracture energy of the interface, G_f is the mode I fracture energy of the fiber, and α is a measure of elastic anisotropy. [After He and Hutchinson (1989).] $E' = E/(1 - \nu^2)$.

mismatch, i.e., for $\alpha = 0$, the fiber/matrix interface will debond for G_i/G_f less than about 0.25. Conversely, for G_i/G_f greater than 0.25, the crack will propagate across the fiber. In general, for the elastic mismatch, with α greater than zero, the minimum interfacial toughness required for interface debonding increases, i.e., high modulus fiber tends to favor debonding. One shortcoming of this analysis is that it treats the fiber and matrix as isotropic materials. This is not always true, especially for the fiber. Gupta et al. (1993) have derived strength and energy criteria for crack deflection at a fiber/matrix interface for several composite systems, taking due account of the anisotropic nature of the fiber. Their experimental technique, laser spallation experiment using a laser Doppler displacement interferometer, was described in Chapter 4. By this technique they can measure the tensile strength of a planar interface. They have tabulated the required values of the interface strength and fracture toughness for delamination in number of ceramic, metal, intermetallic, and polymer matrix composites.

7.5 Thermal shock resistance

Thermal shock resistance is a very important characteristic with ceramics and ceramic composites that are meant to be used at high temperatures and must undergo thermal cycling. We define a thermal shock resistance (TSR) parameter as

$$Thermal\ Shock\ Resistance = \frac{\sigma k}{E\alpha}$$

where σ is the fracture strength, k is the thermal conductivity, E is the Young's modulus, and α is the coefficient of thermal expansion. Most ceramics have low thermal conductivity which is one problem. But high a coefficient of thermal expansion, α, will aggravate the situation further. Thus, common soda-lime glass and alumina have a high α, about 9.10^{-6} K^{-1}. This makes them very poor in TSR. If we reduce CaO and Na_2O in the common glass, and add B_2O_3 to form a borosilicate, we get a special glass that has an $\alpha = 3.10^{-6}$ K^{-1}. Such a glass will show superior TSR and is commercially available under the trade names, Pyrex or Duran glass. Glass-ceramics such as lithium aluminosilicates (LAS) have an α close to zero and thus excellent TSR. LAS are used to make the Corningware which can be take directly from the freezer to the cooking range or oven without causing it to shatter!

Boccaccini et al. (1997a, 1997b) studied the cyclic thermal shock behavior of Nicalon fiber reinforced Duran glass matrix composites. The thermal mismatch between the fiber and the matrix in this system is almost nil. A decrease in Young's modulus and a simultaneous increase in internal friction as a function of thermal cycles were observed. The magnitude of internal friction was more sensitive to microstructural damage than Young's modulus. An interesting finding of theirs involved the phenomenon of crack healing when the glass matrix composite was cycled above the glass transition temperature of the matrix. We discuss this topic further in Chapter 13.

7.6 Applications of CMCs

Ceramic matrix composites find applications in many areas. A convenient classification of the applications of CMCs is aerospace and nonaerospace. Materials-related drivers for applications of CMCs in the aerospace field are:

- High specific stiffness and strength leading to a weight reduction, and, consequently, decreased fuel consumption.
- Reduction in fabrication and maintenance cost.
- higher operating temperatures leading to a greater thermal efficiency.
- longer service life.
- signature reduction.

CMCs can lead to improvements in aerospace vehicles including aircraft, helicopters, missiles, and reentry vehicles. Projected skin temperatures in future hypersonic aircraft are higher than 1600 °C. Other parts, such as radomes, nose tips, leading edges, and control surfaces, will experience only slightly lower temperatures. Currently, one uses sacrifical, nonload-bearing thermal-protection materials on load-bearing components. With the use of CMCs, one can have load bearing at the operating temperatures that are reusable.

Next we give a deseription of some nonaerospace applications of CMCs.

Cutting Tool Inserts

An important area of CMC applications is that of cutting tool inserts. Silicon carbide whisker reinforced alumina (SiC_w/Al_2O_3) is used as cutting tool inserts for high-speed cutting of superalloys. For example, in the cutting of Inconel 718, SiC_w/Al_2O_3 composite tools show performance that is three times better than conventional ceramic tools, and eight times better than cemented carbides. Among the characteristics that make CMCs good candidates for cutting tool inserts are:

- Abrasion resistance
- Thermal shock resistance
- Strength
- Fracture toughness
- Thermal conductivity

Commonly, the volume fraction of SiC_w is 30–45%, and they are made by hot pressing. Figure 7.24 shows the microstructure of such a composite.

Ceramic Composite Filters

Candle-type filters consisting of Nextel™ 312 ceramic fibers in a silicon carbide matrix (see Fig. 7.25) are used to remove particulate matter from high-temperature gas streams up to 1000 °C. The collected particles are removed by reverse pulse jet cleaning.

The high-temperature capability of this filter can eliminate the need to cool the gas stream prior to filtration, which may increase process efficiency and eliminate the cost and complexity of gas dilution, air scrubbers, or heat exchangers. The ceramic fibers toughen the composite construction and result in a filter with excellent resistance to thermal shock and catastrophic failure. The light weight (900 g) of such a filter reduces the strength requirements of the tube sheet, and the excellent thermal shock resistance provides protection during upset conditions. The 3M ceramic composite filter is designed for advanced coal-fired power-generation systems, such as pressurized fluidized bed combustion (PFBC), and integrated gasification combined cycle (IGCC). Among its features are:

- High-temperature capability
- Resistance to thermal shock
- Lightweight
- Resistance to catastrophic failure

Fig. 7.24. Microstructure of SiC whisker–reinforced alumina composite tool insert made by hot pressing.

Potential applications of such filters include:

• Pressurized fluidized bed combustion (PFBC)
• Integrated gasification combined cycle (IGCC)
• Incineration

Supports made of Nicalon fiber reinforced glass matrix composites are used as pad inserts and takeout paddles in direct hot glass-contact equipment (Beier and Markmann, 1997). Figure 7.26 shows an example of such a support made of fiber reinforced glass.

Potentially, CMCs can find applications in heat engines, components requiring resistance to aggressive environments, special electronic/electrical applications, energy conversion, and military systems (Schioler and Stiglich, 1986). Among the barriers that need to be overcome for large-scale applications of CMCs are the high production costs, accepted design philosophy, and the lack of models for strength and toughness. Complicated shapes are difficult to make economically by hot pressing. Sintering or sintering

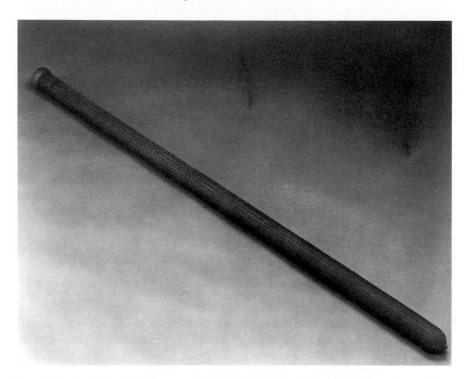

Fig. 7.25. A candle-type filter consisting of Nextel™ 312 ceramic fibers in a silicon carbide matrix. The filter is 1.3 m long. Such filters are used to remove particulate matter from high-temperature gas streams up to 1000 °C. (Courtesy of 3M Co.)

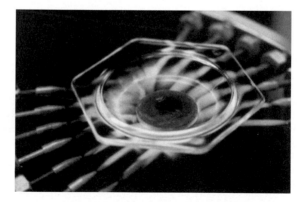

Fig. 7.26. An example of a support made of fiber reinforced glass. (Courtesy of Schott Glaswerke.)

followed by hot isostatic pressing (HIP) are the alternate routes for non-glassy matrices.

References

J. Aveston, G.A. Cooper, and A. Kelly (1971). In *The Properties of Fibre Composites*, IPC Science & Technology Press, Guildford, UK, p. 15.

S.J. Barclay, J.R. Fox, and H.K. Bowen (1987). *J. Mater Sci.*, **22**, 4403.

P.F. Becher and G.C. Wei (1984). *Comm. Am. Ceram. Soc.*, **67**, 259.

W. Beier and S. Markmann (Dec., 1997). *Adv. Mater. & Processes*, **152**, 37.

R.T. Bhatt (1986). *NASA TN-88814*.

R.T. Bhatt (1990). *J. Mater. Sci.*, **25**, 3401.

A.R. Boccaccini, D.H. Pearce, J. Janczak, W. Beier and C.B. Ponton (1997a). *Materials Science and Technology*, **13**, 852.

A.R. Boccaccini, C.B. Ponton and K.K. Chawla (1997b). *Mat. Sci. Eng.* **A241**, 142.

R.K. Bordia and R. Raj (1988). *J. Am. Ceram. Soc.*, **71**, 302.

J.J. Brennan and K.M. Prewo (1982). *J. Mater. Sci.*, **17**, 2371.

C.V. Burkland, W.F. Bustamante, R. Klacka, and J.-M. Yang (1988). In *Whisker- and Fiber-Toughened Ceramics*, ASM Intl., Materials Park, OH, p. 225.

W.C. Carter, E.P. Butler, and E.R. Fuller, Jr. (1991). *Scripta Metall. et Mater.*, **25**, 579–584.

K.K. Chawla (1993). *Ceramic Matrix Composites*, Chapman & Hall, London.

K.K. Chawla, M.K. Ferber, Z.R. Xu, and R. Venkatesh (1993a). *Mater. Sci. & Eng.*, **A162**, 35–44.

K.K. Chawla, Z.R. Xu, A. Hlinak, and Y.-W. Chung (1993b). In *Advances in Ceramic-Matrix Composites*, Am. Ceram. Soc., p. 725–736.

N. Chawla, P.K. Liaw, E. Lara-Curzio, R.A. Lowden, and M.K. Ferber (1994). In *High Performance Composites: Commonalty of Phenomena*, The Minerals, Metals & Materials Society, Warrendale, PA, p. 291.

A.H. Chokshi and J.R. Porter (1985). *J. Am. Ceram. Soc.*, **68**, c144.

J. Cook and J.E. Gordon (1964). *Proc. R. Soc., London*, **A228**, 508.

J.A. Cornie, Y.-M. Chiang, D.R. Uhlmann, A. Mortensen, and J.M. Collins (1986). *Am. Ceram. Soc. Bull.*, **65**, 293.

R.W. Davidge (1979). *Mechanical Behavior of Ceramics*, Cambridge University Press, Cambridge, p. 116.

L.C. De Jonghe, M.N. Rahaman, C.H. Hseuh (1986). *Acta Met.*, **39**, 1467.

J.A. DiCarlo (June 1985). *J. Met.*, **37**, 44.

A.G. Evans, (1985). *Mater. Sci. Eng.*, **71**, 3.

A.G. Evans and D.B. Marshall (1989). *Acta Met.*, **37**, 2567.

E. Fitzer and R. Gadow (1986). *Am. Ceram. Soc. Bull.*, **65**, 326.

E. Fitzer and D. Hegen (1979). *Angew. Chem.*, **91**, 316.

E. Fitzer and J. Schlichting (1980). *Z. Werkstofftech.*, **11**, 330.

C.W. Forrest, P. Kennedy, and J.V. Shennan (1972). *Special Ceramics*, British Ceramic Research Association, Stoke-on-Trent, UK, **Vol. 5**, p. 99.

V. Gupta (1991). *MRS Bulletin*, **XVI-4**, 39.

V. Gupta, J. Yuan, and D. Martinez (1993), *J. Amer. Ceram. Soc.*, **76**, 305.

B. Harris (1980). *Met. Sci.*, **14**, 351.

M.Y. He and J.W. Hutchinson (1989). *J. App. Mech.*, **56**, 270.

M. Herron and S.H. Risbud (1986). *Am. Ceram. Soc. Bull.*, **65**, 342.

W.B. Hillig (1988). *J. Amer. Ceram. Soc.*, **71**, C-96.

J. Homeny, W.L. Vaughn, and M.K. Ferber (1987). *Amer. Cer. Soc. Bull.*, **67**, 333.

J.W. Hutchinson and H.M. Jensen (1990). *Mech. Matls.*, **9**, 139–163.

P.D. Jero (1990). *Am. Ceram. Soc. Bull.*, **69**, 484.

P.D. Jero and R.J. Kerans (1990). *Scripta. Metall.*, **24**, 2315–2318.

P.D. Jero, R.J. Kerans, and T.A. Parthasarathy (1991). *J. Am. Ceram. Soc.*, **74**, 2793–2801.

B. Kellett and F.F. Lange (1989). *J. Am. Ceram. Soc.*, **67**, 369.

R.J. Kerans and T.A. Parthasarathy (1991). *J. Am. Ceram. Soc.*, **74**, 1585–1596.

D. Lewis (1991). In *Metal Matrix Composites: Processing and Interfaces*, Academic Press, Boston, p. 121.

H.Y. Liu, N. Claussen, M.J. Hoffmann, and G. Petzow (1991). *J. Eur. Ceram. Soc.*, **7**, 41.

T.J. Mackin, P.D. Warren, and A.G. Evans (1992). *Acta Metall. Mater.*, **40**, 1251–1257.

D.R. Mumm and K.T. Faber (1992). *Ceram. Eng. & Sci. Proc.*, **7–8**, 70–77.

S. Nourbakhsh, F.L. Liang, and H. Margolin (1990). *Met. Trans. A*, **21A**, 213.

S. Nourbakhsh and H. Margolin (1990). *Met. Trans. A*, **20A**, 2159.

D.C. Phillips (1983). In *Fabrication of Composites*, North-Holland, Amsterdam, p. 373.

D.C. Phillips, R.A.J. Sambell, and D.H. Bowen (1972). *J. Mater. Sci.*, **7**, 1454.

K.M. Prewo (1982). *J. Mater. Sci.*, **17**, 3549.

K.M. Prewo (1986). In *Tailoring Multiphase and Composite Ceramics*, Vol. 20, Materials Science Research, Plenum Press, New York, p. 529.

K.M. Prewo and J.J. Brennan (1980). *J. Mater. Sci.*, **15**, 463.

K.M. Prewo, J.J. Brennan, and G.K. Layden (1986). *Am. Ceram. Soc. Bull.*, **65**, 305.

M.N. Rahaman and L.C. De Jonghe (1987). *J. Am. Ceram. Soc.*, **70**, C-348.

R. Raj and R.K Bordia (1989). *Acta Met.*, **32**, 1003.

M. Ruhle and A.G. Evans (1988). *Mater. Sci. and Eng.*, **A107**, 187.

M.D. Sacks, H.W. Lee, and O.E. Rojas (1987). *J. Am. Ceram. Soc.*, **70**, C-348.

R.A.J. Sambell, D.H. Bowen, and D.C. Phillips (1972). *J. Mater. Sci.*, **7**, 773.

R.A.J. Sambell, D.C. Phillips, and D.H. Bowen (1974). In *Carbon Fibres: Their Place in Modern Technology*, the Plastics Institute, London, p. 16/9.

L.J. Schioler and J.J. Stiglich (1986). *Am. Ceram. Soc. Bull.*, **65**, 289.

P.D. Shalek, J.J. Petrovic, G.F. Hurley, and F.D. Gac (1986). *Amer. Ceram. Soc. Bull.*, **65**, 351.

B.F. Sorensen (1993). *Scripta Metall. et Mater.*, **28**, 435–439.

D.P. Stinton, A.J. Caputo, and R.A. Lowden (1986). *Am. Ceram. Soc. Bull.*, **65**, 347.

T.N. Tiegs and P.F. Becher (1986). In *Tailoring Multiphase and Composite Ceramics*, Plenum Press, New York, p. 639.

A.W. Urquhart (1991). *Mater. Sci. Eng.*, **A144**, 75.

R. Venkatesh and K.K. Chawla (1992). *J. Mater. Sci.*, **11**, 650–652.

G.C. Wei and P.F. Becher (1984). *Am. Ceram. Soc. Bull.*, **64**, 298.

P.A. Willermet, R.A. Pett, and T.J. Whalen (1978). "Development and Processing of Injection-Moldable Reaction-Sintered SiC Compositions," *Amer. Ceram. Soc. Bull.*, **57**, 744.

M. Yang and R. Stevens (1990). *J. Mater. Sci.*, **25**, 4658.

Suggested Reading

K.K. Chawla (1993). *Ceramic Matrix Composites*, Chapman & Hall, London.

K.S. Mazdiyasni (Ed.) (1990). *Ceramic Fiber Reinforced Composites*, Noyes Pub., Park Ridge, NJ.

D.C. Phillips (1983). "Fiber Reinforced Ceramics," In *Fabrication of Ceramics*, vol. 4 of *Handbook of Composites*, North-Holland, Amsterdam, p. 373.

R. Warren (Ed.) (1991). *Ceramic Matrix Composites*, Blackie & Sons, Glasgow, UK.

CHAPTER 8

Carbon Fiber Composites

Carbon fiber reinforced polymer matrix composites can be said to have had their beginning in the 1950s and to have attained the status of a mature structural material in the 1980s. Not unexpectedly, applications in the defense-related aerospace industry were the main driving force for the carbon fiber reinforced polymer matrix composites, followed by the sporting goods industry. The availability of a large variety of carbon fibers (Chap. 2), coupled with a steady decline in their prices over the years, and an equally large variety of polymer matrix materials (Chap. 3) made it easier for carbon fiber polymer composites to assume the important position that they have. This is the reason we devote a separate chapter to this class of composites. Epoxy is the most commonly used polymer matrix with carbon fibers. Polyester, polysulfone, polyimide, and thermoplastic resins are also used. Carbon fibers are the major load-bearing components in most such composites. There is, however, a class of carbon fiber composites wherein the excellent thermal and, to some extent, electrical conductivity characteristics of carbon fibers are exploited; for example, in situations where static electric charge accumulation occurs, parts made of thermoplastics containing short carbon fibers are frequently used. Carbon fibers coated with a metal, e.g., nickel, are used for shielding against electromagnetic interference. Mesophase pitch-based carbon composites show excellent thermal conductivity and thus find applications in thermal management systems. Composites of carbon fibers with metallic and ceramic and glass matrix materials have also found applications. Special mention should be made of carbon/carbon composites, which can be treated as an important subclass of ceramic matrix composites. As we did for other composite systems, we describe the fabrication, properties, interfaces, and applications of carbon fiber reinforced composites.

8.1 Processing of Carbon Fiber Reinforced Composites

Most fabrication methods described in Chapter 5 for polymer matrix composites (PMCs), such as pultrusion, vacuum molding, filament winding, and

Fig. 8.1. A helicopter windshield post made of carbon fibers/vinyl ester resin by pultrusion. The post is 1.5 m long. (Courtesy of Morrison Molded Fiber Glass Co.)

laminated composites starting from prepregs, are also used for carbon fiber reinforced PMCs. An example of a product obtained by pultrusion is shown in Figure 8.1. The hollow trapezoidal-shaped product shown is a helicopter windshield post made of carbon fiber mat and tows in a high-temperature vinyl ester resin matrix. Injection or compression molding of chopped carbon fibers in a thermoplastic or thermoset resin is an economical technique, especially where mechanical property requirements are not very critical. In the aerospace industry, carbon fiber composites are generally made from prepreg sheets or tapes that allow a high fiber volume fraction to be attained. As we described in Chapter 5, prepregs are rolls of sheets of unidirectionally oriented fibers preimpregnated with a partially cured resin matrix. A removable backing sheet is used to prevent sticking of prepregs. Prepregs are stacked in an appropriate sequence (see Chap. 11) and consolidated into a composite component in an autoclave (see Chap. 5). Figure 8.2 shows an optical micrograph of a laminated composite made from carbon fiber/epoxy prepregs. Note the different fiber orientation in different layers. Carbon fiber reinforced carbon matrix composites, or the so-called carbon/carbon composites (C/C), do require different processing techniques. These are actually closer to ceramic matrix composites than polymer matrix composites. Carbon is an excellent high-temperature material when used in an inert or nonoxidizing atmosphere. In carbon fiber reinforced ceramics, the matrix may be carbon or some other glass or ceramic. Unlike other nonoxide ceramics, carbon powder is nonsinterable. Thus, the carbon matrix is generally obtained from pitch or phenolic resins. A heat treatment decomposes the pitch or phenolic to carbon. Many pores are formed during this conversion from a hydrocarbon to carbon. This makes it difficult to fabricate a

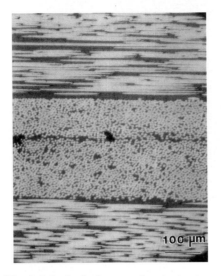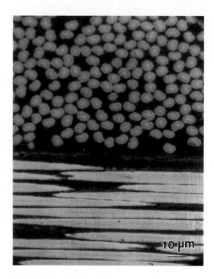

Fig. 8.2 Optical micrographs of a laminated composite made from carbon fiber/epoxy prepregs. Note the different fiber orientation in different layers.

dense and strong pore-free carbon/carbon composite. Thermal transport properties of carbon/carbon composites are of special interest. According to Whittaker and Taylor (1990), it is the matrix thermal conductivity that dominates overall thermal conductivity of a carbon/carbon composite. Thus, one should be able to improve the matrix conductivity by improving its density and orientation.

Carbon/carbon composites can be made by one of the following three methods:

1. A woven carbon fiber preform is impregnated under heat and pressure with pitch from coal tar or petroleum sources. This is followed by pyrolysis of the pitch to obtain a carbonaceous matrix. Generally, the matrix after the first such cycle is highly porous. The cycle may be repeated to obtain the desired amount of densification.
2. A carbon fiber polymeric matrix composite is made by one of the PMC fabrication techniques, followed by pyrolysis of the resin to get a carbonaceous matrix. Commonly, phenolic is used as the polymeric matrix because it gives a high char yield. Multiple cycles involving reimpregnation and repyrolysis of the resin matrix are done to obtain the desired amount of densification.
3. Carbon matrix is obtained by chemical vapor deposition from a gaseous phase (methane + nitrogen + hydrogen) onto and between carbon fibers in a preform.

A schematic of the phenolic or pitch route for obtaining carbon/carbon composites is shown in Figure 8.3. The starting material may be a carbon/

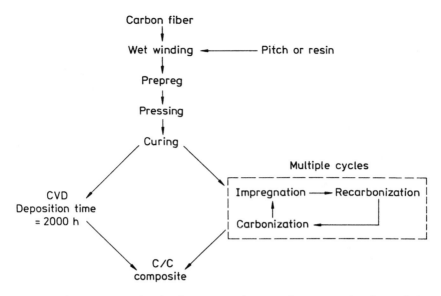

Fig. 8.3. Schematic of carbon/carbon composite manufacture starting from pitch or phenolic resin. [From Fritz et al. (1979), used with permission.]

fiber phenolic prepreg or a three-dimensional woven structure, say a cone, which can be impregnated under pressure with resin or pitch. In the case of prepregs, they are cut and laid up in a form appropriate for the final desired shape. Application of heat and pressure results in curing of the matrix and fixing of the carbon fibers in the desired shape. This is followed by a densification cycle involving carbonization in an inert atmosphere, with an optional high-temperature graphitization treatment, and reimpregnation by resin or pitch and repyrolysis to obtain the desired density. The densification cycle, indicated by the dashed lines in Figure 8.3, may be repeated several times to reduce the porosity to an acceptable level. This densification may also be carried out by CVD of carbon from a hydrocarbon gas (e.g., methane, acetylene, or benzene). A characteristic feature of pyrolysis of the polymer or pitch infiltrates is the rather heavy weight loss, anywhere between 10 and 60%, which is inevitably accompanied by rather large shrinkage porosity (Fritz et al., 1979; McAllister and Lachman, 1983; Chawla, 1993).

8.2 Properties

Since a variety of carbon fibers is available, one can produce carbon fiber reinforced composites showing a range of properties. Two major types of carbon fiber are the high-strength and high-modulus types (Sect. 2.4). Accordingly, we summarize some of the mechanical characteristics of uni-

Table 8.1. Typical mechanical properties of some carbon fiber/epoxy composites[a]

Property	AS	HMS	Celion 6000	GY 70
Tensile strength (MPa)	1850	1150	1650	780
Tensile mudulus (GPa)	145	210	150	290–325
Tensile strain to fracture (%)	1.2	0.5	1.1	0.2
Compressive strength (MPa)	1800	380	1470	620–700
Compressive modulus (GPa)	140	110	140	310
Compressive strain to fracture (%)	—	0.4	1.7	—
Flexural strength (4 points) (MPa)	1800	950	1750	790
Flexural modulus (GPa)	120	170	135	255
Interlaminar shear strength (MPa)	125	55	125	60

[a] Values given are indicative only and are for a unidirectional composite (62% V_f) in the longitudinal direction.
Source: Adapted with permission from Riggs et al. (1982).

directionally aligned, 62% volume fraction carbon fibers from different sources in an epoxy matrix in Table 8.1 (Riggs et al., 1982). Carbon fiber reinforced composites also show a linear relationship between the longitudinal elastic modulus and fiber volume fraction. Thus we can write the longitudinal Young's modulus, E_{cl}, of the composite

$$E_{cl} = E_f V_f + E_m(1 - V_f)$$

where E is the Young's modulus, V is the volume fraction, and the subscripts f and m designate fiber and matrix, respectively. Such a relationship is generally valid in tension and compression. It will be valid for flexure if shear effects can be neglected.

Carbon fibers find a major outlet in the aerospace industry. Because aerospace structures are exposed to a range of environments and temperatures, for example, oils, fuels, moisture, acids, and hot gases, the excellent corrosion-resistance characteristics of carbon/epoxy composites are of great value under such conditions. Commonly encountered damage to polymers by ultraviolet rays is minimized by properly painting the exterior of the composite. Moisture is a major damaging agent. Epoxy matrices can absorb water to as much as 1% of the composite weight; however, unlike glass fiber, which is attacked by moisture, the carbon fiber itself is unaffected by moisture. Thus, moisture absorbed in carbon fiber PMCs opens up the polymer structure and reduces its glass transition temperature; that is, the moisture acts as a plasticizer for the polymeric matrix. Moisture absorption in polymers occurs according to Fick's law; that is, the weight gain owing to moisture intake varies as the square root of the exposure time. This Fickian moisture absorption in a 16-ply carbon epoxy laminate is shown in Figure 8.4a (Shirrel and Sandow, 1980). An example of moisture-induced degradation in a model composite consisting of a single carbon fiber in an epoxy matrix is shown in Figures 8.4b, c, and d. A photoelastic technique that shows stress-induced birefringence was used to analyze the stress patterns obtained as a function of immersion time in water. Figure 8.4b shows the initial condition. After an immersion for 10.5 hours, a net tensile stress

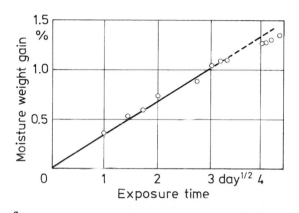

a

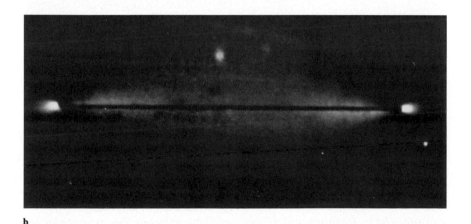

b

Fig. 8.4. a Fickian moisture absorption in a carbon/epoxy laminate. [From Shirrel and Sandow (1980), used with permission.] **b, c,** and **d** Moisture-induced degradation in a model composite consisting of a single carbon fiber in an epoxy. The immersion times are 0, 10.5, and 151 hours, respectively. After an immersion for 10.5 hours, a net tensile stress birefringence can be seen, while after 151 hours, large swelling of the epoxy matrix occurred, causing an increase in the axial tensile stress in the fiber and eventually fiber fracture. (Courtesy of Z.R. Xu.)

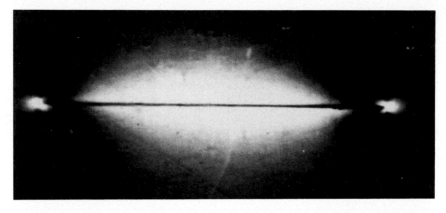

c

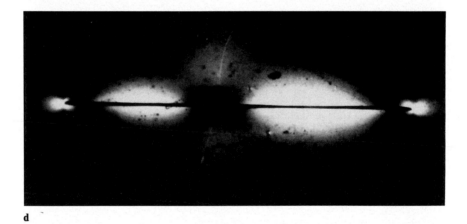

d

Fig. 8.4 (*continued*)

birefringence was obtained (Fig. 8.4c). With continued moisture absorption, more swelling of the epoxy matrix occurred, causing an increase in the axial tensile stress in the fiber and eventually leading to fiber fracture after 151 hours, as shown in Figure 8.4d (Xu and Ashbee, 1994). The moisture absorption problem in PMCs is analogous to that of degradation by temperature effects. Collings and Stone (1985) present a theoretical analysis of strains developed in longitudinal and transverse plies of a carbon/epoxy laminate owing to hygrothermal effects. An interesting finding of theirs is that tensile thermal strains that develop in the matrix after curing are reduced by compressive strains generated in the matrix by swelling resulting from water absorption. It is worth pointing out that moisture absorption causes compressive stresses in the resin and tensile stresses in the fibers (see Fig. 8.4). Also, a temperature increase ΔT generates strains of the same sign

Fig. 8.5. Comparison of flexural strength of carbon/polyimide (in air) and carbon/carbon and carbon/aluminum (insert atmosphere) as a function of temperature. [From Fitzer and Heym (1976), used with permission.]

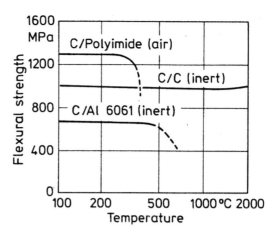

as those caused by an increase in the moisture content (Collings and Stone, 1985). Thus, it is understandable that moisture absorption should reduce the residual strains after curing.

In view of the high-temperature-strength requirements, it is instructive to look at the high-temperature strength of carbon fiber reinforced composites. Figure 8.5 shows flexural strengths of carbon/polyimide in air, and those of carbon/carbon and carbon/aluminum in an inert atmosphere, as a function of temperature (Fitzer and Heym, 1976). While carbon/polyimide shows a superior high-temperature strength than most carbon fiber PMCs, we note that carbon/aluminum shows good strength up to about 500 °C and that carbon/carbon can withstand temperatures as high as 2000 °C. It is worth pointing out that the excellent high-temperature characteristics of carbon/aluminum and carbon/carbon are obtainable in inert atmospheres only. In air, carbon fibers are readily oxidized and protective coatings are needed.

Carbon fiber/epoxy composites exhibit superior properties in creep compared to aramid/epoxy. This is because aramid fibers, similar to other polymeric fibers, creep significantly even at quite low stresses (Eriksen, 1976). The ply stacking sequence can affect the composite properties. Figure 8.6 shows tensile creep strain at ambient temperature as a function of time for two different stacking sequences (Sturgeon, 1978). Note that a laminate with carbon fibers at $\pm 45°$ shows more creep strain than one containing plies at $0°/90°/\pm 45°$. The reason for this is that in the $\pm 45°$ sequence, the epoxy matrix undergoes creep strain by tension in the loading direction, shear in the $\pm 45°$ directions, and rotation of the plies in a scissor-like action. As we shall see in Chapter 11, $0°$ and $90°$ plies do not show the scissor-like rotation characteristic of $45°$ plies. Thus, the addition of $0°$ and $90°$ plies reduces the matrix shear deformation. Consequently, the creep resistance of $0°/90°/\pm 45°$ sequence is better than that of the $\pm 45°$ sequence.

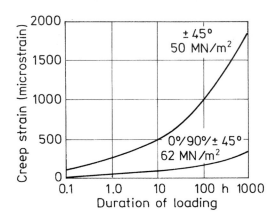

Fig. 8.6. Creep strains at ambient temperature for a +45 and for a 0/90~/+45 carbon/epoxy laminate. [From Sturgeon (1978), used with permission.]

Carbon fiber reinforced PMCs generally show excellent fatigue strength. Depending on the ply stacking arrangement, their fatigue strength (tension-tension) may vary from 60 to 80% of the ultimate tensile strength for lives over 10^7 cycles (Baker, 1983). The higher fatigue strength levels pertain to composites having more than 50% of the fibers in the longitudinal direction (0°), which leads to high longitudinal stiffness and low strains. Pipes and Pagano (1970) showed that certain stacking sequences can result in tensile stresses at the free edges, which can lead to early local delamination effects in fatigue and consequently to lower fatigue lives. Further discussion of these topics can be found in Chapters 11 through 14.

One of the areas where extensive efforts have been made is that of toughening the carbon fiber PMCs. This has involved modifying epoxies and using polymeric matrix materials other than epoxies. Among the latter are modified bismaleimides and some new thermoplastic materials (see also Chap. 3). The latter category includes poly (phenylene sulfide) (PPS), poly-sulfones (PS), and polyetheretherketone (PEEK), among others. PEEK, a semicrystalline polyether, combines excellent toughness with chemical inertness. Typically, PEEK-based carbon fiber composites are equivalent to high-performance epoxy-based carbon fiber composites, with the big advantage that the former have an order-of-magnitude higher toughness than the latter. PPS is a semicrystalline aromatic sulfide that has excellent properties. A special process has been developed to make continuous fiber prepregs with PPS for use in making composite laminates. It should be pointed out that incorporation of fibers in a thermoplastic polymer can result in alteration of crystallization kinetics of the thermoplastic matrix (Waddon et al., 1987; Hull, 1994). Figure 8.7 shows a schematic representation of two possible spherulitic morphologies resulting from constrained growth in PEEK, a thermoplastic matrix, because of the presence of carbon fibers. The nucleation density of spherulites on the fibers is the same in the two cases but the fiber volume fraction is different, i.e., the interfiber spacing is smaller for the high fiber volume fraction composite. Hence, the constraint on the growth of

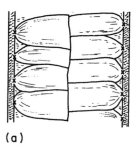 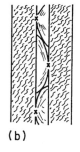

☒ Carbon fiber

× Nucleation site

(a)   (b)

Fig. 8.7. A schematic representation of two possible spherulitic morphologies resulting from constrained growth in PEEK, a thermoplastic matrix, because of the presence of carbon fibers. (a) low fiber volume fraction (b) high fiber volume fraction. The nucleation density of spherulites on the fibers is the same in the two cases but the fiber volume fraction is different, i.e., the interfiber spacing is smaller for the high fiber volume fraction composite. Hence, the constraint on the growth of spherulites is greater in the high fiber volume fraction case. [After Waddon et al. (1987).]

spherulites is greater in the high fiber volume fraction case. This is really a general issue in all kinds of composites, namely, the microstructure and properties of matrix change during processing of the composite. See Chapter 12 for a further discussion of the importance of such structure-sensitive issues in composites.

Properties of carbon/carbon composites depend on the type of carbon fiber used (high-modulus or high-strength type), fiber volume fraction, the fiber distribution. One-, two-, and three-dimensionally woven carbon fibers may be used. Figure 8.8 shows a two-dimensional (2-D) plain weave and a five-harness satin weave (McAllister and Lachman, 1983). Modifications of the basic 3-D orthogonal weave involving 5, 7, or 11 fiber directions are possible, which give a highly isotropic final composite (McAllister and Lachman, 1983). Carbon fiber can be woven in a variety of weaves for reinforcement in two or more dimensions. Two main categories of weave are plain and satin. A *plain weave* has one warp yarn running over and under one fill yarn and is the simplest weave. *Satin-type weaves* are more flexible; that is, they can conform to complicated shapes easily. A five-harness satin weave, shown in Figure 8.8b, has one warp yarn running over four fill yarns and under one fill yarn. The precursor, processing, and high temperature involved all influence the carbonaceous matrix significantly. Table 8.2 summarizes the room-temperature properties of some carbon/carbon composites. The values refer to a high-modulus carbon fiber, a final heat treatment at 1000 °C, four to six densification cycles, and a fiber volume fraction of 55. The properties of a 2-D weave depend on the type of weave of carbon cloth. The properties depend strongly on the weave pattern and the amount of fibers in the x, y, and z directions. Figure 8.9 shows schematically the stress-strain curves, at room temperature, of 1-D, 2-D, and 3-D carbon/carbon

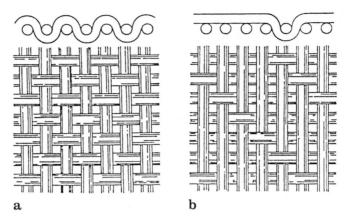

a   b

Fig. 8.8. Two-dimensional fabric: **a** plain weave; **b** five-harness satin. (Courtesy of Fiber Materials Inc.)

composites (Fritz et al., 1979). Note the change in fracture mode from semi-brittle (1-D) to nonbrittle (3-D). The latter is due to the existence of the continuous crack pattern in the composites. Because carbon/carbon composites are meant for high-temperature applications, their thermal expansion behavior is of great importance. As expected, fibers control the thermal expansion behavior parallel to the fibers, while perpendicular to the fiber axis the carbonaceous matrix controls the expansion behavior. The amount of porosity in the matrix will also influence the thermal expansion behavior.

8.3   Interface

As in any other composite, the carbon fiber/polymer matrix interface is very important in determining the final properties of the composite. Carbon fiber is a highly inert material. This makes it difficult to have a strong adhesion between carbon fiber and a polymer matrix. One solution is to make the fiber surface rough by oxidation or etching in an acid. This results in an increased

Table 8.2. Mechanical properties of carbon/carbon composites at room temperature

Weave	Flexural Strength (MPa)	Young's Modulus (GPa)	Interlaminar Shear Strength (MPa)
1-D, 55% V_f	1200–1400	150–200	20–40
2-D, 8H/S weave, 35% V_f	300	60	20–40
3-D, felt, 35% V_f	170	15–20	20–30

Source: Adapted with permission from Fritz et al., 1979.

Fig. 8.9. Stress—strain curves (schematic) of 1-D, 2-D, and 3-D carbon/carbon composites. [From Fritz et al. (1979), used with permission.]

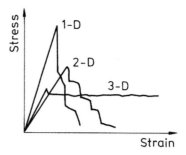

specific surface area and improved wetting, which in turn results in an improved interlaminar shear strength because of the mechanical keying effect at the fiber/matrix interface. Recall that the carbon fiber microstructure is inhomogeneous through its cross section. Specifically, in the surface layer, the basal planes are aligned parallel to the surface. This graphitic layer is very smooth and weak in shear and hard to bind with a matrix. That is the reason that a surface treatment is given to carbon fibers. A variety of surface treatments can be used to accomplish the following:

a. increase the surface roughness
b. increase the surface reactivity

The interface region between a carbon fiber and the polymer matrix is quite complex. It is therefore not surprising that a unified view of the interface in such composites does not exist. As pointed out in Chapter 2, the carbon fiber structure, at submicrometer level, is not homogeneous through its cross section. The orientation of the basal planes depends on the precursor fiber and processing conditions. In particular, the so-called onion-skin structure is frequently observed in PAN-based fibers, wherein basal planes in a thin surface layer are aligned parallel to the surface while the basal planes in the core are less well-aligned. Figure 8.10 shows the structure of a carbon fiber/epoxy composite (Diefendorf, 1985). The onion-skin zone (C in Fig. 8.10) has a very graphitic structure and is quite weak in shear. Thus, failure is likely to occur in this thin zone. Additionally, the skin can be hard to bind with a polymeric matrix because of the high degree of preferred orientation of the basal planes, thus, facilitating interfacial failure (zone D). The matrix properties in zone E (close to the interface) may be different from those of the bulk epoxy (zone B). Carbon fibers meant for polymer reinforcement invariably receive some form of surface treatment from the manufacturer to improve their compatibility with the polymer matrix and their handleability. Organic *sizes* are commonly applied by passing the fibers through a sizing bath. Common sizes include polyvinyl alcohol, epoxy, polyimide, and titanate coupling agents.

Carbon fibers, especially high-modulus carbon fibers that have undergone a high-temperature graphitization, are quite smooth. They have a rather low

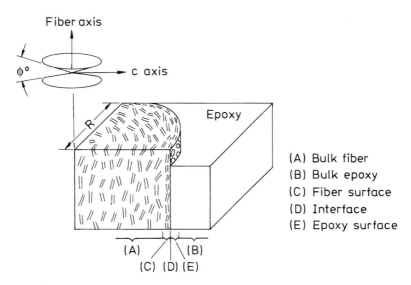

Fig. 8.10. Structure of carbon fiber, interface, and epoxy matrix. [From Diefendorf (1985), used with permission.]

specific area, varying from 0.1 to 2 m^2 g^{-1}. Invariably, there is a microscopic scale of roughness, mostly as longitudinal striations (see Fig. 2.20). Carbon fibers are also generally chemically inert; that is, interfacial interactions in carbon fiber–based composites would be rather weak. Generally, a short beam bending test is conducted to measure what is called *interlaminar shear strength* (ILSS). Admittedly, such a test is not entirely satisfactory, but for lack of any better, quicker, or more convenient test, the ILSS test value is taken as a measure of bond strength.

Ehrburger and Donnet (1980) point out that there are two principal ways of improving interfacial bonding in carbon fiber composites: increase the fiber surface roughness, and thus the interfacial area, and increase the surface reactivity. Many surface treatments have been developed to obtain improved interfacial bonding between carbon fibers and the polymer matrix (Donnet and Bansal, 1984; McKee and Mimeault, 1973). A brief description of these follows.

8.3.1 Chemical Vapor Deposition (CVD)

Silicon carbide and pyrocarbon have been deposited on carbon fibers by CVD. Growing whiskers on the carbon fiber surface, called *whiskerization* in the literature, can result in a two- to threefold increase of ILSS. SiC whiskers ($\sim 2\,\mu$m in diameter) are grown on the fiber surface. The improvement in ILSS is mainly due to the increase in surface area. Whiskerization involves the growth of single-crystal SiC whiskers perpendicular to the fiber, which

Fig. 8.11. A linear relationship between the carbon fiber surface roughness as measured by AFM and interlaminar shear strength in a series of carbon fiber/epoxy composites [Jangehud et al., (1993).]

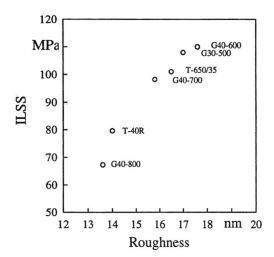

results in an efficient mechanical keying effect with the polymer matrix. The CVD method is expensive, and handling of whiskerized fibers is difficult. Although this treatment results in an increase in ILSS, there also occurs a weakening of the fibers.

8.3.2 Oxidative Etching

Treating carbon fibers with several surface-oxidation agents leads to significant increases in the ILSS of composites. This is because the oxidation treatment increases the fiber surface area and the number of surface groups (Ehrburger and Donnet, 1980). Figure 8.11 shows a linear relationship between the fiber surface roughness as measured by AFM and interlaminar shear strength in a series of carbon fiber/epoxy composites (Jangehud et al., 1993). Yet another reason for increased ILSS may be the removal of surface defects, such as pores, weakly bonded carbon debris, and impurities (Donnet and Bansal, 1984). Oxidation treatments can be carried out by a gaseous or a liquid phase. Gas phase oxidation can be done with air or oxygen diluted with an inert gas (Clark et al., 1974). Gas flow rates and the temperature are the important parameters in this process. Oxidation results in an increase in the fiber surface roughness by pitting and increased longitudinal striations (Donnet and Bansal, 1984). Too high an oxidation rate will result in non-uniform etching of the carbon fibers and a loss of fiber tensile strength. Since oxidation results in a weight loss, we can conveniently take the amount of weight loss as an indication of the degree of oxidation. Figure 8.12 shows ILSS versus weight loss for high-strength and high-modulus carbon fiber composites (Clark et al., 1974). The maximum ILSS corresponds to less than a 10% weight loss in both cases. Overoxidation results in a loss of fiber strength and lower ILSS.

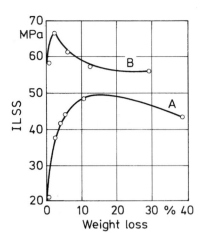

Fig. 8.12. Interlaminar shear strength (ILSS) as a function of weight loss on oxidation. A high-weight loss corresponds to a high degree of oxidation. Note that the maximum ILSS corresponds to less than 10% weight loss in both cases. A High-modulus-type carbon fiber; B high-strength type. [From Clark et al. (1974), used with permission.]

8.3.3   Liquid Phase Oxidation

Liquid phase oxidation involves treatments in nitric acid, sodium hypo-chlorite, potassium permanganate, and anodic etching (Donnet and Bansal, 1984). Liquid phase oxidation by nitric acid and sodium hypochlorite results in an increase in the interfacial area and formation of oxygenated surface groups due to fiber etching. Wetting of the carbon fibers by the polymer is enhanced by these changes.

Graphitic oxides are lamellar compounds having large amounts of hydroxylic and carboxylic groups. The formation of a graphitic oxide layer increases the number of acidic groups on the carbon fibers.

Anodic etching or electrochemical oxidation using dilute nitric acid or dilute sodium hydroxide solutions results in no significant decrease in tensile strength of the carbon fibers, according to Ehrburger and Donnet (1980). The fiber weight loss is less than 2% and no great change in surface area or fiber roughness occurs, the major change being an increase in the acidic surface groups. Oxidative treatments produce functional groups ($-CO_2H$, $-C-OH$, and $-C=O$) on carbon fiber surfaces. They form at the edges of basal planes and at defects. These functional groups form chemical bonds with unsaturated resins.

According to Drzal et al. (1983a, 1983b) various carbon fiber surface treatments promote adhesion to epoxy materials through a two-pronged mechanism: (a) the surface treatment removes a weak outer layer that is present initially on the fiber and (b) chemical groups are added to the surface which increase interaction with the matrix. When a fiber finish is applied, according to Drzal and coworkers, the effect is to produce a brittle but high modulus interphase layer between the fiber and the matrix. As amine content is reduced, the modulus of epoxy goes up accompanied by lower fracture strength and strain. That is, the interphase being created between the fiber and the matrix has high modulus but low toughness. This promotes matrix fracture as opposed to interfacial fracture.

8.4 Applications

The aerospace industry is the major user of lightweight structures made of carbon fiber reinforced polymer matrix composites. Carbon fiber PMCs started entering the aircraft industry as early as the 1970s. In commercial aircraft, such as the Boeing 757, 767, and 777 and Airbus 320 planes, carbon fiber composites are used in a number of locations. The weight savings achieved in the Boeing 767 through the use of composites amounted to about 1000 kg over conventional metallic structures. An aircraft louvered door made of carbon fiber (40%) in a nylon 6/12 matrix that is used on the engine nacelle of the Boeing 757 aircraft is shown in Fig. 8.13. The helicopter industry in particular has been an enthusiastic user of carbon fiber PMCs. Figure 8.1 shows a helicopter windshield post made of carbon fiber/vinyl ester. It is claimed that composite helicopter rotor blades have lower direct operating costs than aluminum blades (Mayer, 1974), and this is with-

Fig. 8.13. A louvered door made of carbon fiber (40% V_f)/nylon thermoplastic used on the engine nacelle of the Boeing 757 aircraft. (Courtesy of LNP Corporation.)

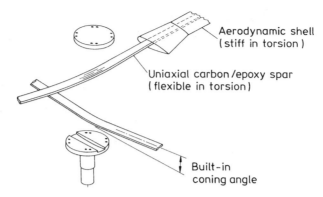

Fig. 8.14. Schematic carbon/epoxy helicopter rotor blades. [From Mayer (1974), used with permission.]

out taking into account the intangibles such as longer fatigue life, reduced maintenance, and lower manufacturing costs. Boeing Vertol, Bell Aircraft, and Sikorsky, among others, use composite rotor blades in their helicopters. Figure 8.14 shows a schematic of a carbon/epoxy helicopter rotor blade. Cargo bay doors, maneuvering engine pods, arm booms, and booster rocket casings in the U.S. Space Shuttle Orbiter are made of carbon fiber/epoxy composites. Figure 8.15 shows the primary tower structure and several

Fig. 8.15. Intelsat 5 has a primary structure and several antennas made of carbon fiber PMCs. (Courtesy of Fiberite Co.)

antennas made of carbon fiber/epoxy for use in Intelsat 5. The main attractions for their use are lightness and dimensional stability.

In the road transport industry carbon fiber/epoxy composites have been tried in leaf springs, drive shafts, and various chassis parts. Carbon fiber composites allow a one-piece design to replace a two-piece drive shaft, resulting in about 50–60% weight reduction. Blades in turbines, compressors, and windmills, as well as in ultracentrifuges and flywheels, have been made of carbon fiber PMCs. Other applications include cargo shipping containers for rail, sea, and land transportation. Pressurized tanks, especially compressed natural gas tanks (CNG) for bus and other vehicle fleet market are other applications of carbon fiber reinforced composites (see Chapter 5). Use of carbon fiber composites leads to a reduction in weight and increased payload.

Carbon fibers are used as an additive in the automotive airbag propellant mix in the pyrotechnic type of airbags in automobiles. Carbon fibers help improve the performance of the inflation process.

Medical applications of carbon fiber composites have involved their use as prosthetic devices, as well as in the manufacture of medical equipment, for example, X-ray film holders and tables for X-ray equipment. The ability to tailor the required stiffness and strength properties of internal bone plates made of carbon fiber composites gives them a definite advantage over metallic parts.

The leisure and sporting goods industry is another big market for carbon fiber PMCs. Golf clubs, archery bows, fishing rods, tennis rackets, cricket bats, and skis are commonplace items in which carbon fiber PMCs are used. Figure 8.16a shows a tennis racket, a pair of skis, and a fishing rod while Fig. 8.16b shows a bicycle frame made of carbon fiber/epoxy. Note the sleek lines that have become a hallmark of composite construction. In addition to attractive mechanical characteristics, carbon fiber reinforced thermoplastic composites have excellent electrical properties. This is exploited in situations where a static charge builds up easily, for example, in high-speed computer parts and in musical instruments where rubbing, sliding, or separation of an insulating material results in electrostatic voltages. In parts made of insulating polymeric resins, this charge stays localized until the polymer comes in contact with a body at a different potential and the electrostatic voltage discharges via an arc or spark. Voltages as high as 30–40 kV can build up. This electrostatic charge can be painful to a human being—or even fatal in extreme circumstances. Some very sensitive microelectronic parts can be damaged by an electrostatic discharge of a mere 20 V. Thus, it is not surprising that carbon fibers dispersed in thermoplastic resins find applications where dissipation of a static charge is important. Of course, conductive fillers of one kind or another can be used to overcome the problem of static electricity, but carbon fibers also serve to reinforce mechanically in situations requiring high strength and wear resistance. Figures 8.17a and b show a microphone and head shell, the unit that holds the stylus at the end of the

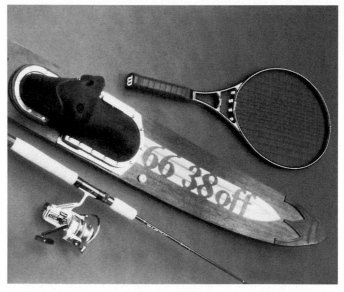

a

b

Fig. 8.16. Examples of sporting goods made of carbon fiber composites where lightness, good mechanical characteristics, and sleek lines make the items very attractive: **a** tennis racket, a pair of skis, and a fishing rod. (Courtesy of Fiberite Co.), **b** a carbon fiber, a bicycle frame made of carbon fiber/epoxy. (Courtesy of Trek Co.)

a

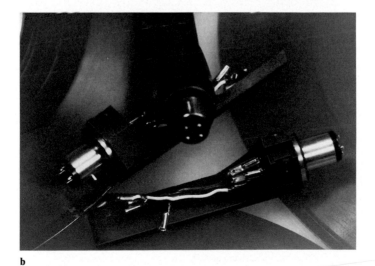

b

Fig. 8.17. Use of carbon fiber/thermoplastic matrix composites in situations involving static charge: **a** microphone, **b** a head shell unit of a turntable arm. (Courtesy of LNP Corporation.)

turntable arm. Carbon fibers dissipate the static directly, thus eliminating the need for copper conductors. This is coupled with a high stiffness-to-weight ratio of these composites, which allows a weight reduction of the part as well. We should also mention that quite a few companies are making musical instruments such as guitars, violins, and violas as well as bows out of carbon fiber reinforced composites.

Electromagnetic Shielding

Shielding against electromagnetic interference (EMI) is another area where highly conductive composites based on carbon fibers are finding applications. EMI is nothing but electronic pollution or noise caused by rapidly changing voltages. Examples include avionic housings, computer enclosures, and any other electronic device that needs protection against stray EMI. Nickel-coated carbon fibers are used in shielding against electromagnetic and radio frequency interference. Nickel confers excellent conductivity while retaining the flexibility of carbon fibers. In aircraft, such a composite can provide protection against lightning strikes. Radar-absorbing materials are used to reduce EMI. Navy ships carry a large number of antennas, computers, and telecommunications equipment. A myriad of EMI problems can arise under such circumstances. Nickel is deposited by chemical vapor deposition on carbon fibers. Nickel-coated carbon fibers can be incorporated into a polymeric matrix by any of the methods described earlier. Finely chopped nickel-coated carbon fibers can be incorporated in adhesives, sealants, gaskets, and battery electrodes for use in aerospace and for electronic applications requiring EMI shielding.

It is convenient to measure shielding effectiveness or attenuation in decibels (dB), which is nothing but a logarithmic scale. The decibel scale is preferred when a quantity can vary over several orders of magnitude. Electromagnetic shielding effectiveness (SE) can be defined as follows:

$$Shielding\ effectiveness\ (SE) = 20 \log\left(\frac{incident\ field\ strength}{transmitted\ field\ strength}\right)$$

Table 8.3 shows the correspondence between shielding effectiveness in dB and signal attenuation in percentage. A shielding level of 80 dB implies

Table 8.3. Correspondence between shielding effectiveness in dB and % signal attenuation

Shielding Effectiveness (dB)	Attenuation (%)
20	90
40	99
60	99.9
80	99.99

99.99% attenuation of the incident electromagnetic radiation. For most business electronic equipment with 30–1000 MHz frequency, 35–45 dB attenuation is adequate. Requirements of military applications are stringent, and 60–80 dB attenuation is not uncommon.

While there are applications involving the use of the conductive properties of carbon fibers, there are also certain problems associated with this characteristic of carbon fibers. Carbon fibers are extremely fine and light, and if they become accidently airborne, for any reason, and settle on electrical equipment, short circuiting can occur. The conclusion of extensive studies conducted at NASA on this problem was that despite the risk, there was no reason to prohibit the use of carbon fibers in structures; see NASA (1980).

Civil Infrastructure Applications

A major development in the 1990s has been the use of fiber reinforced composites in civil infrastructure. Among the major drivers for this are: reduced cost, corrosion resistance, improved life and reduced maintenance, and possible reduction in seismic problems. Major effort has been in the area of bridges, new as well repair and retrofit. As an example, concrete and masonry structures can be strengthened by composites rather than steel plates. One such process involves surface preparation and application of an appropriate primer to the concrete structure, application of a resin coat, followed by a sheet of composite and a second coat of resin. Both glass fiber– and carbon fiber–reinforced composites are used, although carbon fiber–reinforced polymer composites give superior results. As an illustrative example, we show in Figure. 8.18a a concrete column wrapped around with a jacket of carbon fiber/epoxy composite, while Figure 8.18b shows the results after the application of a compressive load. In the part of the concrete column without the composite jacket, the spalling of concrete and buckling of the steel reinforcement bars can be seen. In the portion of the column protected with a composite jacket, there is no visible effect. Potentially, composite wrapping of structural columns for seismic reinforcement would appear to be a huge market. The reader should not get the impression that everything is perfect in this area of use of composites in the civil infrastructure. There are many problems, such as durability, moisture absorption by the polymer matrix and other environmental effects. In addition, there is a lack of design guidelines and database of material specifications, which makes it difficult for civil engineers to accept the composites.

Carbon fiber is also used to reinforce cement. The resultant brittle composite shows an improved tensile and flexural strength, high impact strength, dimensional stability, and a high resistance to wear. Another application of carbon fiber PMCs in civil infrastructure has to do with the ability of bridges and other structures to withstand earthquakes. Earthquake-proofing of bridge columns is done by wrapping columns with carbon fiber reinforced composites as described above.

a   b

Fig. 8.18. a A concrete column wrapped around with a jacket of carbon fiber/epoxy composite while **b** On application of a compressive load, spalling of concrete and buckling of the steel reinforcement bars can be seen in the part of the concrete column without the composite jacket. In the portion of the column protected with a composite jacket, there is no visible effect. (Courtesy of B. Kad.)

Applications of Carbon/Carbon Composites

Major applications of carbon/carbon composites involve uses at high temperatures, for example, as heat shields for reentry vehicles, aircraft brakes, hot-pressing dies, and high-temperature parts such as nozzles. Figure 8.19 shows a fully processed three-dimensional carbon/carbon frustum. Hot-pressing dies made of carbon/carbon composites are commercially available, while brake disks are used in some subsonic aircraft, the Concorde supersonic aircraft, and racing cars. Heat shields and nozzles are generally made of multidirectionally reinforced carbon/carbon composites. Consider the comparison of the flexural strength of various carbon fiber composites shown in Figure 8.5. Note that carbon/carbon composites can withstand high temperatures in an inert atmosphere. Lack of oxidation resistance is a major problem, and a great deal of effort has been put into the development of oxidation-resistant coating for carbon fibers, with SiC coating being a primary candidate.

Potential applications of carbon/carbon composites may call for use in service at temperatures exceeding 1000 °C and even approaching 2200 °C for

Fig. 8.19. A fully processed three-dimensional carbon/carbon frustum. (Courtesy of Fiber Materials Inc.)

times ranging from 10 hours to a few thousand hours. Thus, the major drawback of carbon/carbon composites is the formation of gaseous oxides of carbon upon reaction with oxygen. There are two main approaches to protecting carbon/carbon composites against oxidation. One is to use inhibitors to slow down the rate of reaction between carbon and oxygen. The other is to use diffusion barriers to prevent oxygen from reaching the carbon and reacting with it. Commonly, silicon-based ceramics, such as SiC and Si_3N_4, are used as the primary oxygen barriers. The two approaches are frequently combined, for examples, using diffusion barriers and internal glass-forming inhibitors and sealants. Sealants have the added advantage that they seal thermal stress cracks.

Carbon/carbon brakes are used in some aircraft and automobiles. Let us examine this important application of carbon/carbon composites. Brakes, in general, have the following general requisites (Awasthi and Wood, 1988):

- Oxidation resistance
- High thermal capacity

- Good strength, impact resistance, strain to failure
- Adequate and consistent friction characteristics
- High thermal conductivity

For use as a braking material, carbon/carbon composites are superior to high strength bulk graphite. Carbon/carbon composites can obtain much higher strength at the same density. The brake design involves the following items. Friction members must generate stopping torque (over a range of environmental conditions), heat sinks must absorb the kinetic energy of the aircraft, and the structural elements should be able to transfer torque to the tires. Multiple-disk brakes have alternating rotors and stators forced against adjacent members by hydraulic pressure. Friction between rotating and stationary disks causes them to heat up to 1500 °C (surface temperature can be as high as 3000 °C), so good thermal shock resistance is required. In view of the requisites listed earlier, any braking material must be a good structural material, an efficient heat sink, and have excellent abrasion resistance. Consider the demands made of a braking material in a Boeing 767 aircraft. It has a mass of about 170,000 kg. Let us say it has a take-off velocity of about 320 km/h. This will give us a kinetic energy at take-off of 670 MJ. In the event of an aborted take-off, this energy must be dissipated in about 30 seconds by the eight brakes on the aircraft. An aborted take-off is, indeed, the worst case scenario, but then the braking material must be able to meet such requirements. Let us consider the weight savings that result from replacement of conventional brakes by carbon/carbon brakes. In a large aircraft, a multiple stator and rotor arrangement (a sintered high-friction material sliding against a high-temperature steel) weighs about 1100 kg. Carbon/carbon brakes (both the stator and the rotor are made of carbon/carbon composite) weigh about 700 kg, resulting in a weight savings of 400 kg.

Other applications of carbon/carbon composites include their use as implants and internal repair of bone fractures because of their excellent biocompatibility. They are also used to make molds for hot pressing. Carbon/carbon molds can withstand higher pressures and offer a longer use life than polycrystalline graphite.

References

S. Awasthi and J.L. Wood (1988). *Adv. Ceramic Materials*, **3**, 449.

A. Baker (1983). *Met. Forum*, **6**, 81.

K.K. Chawla (1993). *Ceramic Matrix Composites*, Chapman & Hall, London.

D. Clark, N.J. Wadsworth, and W. Watt (1974). In *Carbon Fibres, Their Place in Modern Technology*, the Plastics Institute, London, p. 44.

T.A. Collings and D.E.W. Stone (1985). *Composites*, **16**, 307.

R.J. Diefendorf (1985). In *Tough Composite Materials*, Noyes Publishing, Park Ridge, NJ, p. 191.

J.-B. Donnet and R.C. Bansal (1984). *Carbon Fibers*, Marcel Dekker, New York, p. 109.

L.T. Drzal, M.J. Rich, and P.F. Lloyd (1983a). *J. Adhesion*, **16**.

L.T. Drzal, M.J. Rich, M.F. Koenig, and P.F. Lloyd (1983b). *J. Adhesion*, **16**, 133.

P. Ehrburger and J.B. Donnet (1980). *Philos. Trans. R. Soc. London*, **A294**, 495.

R.H. Eriksen (1976). *Composites*, **7**, 189.

E. Fitzer and M. Heym (1976). *Chem. Ind*, 663.

W. Fritz, W. Hüttner, and G. Hartwig (1979). In *Nonmetallic Materials and Composites at Low Temperatures*, Plenum Press, New York, p. 245.

D. Hull (1994). *Mater. Sci. & Eng.*, **A184**, 173.

I. Jangehud, A.M. Serrano, R.K. Eby, and M.A. Meador (1993). In *Proc. 21st Biennial Conf. on Carbon*, Buffalo, NY, June 13–18.

N.J. Mayer (1974). In *Engineering Applications of Composites*, Academic Press, New York, p. 24.

L.E. McAllister and W.L. Lachman (1983). In *Fabrication of Composites*, North-Holland, Amsterdam, p. 109.

D. McKee and V. Mimeault (1973). In *Chemistry and Physics of Carbon*, Vol. 8, Marcel Dekker, New York, p. 151.

F. Molleyre and M. Bastick (1977). *High Temp. High Pressure*, **9**, 237.

NASA (1980). *Risk to the Public from Carbon Fibers Released in Civil Aircraft Accidents*, SP-448, NASA, Washington DC.

R.B. Pipes and N.J. Pagano (1970). *J. Composite Mater.*, **1**, 538.

D.M. Riggs, R.J. Shuford, and R.W. Lewis (1982). In *Handbook of Composites*, Van Nostrand Reinhold, New York, p. 196.

C.D. Shirrel and F.A. Sandow (1980). In *Fibrous Composites in Structural Design*, Plenum Press, New York, p. 795.

J.B. Sturgeon (1978). In *Creep of Engineering Materials*, a *Journal of Strain Analysis* Monograph, p. 175.

A.J. Waddon, M.J. Hill, A. Keller, and D.J. Blundell (1987). *J. Mater. Sci.*, **27**, 1773.

A. Whittaker and R. Taylor (1990). *Proc. Roy. Soc. Lond.*, **430A**, 167.

Z.R. Xu and K.H.G. Ashbee (1994). *J. Materials Sci.*, **29**, 394.

Suggested Reading

K.K. Chawla (1993). *Ceramic Matrix Composites*, Chapman & Hall, London.

J.-B. Donnet and R.C. Bansal (1984). *Carbon Fibers*, second ed. Marcel Dekker, New York.

E. Fitzer (1985). *Carbon Fibres and Their Composites*, Springer-Verlag, Berlin.

L.H. Peebles (1995). *Carbon Fibers*, CRC Press, Boca Raton, FL.

G. Savage (1992). *Carbon-Carbon Composites*, Chapman & Hall, London.

C.R. Thomas (Ed.) (1993). *Essentials of Carbon-Carbon Composites*, Royal Society of Chemistry, Cambridge, UK.

CHAPTER 9

Multifilamentary Superconducting Composites

9.1 Introduction

Certain materials lose all resistance to the flow of electricity when cooled to within a few degrees of absolute zero. The phenomenon is called *superconductivity*, and the materials exhibiting this phenomenon are called *superconductors*. Superconductors can carry a high current density without any electrical resistance; thus, they can generate the very high magnetic fields that are common in high-energy physics and fusion energy programs. Other fields of application include magnetic levitation vehicles, magneto hydrodynamic generators, rotating machines, and magnets in general. Kammerlingh Onnes discovered the phenomenon of superconductivity in 1911. Since then, some 27 elements and hundreds of solid solutions or compounds have been discovered that show this phenomenon of total disappearance of electrical resistance below a critical temperature, T_c. Figure 9.1 shows the variation of electrical resistivity with temperature of a normal metal and that of a superconducting material, Nb_3Sn. The critical temperature is a characteristic constant of each material. Kunzler et al. (1961) discovered the high critical field capability of Nb_3Sn and thus opened up the field of practical, high-field superconducting magnets. It turns out that most of the superconductors came into the realm of economic viability when techniques were developed to put the superconducting species in the form of ultra-thin filaments in a copper matrix; see the following discussion. A similar development route is used for the newer oxide superconductors.

Multifilamentary composite superconductors started becoming available in the 1970s. These are niobium-based (Nb-Ti and Nb_3Sn) superconductors, also referred to as *conventional* superconductors. The record critical temperature at which a conventional material became a superconductor was 23 K and was set in 1974. In 1986, there began a new era in the field of superconductivity, called high-temperature superconductivity (HTS), which started with the now well-known work of Bednorz and Muller. They reported of superconductivity at 30 K in a ceramic containing lanthanum, copper, oxygen, and barium. This original discovery, for which Bednorz and Muller

Fig. 9.1. Variation of electrical resistivity with temperature for a normal metal and a superconducting material, Nb_3Sn.

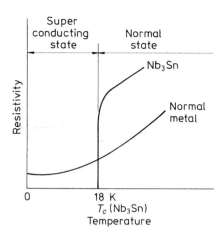

received a Nobel prize, set off a storm of activity. In 1987, Chu and Wu found a related oxide superconductor with a critical temperature above 77 K, the boiling point of liquid nitrogen. This had the chemical composition of $YBa_2Cu_3O_{7-x}$, and is commonly referred to as the *1-2-3 superconductor* because of the Y:Ba:Cu ratio. Since then other ceramic compounds with critical temperatures above 77 K have been discovered.

The oxide superconductors have the great advantage of having a T_c around 90 K, i.e., above the liquid nitrogen temperature (77 K). Thus, potentially, liquid nitrogen, a cheap and easily available cooling medium, could replace liquid helium, which is expensive and limited in supply. There are, however, many problems with these superconductors. These new ceramic superconductors have layered perovskite body-centered tetragonal structure and, not surprisingly, are very brittle. This problem of brittleness, as was pointed out in 1987 by some researchers including this author, makes it very difficult to make thin filaments of these oxide superconductors for use in, for example, magnet windings. Besides the inherent brittleness of ceramics, these superconductors carry very low current densities. It is true that the electrical resistance goes to zero around 90 K, but the troublesome fact is that these materials lose their superconductivity at very modest current densities, of the order of a few hundred A cm^{-2}. This has been attributed to impurities, misaligned grains, and the like. It turns out that the layered perovskite cuprates are all anisotropic, difficult to prepare in desirable shapes such as long continuous wires, and very poor at carrying current. The grain boundaries in a polycrystalline sample block the current. This is attributed to the extremely short (a few nm) coherence length in these oxides (Larbalestier, 1996). Yet another problem is that the performance of these oxide superconductors deteriorates drastically as the applied magnetic field increases. At a magnetic field of 1 T, the critical current density, J_c, is about 1–10 A cm^{-2}. For any reasonable commercial application, one needs a J_c of about 10^5 A cm^{-2}. The critical current density, J_c, is a function of the processing and, consequently, of the microstructure of the superconductor. This is where some innovative processing can be of great value. Heine et al. (1989) took just such a step when they partially melted Bi-Sr-Ca-Cu-O (BSCCO) in a silver sheath and observed that the resultant superconductor showed high J_c in the grains. There remains a large gap to be bridged between producing a small sample for testing in the laboratory and making a viable commercial product. As Larbalestier (1996) says, we need superconductors that "... must be electrically continuous, otherwise there can be no applications." They must also be strong and tough, amenable to being economically made into long lengths, and possess high overall J_c values. The term *overall* implies that all of the space required to make the superconductor must be taken into account when critical current density is computed (Larbalestier, 1996). It should be pointed out that the Nb-Ti system took 15–20 years between its discovery and commercial availability. The new high-temperature oxide superconductors hold great promise, and it would appear that eventually some kind of composite superconductors such as the oxide superconductor filaments in a silver matrix will be used commercially as are the Nb-based conventional superconductors. Thus, it is quite instructive to review the composite material aspects of niobium-based superconductors. In this chapter we describe the processing, structure, and properties of the conventional and high-T_c superconductors. Both are truly multifilamentary metal matrix composites. But first let us describe why we need these superconductors in the form of multifilamentary metal matrix composites.

9.1.1 The Problem of Flux Pinning

For a variety of applications, we need long-length superconductors of uniform properties, for example, large solenoids and coils for rotating machinery and magnets for plasma confinement in a fusion reactor. In conventional superconductors, ultra-thin superconducting filaments are incorporated in a copper matrix to form a filamentary fibrous composite. In the high-T_c superconductors, we have ceramic superconducting material in the of thin filaments in a silver matrix. Superconducting filaments have a micrometer-size diameter that helps to reduce the risk of flux jump in any given filament. If a superconductor is perturbed, say, by motion or a change in the applied field, it leads to a rearrangement of magnetic flux lines in the superconductor. This phenomenon is called *flux motion* and is an energy-dissipative or heat-producing process. When a current density J flows in a superconductor under a magnetic field, B, it experiences a Lorentz force per unit volume F_L, given by

$$F_L = J \times B$$

The Lorentz force acts in a direction perpendicular to both J and B. This force can result in motion of flux lines. Whenever the flux lines move (no

Fig. 9.2. Flux movement in a superconductor is caused by any change in current of field or by any mechanical disturbance. The end result of flux movement is that the superconductor is quenched to its normal state. Filamentary composite superconductors help pin the flux.

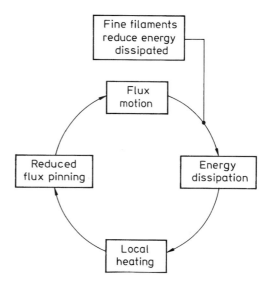

matter what the source of this motion), they produce resistance. Hence, we need to pin the flux lines in a superconductor, which is done by appropriate microstructural control. The critical current density, J_c, corresponds to that Lorentz force, which unpins the flux lines and makes them move. Thus, the pinning force $= J_c \times B$. Any heat generated by flux motion will result in a temperature increase, which in turn will lead to a reduced critical current and more flux motion results. The net result is that the superconductor is heated above T_c and reverts to the normal state. A practical solution to this problem is to make the superconductor in the form of ultra-thin filaments so that the amount of energy (heat) dissipated by flux motion is too small to cause this runaway behavior; see Figure 9.2. The high-purity copper or silver metal provides a high-conductivity alternate path for the current. In the case of a *quench*, that is, superconductor reverting to the normal state, the metal matrix carries the current without getting excessively hot. The superconductor is cooled again below its T_c and carries the electric current again. This is the so-called *cryogenic stability* or *cryostabilization* design concept; namely, the superconductor is embedded in a large volume of low-resistivity metal and a coolant (liquid helium or liquid nitrogen) is in intimate contact with all windings.

9.2 Types of Superconductor

The technologically most important property of a superconductor is its capacity to carry an electric current without normal I^2R losses up to a critical current density, J_c. Here, I and R are the current and resistance,

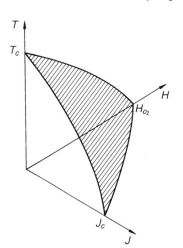

Fig. 9.3. The state of superconductivity in a material is described by three critical parameters: magnetic field (H), temperature (T), and current density (J). The material will remain in a superconducting state as long as it is below the shaded portion.

respectively. The critical current density, J_c, is a function of the applied field and temperature. The commercially available superconductors in the 1980s could demonstrate critical current densities of $J_c > 10^6$ A cm^{-2} at 4.2 K and an applied field of 5 T.

There are three parameters that limit the properties of a superconductor, namely, the critical temperature (T_c), the critical electrical current density (J_c), and the critical magnetic field (H_c); see Figure 9.3. As long as the material stays below the shaded area indicated in Figure 9.3, it will behave as a superconductor.

There are two types of superconductor:

Type I: These are characterized by low T_c values, and they lose their superconductivity abruptly at H_c.

Type II: These behave as diamagnetic materials up to a field H_{c1}. Above this field, the magnetic field penetrates gradually into the material, and concomitantly the superconductivity is gradually lost, until at the critical magnetic field H_{c2} the material reverts to the normal state.

All major applications of superconductivity involve the use of these type II superconductors. The magnetic field penetrates a type II superconductor in the form of thin filaments called *magnetic vortices*, which are fixed by what are called *pinning centers*. When an electric current is applied, the Lorentz force exerted by current tends to move these vortices. If there is no pinning or if the pinning is weak such that the Lorentz force moves the vortices, then the motion of these vortices generates voltage and the resultant heat will destroy the superconducting state. An interesting demonstration of such a dynamic interaction of vortices with pinning centers was done by Matsuda et al. (1996) by means of a technique called *Lorentz microscopy*. They studied a niobium film sample in which they had made a square lattice of defects by ion irradiation. The sample was placed in a low-temperature specimen stage of a 300-kV field emission transmission electron microscope. The objective lens of TEM was replaced by an intermediate lens to make an out-of-focus image (i.e., a Lorentz micrograph) in which vortices appear as dark spots.

In the following sections, we describe processing, microstructure, and properties of some important superconducting composite systems. Finally, we provide a summary of their applications.

9.3 Processing and Structure of Multifilamentary Superconductors

There are three main categories of multifilamentary superconductors: Nb-Ti-based ductile alloy superconductors, Nb$_3$Sn-type brittle superconductors, and ceramic oxide brittle superconductors.

9.3.1 Niobium-Titanium Alloys

Niobium-titanium alloys provide a good combination of superconducting and mechanical properties. A range of compositions is available commercially: Nb—44% Ti in the United Kingdom, Nb—46.5% Ti in the United States, and Nb—50% Ti in Germany.

In all these alloys, a $J_c > 10^6$ A cm^{-2} at 4.2 K and an applied field of 7 T can be obtained by a suitable combination of mechanical working and annealing treatments. Strong flux pinning and therefore high J_c are obtained in these alloys by means of dislocation of cell walls and precipitates. The flux pinning by precipitates becomes important in high-Ti alloys because the Nb-Ti phase diagram indicates precipitation of α-Ti in these alloys; see Figure 9.4. As pointed out earlier, the condition of stability against flux motion requires that the superconductor be manufactured in the form of a composite system: extremely fine superconducting filaments embedded in a copper

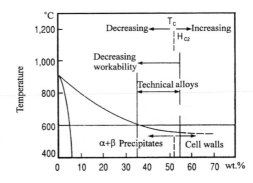

Fig. 9.4. Nb-Ti phase diagram showing the range of compsitions used to make Nb-Ti superconductors. [After Hillmann (1981).]

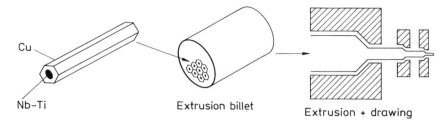

Cu

Nb–Ti   Extrusion billet

Extrusion + drawing

Fig. 9.5. Fabrication route for Nb-Ti/Cu composite superconductors.

matrix provide flux stability and reduced losses caused by varying magnetic fields. Fortunately, Nb-Ti and copper are compatible, chemically and otherwise, and amenable to making filamentary composites. Figure 9.5 shows the essential steps in the fabrication of Nb-Ti composite superconductors. Annealed Nb-Ti rods are inserted into hexagonal-shaped high-purity copper tubes. These rods are next loaded into an extrusion billet of copper, then evacuated, sealed, and extruded. The extruded rods are cold drawn to an intermediate size and annealed to provide the necessary dislocation cell walls and precipitates for the flux pinning. This is followed by more cold drawing passes to the proper final size and a final anneal to get back the high conductivity of the copper matrix. Consider the specific case of Nb—50% Ti alloy. Its initial microstructure consists of a β solid solution. The necessary cell structure and dislocation density for flux pinning purposes depend on the purity of the alloy and the size and distribution of the α particles after precipitation heat treatment. The α phase is nonsuperconducting; its main function is to aid dislocation cell structure formation. Dislocations and α particles are responsible for flux pinning and thus contribute to high J_c values. The amount and distribution of α particles depend on the alloy chemical composition, processing, and annealing temperature and time. For a Nb—50% Ti alloy, a heat treatment of 48 hours at 375 °C is generally used and results in about 11% of α particles. A greater amount of α will reduce the ductility of the alloy. Higher annealing temperatures result in excessive softening, fewer dislocations, and lower J_c. The precipitation treatment is generally followed by some cold working to refine the structure and obtain a high J_c. The precise amount of strain given in this last step is a function of the superconductor design, that is, the distribution of Nb-Ti filaments and the ratio of Nb-Ti/Cu in the cross section. Hillmann (1981) points out that undesirable intermetallic compounds can form from any scratches. At a scratch on the surface of a NbTi rod, a hard ball of oxidized NbTi about 1 μm in diameter can appear. During extrusion, there is a snowball-like effect causing mechanical alloying and formation of a NbTi + Cu mixture. During subsequent annealing treatment $(NbTi)Cu_2$ and $(NbTi)_2Cu$ intermetallic compounds can form. These brittle compounds can eventually cause rupture of the superconducting filament.

Figure 9.6a shows a compacted strand cable-type superconductor made from 15 fine multifilamentary strands (strand diameter = 2.3 mm). Each strand consists of 1060 filaments of Nb-Ti (diameter = 50 μm) embedded in a copper matrix. A magnified picture of one of the strands is shown in Figure 9.6b, while a scanning electron microscope picture is shown in Figure 9.7. The Cu-Ni layers seen in Figure 9.7 provide the low-conductivity barriers to prevent eddy current losses in alternating fields.

9.3.2 A-15 Superconductors

For applications involving fields greater than 12 T and temperatures higher than 4.2 K, the ordered intermetallic compounds having an A-15 crystal structure (Nb_3Sn, V_3Ga) are better suited than the Nb-Ti type. In particular, Nb_3Sn has a T_c of 18 K. It is easy to appreciate that the higher the T_c the lower the refrigeration costs will be. Hence, the tremendous interest in high T_c oxide superconductors. Superconducting magnets form an integral part of

a

Fig. 9.6. a Compacted strand cable superconductor made from 15 multifilamentary strands each of diameter = 2.3 mm. **b** Magnified picture of one of the strands in **a** containing 1060 filaments (diameter = 50 μm). (Courtesy of Hitachi Cable Co.)

b

Fig. 9.6 (*continued*)

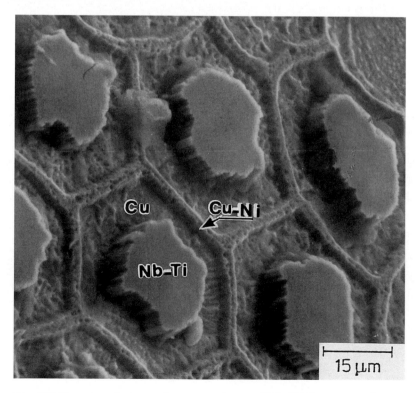

Fig. 9.7. Scanning electron microscope picture of Nb-Ti/Cu superconductor.

any thermonuclear fusion reactor for producing plasma-confining magnetic fields (see Sect. 9.6). Nb_3Sn is the most widely used superconductor for high fields and high temperatures. A characteristic feature of these intermetallic compounds having an A-15 crystal structure is their extreme brittleness (typically, a strain to fracture of 0.2% with no plasticity). Compare this to Nb-Ti, which can be cold worked to a reduction in area over 90%. Initially, the compound Nb_3Sn was made in the form of wires or ribbons either by diffusion of tin into niobium substrate in the form of a ribbon or by chemical vapor deposition. V_3Ga on a vanadium ribbon was also produced in this manner. The main disadvantages of these ribbon-type superconductors were: (1) flux instabilities due to one wide dimension in the ribbon geometry, and (2) limited flexibility in the ribbon width direction. Later, with the realization that flux stability could be obtained by the superconductors in the form of extremely fine filaments, the filamentary composite approach to A-15 superconductor fabrication was adopted. This breakthrough involving the composite route came through early in the 1970s. Tachikawa (1970) showed that V_3Ga could be produced on vanadium filaments via a Cu-Ga matrix while Kaufmann and Pickett (1970) demonstrated that Nb_3Sn could be obtained on niobium filaments from a bronze (Cu-Sn) matrix. Figure 9.8 shows schematically the process of producing Nb_3Sn (or V_3Ga). Niobium rods are inserted into a bronze (Cu-13 wt. % Sn) extrusion billet. This billet is sealed and extruded into rods. This is called the *first-stage extrusion*. These first-stage extrusion rods are loaded into a copper can that has a tantalum or niobium barrier layer and extruded again. The second-stage extruded composite is cold drawn and formed into a cable, clad with more copper and compacted to form a monolithic conductor. This is finally given a heat treatment to convert the very fine niobium (vanadium) filaments into Nb_3Sn (V_3Ga). Specifically, for the Nb_3Sn superconductors, the ratio of cross-sectional areas of bronze matrix to that of niobium, $R = $ Cu-Sn/Nb, is an important design parameter that strongly affects the J_c value. It is important that a right amount of tin be available to form the stoichiometric Nb_3Sn phase. Too much of the bronze matrix will reduce the J_c value. Thus, one uses a bronze with about 13 wt. % tin, the limit of solubility of tin in copper, which yields enough tin without affecting the formability. Arriving at an optimum value of the ratio R is difficult because the resultant Nb_3Sn may not be stoichiometric. Superconductors meant for high fields ($\sim 15\,T$) should have a stoichiometric Nb_3Sn. One ensures this by using a high ratio of R (>3) and sufficiently long diffusion times. An example of an Nb_3Sn monolithic superconductor used in high-field magnets is shown in Figure 9.9. Each little dot in this photograph (Fig. 9.9a) is a 4-μm Nb_3Sn filament. Wide strips in the hexagonal form are niobium; they serve as barriers to tin diffusion from the bronze into the copper. Tantalum is also used as a diffusion barrier. Figure 9.9b shows a scanning electron micrograph of a niobium/bronze composite before heat treatment to form Nb_3Sn. The good flux pinning and the consequent high critical currents in these superconductors are

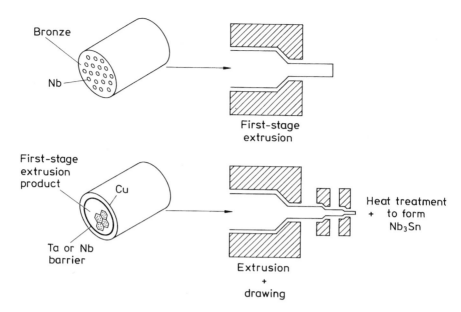

Fig. 9.8. Schematic of the bronze method of fabricating Nb$_3$Sn/Cu superconductor.

due to the grain boundaries (Scanlan et al., 1975; Nembach and Tachikawa, 1979). At a heat treatment temperature of about 650–700 °C, the Nb$_3$Sn reaction layer forms and consists of very fine grains, less than 80 nm (0.08 µm) in diameter. The critical current density for Nb$_3$Sn under these conditions is more than 2×10^5 A cm^{-2} at 10 T and 4.2 K.

The problem of thermal mismatch, i.e., the differential expansion or contraction between the components of a composite (Chawla, 1973a, 1973b), something very much inherent to composites, also exists in superconductors. In the case of A-15-type composite superconductors, their brittle nature makes this thermal mismatch problem of great importance. It turns out that when cooled from the reaction temperature (~ 1000 K) to 4.2 K, the different coefficients of thermal expansion of copper and Nb$_3$Sn lead to rather large compressive strains in the brittle Nb$_3$Sn filament. Luhman and Suenaga (1976) showed that the T_c of Nb$_3$Sn varied with the strain applied to the Nb$_3$Sn filament by the copper matrix. Later, measurements of critical currents as a function of applied strain confirmed these results, and one could explain the strain-critical current behavior in terms of the effects of strain on T_c and H_{c2}. This understanding of strain-critical current behavior is used to good effect in the design of Nb$_3$Sn superconducting magnets. As pointed out, Nb$_3$Sn filaments fail at a tensile strain of a bout 0.2%. Thus, if the superconductor is designed so that the Nb$_3$Sn filaments in the bronze matrix experience compression between 0.4 and 0.6%, then one can expect these

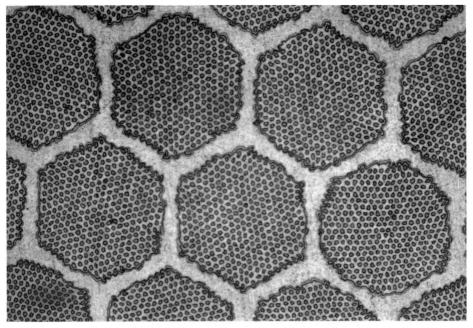

a

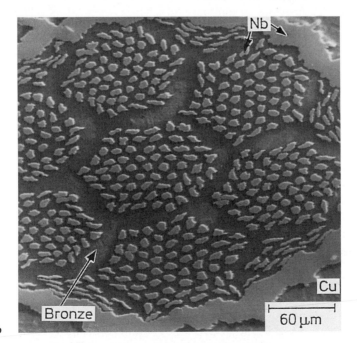

b

Fig. 9.9. a An Nb_3Sn/Cu composite superconductor used in high-field magnets. Each little dot in this picture is a 4-μm diameter Nb_3Sn filament. Hexagonal strips are niobium diffusion barriers for tin. (Courtesy of Hitachi Cable Co.) **b** Scanning electron micrograph of a niobium/bronze composite before the heat treatment to form Nb_3Sn.

composites to withstand applied strains of between 0.6 and 0.8% before fracture ensues in the Nb_3Sn filaments.

Although the bronze route of manufacturing the superconductor is a commercial process now, there are some disadvantages involved. One must have frequent interruptions for annealing the work-hardened bronze matrix. To shorten this process of fabrication, a number of in situ and powder metallurgy techniques have been tried (Roberge and Foner, 1980). These in situ techniques involve melting of copper-rich Cu-Nb mixtures (~ 1800–$1850\,^\circ C$), homogenization, and casting. Niobium is practically insoluble in copper at ambient temperature. Thus, the casting has a microstructure consisting of niobium precipitates in a copper matrix. When this is cold drawn into wire, niobium is converted into fine filaments in the copper matrix. Tin is then plated onto the wire and diffused to form Nb_3Sn. The different melting and cooling techniques used include chill casting, continuous casting, levitation melting, and consumable electrode arc melting. The powder metallurgy techniques involve mixing of copper and niobium powder, pressing, hot or cold extrusion, and drawing to a fine wire that is coated or reacted with tin. The contamination of niobium with oxygen causes embrittlement and prevents its conversion into fine filaments. This has led to the addition of a third element (Al, Zr, Mg, or Ca) to the Cu-Nb mixture. This third element preferentially binds oxygen and leaves the niobium ductile.

Table 9.1 provides a summary of charcteristics of Nb/Ti and Nb_3Sn multifilamentary composites.

Table 9.1. A summary of characteristics of Nb-Ti and Nb_3Sn-type multifilamentary composites (after Hillmann, 1981)

Characteristic	Nb-Ti	Nb_3Sn
T_c	≈ 9 K	18.1 K
B_{c2}	≈ 14.5 K	24.5 K
Stabilization	Simple, coextrusion with high-purity Cu matrix	Complex, coprocessing with bronze matrix, followed by heat treatment to obtain Nb_3Sn in a Cu matrix
Insulation	Conventional; varnish, cotton, and polymeric tape can be used	Complex, in prereaction insulated conductors, high thermal stability of insulation is needed; in post-reaction insulated conductors, careful mechanical handling is required
Advantages	Fabrication process is cheap; insensitive to mechanical handing, simple magnet technology	Applications to high magnetic fields and high temperatures
Disadvantages	Applications limited to 8 T at 4.2 K	Brittleness of the Nb_3Sn, restricted mechanical handling, large number of very fine filaments (diameter $\leq 5\,\mu m$) is required, complicated magnet winding

9.3.3 Ceramic Superconductors

The ceramic oxide superconductors with a transition temperature higher than the conventional superconductors were discovered in late 1986 and early 1987. Since then many oxide superconductors have been discovered, and they have a range of critical transition temperatures. There is a rather complex notation that is commonly used in the ceramic oxide superconductor literature. For example, the one-layer compound $Bi_2Sr_2Cu_1O_y$ is denoted by (2201), the two-layer compound $Bi_2Sr_2Ca_1Cu_2O_y$ is designated by (BSCCO) or (2212), and the three-layer compound $Bi_{2-x}Pb_xSr_2Ca_2Cu_3O_y$ is denoted by (Pb) (2223). Oxide superconductors with a transition temperature around 90 K are available. The reader will recall that such high transition temperatures enable the use of liquid nitrogen (boiling point 77 K), a cheap and easily available coolant. With conventional superconductors, one must use liquid helium as a coolant. There are still many problems with these high-T_c superconductors; one of the major ones is that they carry very low current densities, especially in the presence of magnetic fields. The electrical resistance goes to zero at a T_c of 90 K, but the supercurrents have difficulty going from one grain to another. Various explanations, such as impurities and misaligned grains, have been put forth to explain this behavior. The performance of these new superconductors deteriorates with increasing magnetic field. At a magnetic field of 1 T, the critical current density, J_c, is between 1 and 10 A cm^{-2}. The conventional superconductors, on the other hand, can carry as much as 10^5 A cm^{-2}.

The standard processing of these ceramic superconductors involves cold compaction of the appropriate powders, say, Y, Ba, and Cu, followed by sintering. Such conventionally processed materials do not result in acceptable levels of J_c because the 1-2-3 superconductors are highly anisotropic with respect to their ability to superconduct. A process called *melt-textured growth* results in highly aligned grains in these superconductors (Jin, 1991). This involves melting at 1300 °C followed by cooling down to room temperature via a proprietary regimen. The process results in needle-like crystals (several 100 μm long) aligned along the good conduction direction. The 1-2-3 superconductors processed by this technique have shown a J_c of about 7400 A cm^{-2} at $B = 0$ and a J_c of about 1000 A cm^{-2} at $B = 1$ T. The high critical current density because the J_c of a superconductor is a function of the melt processing. Heine et al. (1989) partially melted Bi-Sr-Ca-Cu-O (BSCCO) in a silver sheath, and the resultant composite showed high J_c values in the grains of HTS.

We alluded earlier to the extreme brittleness of the ceramic superconductors. Very powerful tensile and shear forces develop in dipole magnets due to the phenomenon of Lorentz force, enough to literally explode the ceramic oxides. This also means that processing of such superconductors into long lengths is much more difficult than is the case with a ductile Nb-Ti system. Some of the approaches to go around this difficulty in processing

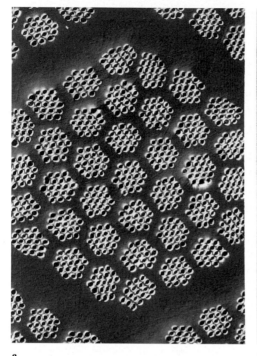

a

b

c

Fig. 9.10. a Low-magnification cross-sectional picture of a cable made by the MP process. **b** longitudinal section **c** transverse section. The superconductor is $Y_{0.9}$ $Ca_{0.1}$ Ba_2 Cu_4 $Ag_{0.65}$ metallic precursor/silver composite containing 962,407 filaments. (Courtesy of American Superconductor Co.)

include the following. In a manner similar to the so-called bronze route of making conventional superconductors, where the superconductor is a very brittle intermetallic, Nb_3Sn, one must avoid the synthesis of the brittle ceramic until the very end. The sintering step should be carried after the formation of the desired shape. This technique, called the *metallic precursor* (MP) method, involves melting the metallic elements (e.g., Y, Ca, Ba, Cu, and Ag), melt spinning, i.e., rapidly solidifying, the molten alloy into a ribbon form, and then pulverizing it to obtain homogeneous alloy powder. This precursor alloy powder is then packed into a silver can, sealed, and extruded into a hexagonal rod. Pieces of the extruded rod are packed into another bundle and further extruded. The final step in this process involves a very large reduction, up to 300:1, and results in round wire or tape (Masur et al., 1994). Figure 9.10a shows a low-magnification cross-sectional picture of a cable made by the MP process. Higher-magnification views of the longitudinal section and transverse cross section of Y-124 (or more precisely, $Y_{0.9}$ $Ca_{0.1}$ Ba_2 Cu_4 $Ag_{0.65}$) metallic precursor/silver composite containing 962,407 filaments are shown in Figures 9.10b and c, respectively. The uniformity of deformation is excellent. The extruded composites are then internally oxidized to form oxides: Y_2O_3, BaO, CuO, and Ba_2Cu_3O. These are then racted to give the Y-124 phase. According to Masur et al. (1994), the precursor oxides are approximately 50% converted to Y-124 after 100 minutes at 700 °C and 80% converted after 300 minutes at 750 °C. The tapes, after the oxidation treatment, are subjected to a thermomechanical treatment (TMT) to obtain suitable textures. The TMT involves multiple heat treatments in the 600–825 °C range, uniaxial pressings at pressure of up to 2 GPa, and the final heat treatment at 750 °C for 100 hours in one-atmosphere oxygen. Recall the similarity between this and the bronze process used for Nb_3Sn.

Another important process for making HTSC filamentary composites is called the *oxide-powder-in-tube* (OPIT) method (Sandhage et al., 1991). In this process, the oxide powder of appropriate composition (stoichiometry, phase content, purity, etc.) is packed in silver tube (4–7 mm inside diameter and 6–11 mm outside diameter). The silver tube is sealed after packing and degassed. The deformation processing techniques commonly used are swaging and drawing for wires and rolling is used for tapes. Heat treatments, intermediate and/or subsequent to deformation is given to form the correct phase, promote grain interconnectivity, crystallographic alignment of the oxide, and to obtain proper oxygenation (Sandhage et al., 1991). Figure 9.11 shows a schematic of this process. A high degree of crystallographic alignment of the superconducting oxide grains results in this process. It is known that the critical current density in HTS is increased with the alignment of oxide grains (Osamura et al., 1990; Uno et al., 1991).

Another method of fabricating HTS, in the same vein as the two mentioned earlier, is called the *continuous tube filling and forming* (CTFF) (Tomsic and Sarkar, 1997). This process is a variant of the process used in the welding industry to make tubular welding wire. One starts with a silver strip, which is formed into a tube and the tube is filled continuously with the

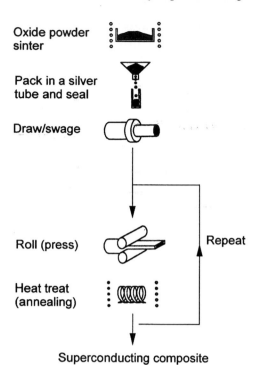

Oxide powder sinter

Pack in a silver tube and seal

Draw/swage

Roll (press)   Repeat

Heat treat (annealing)

Superconducting composite

Fig. 9.11. Schematic of the oxide powder in the tube method of making high-temperature superconducting composite. [After Sandhage et al., (1991).]

precursor powder. The filled tube is about 4 mm in diameter, so there are fewer drawing steps than in the OPIT process. Continuous drawing, rolling, pressing, and heat treatment are done to increase the powder density in the tube.

It should be pointed out that silver tubing material adds to the cost of the composite superconductor. However, all these methods result in a fine filamentary structure, which gives the wire the flexibility. Without this ultra-thin filamentary structure, such wires would be too brittle to be of much use.

9.4   Applications

Filamentary superconducting composites have some very important applications. Examples of applications of metal matrix composite superconducting coils include (Cyrot and Pavuna, 1992):

- High-field magnets for research in high energy and in condensed matter physics.
- Magnetic resonance imaging (MRI), which requires extremely uniform magnetic fields of ~1–2 Tesla).

Table 9.2. Critical current density in a given magnetic field for several applications of superconductors (after Cyrot and Pavuna, 1992)

Application	$B(T)$	$J_c(A/cm^2)$
Interconnects	0.1	5×10^6
AC transmission lines	0.2	10^5
DC transmission lines	0.2	2×10^4
SQUIDs	0.1	2×10^2
Motors and generators	~ 4	$\sim 10^4$
Fault current limiters	>5	$>10^5$

- Coils for windings in motors and generators.
- Magnetic levitating (MAGLEV) coils for high-speed trains.
- Magnetohydrodynamic and electromagnetic thrust systems for propulsion in ships and submarines.

Since the discovery of the so-called high-T_c superconductors, much emphasis seems to have been placed on the high T_c of these superconductors. It is good to remember a rule of thumb for most superconductor applications, which says that critical temperature T_c of a technological superconductor should be about twice the use temperature (Cyrot and Pavuna, 1992), i.e.,

$$T_{use} \sim T_c/2$$

This implies that superconductors that are used with liquid helium (4.2 K) as a coolant should have a $T_c > 8$ K. In fact, due to the heating of the magnet, the actual operational temperature is closer to ~ 7 K, so one needs a $T_c \sim 15$ K. Thus, to use the oxide high-temperature superconductors (HTS) in applications at 77 K one needs a superconductor with a $T_c \sim 150$ K. Such a material has not yet been synthesized, the thallium-compound with $T_c \sim 125$ K is, at present, the best candidate for the "true" technological material at liquid nitrogen temperature. For most electronic-type applications, the rule of thumb is less stringent, and we can write:

$$T_{use} \sim 2/3\, T_c$$

Nb$_3$Sn/Cu superconducting composites are used for magnetic fields greater than 12 T. Such high fields are encountered in thermonuclear fusion reactors, and superconducting composite magnets would represent a sizable fraction of the capital cost of such a fusion power plant. Table 9.2 shows the requirements for critical current density in a given magnetic field for several applications of superconductors. It is estimated that these superconducting magnets will represent a sizable fraction of the capital cost of a fusion power plant. Their performance will affect the plasma as well as the plasma density—hence the importance of materials research aimed at improving the performance limits of these superconducting magnets. The main difference between the magnets used in a fusion reactor and in power transmission is that the former use superconductors at very high magnetic fields while the latter use them at low fields.

A large-scale application of Nb-Ti/Cu superconducting magnets is in magnetically levitated trains. Japan National Railways has already tested such trains over a small stretch at speeds over 500 km h^{-1}. Nb-Ti/Cu superconductor composites are also employed in pulsed magnets for particle accelerators in high-energy physics. Other applications of these superconductor include the fields of magnetohydrodynamics and power. In 1983, researchers at the General Electric Research and Development Center in Schenectady, New York, successfully tested at full load an advanced superconducting electric generator for utility applications. The GE superconducting generator produced 20,600 kilovoltamperes of electricity, twice as much as could be produced by a conventional generator of a comparable size. The superconducting component of the generator was a 4-m-long rotor operating at liquid helium temperature, 4.2 K. The field windings were made from a Nb-Ti/Cu composite. By using superconducting materials, the designers could make a generator that developed a much stronger magnetic field than a conventional generator, permitting a significant reduction in the size of the generator for the same power output. Superconducting magnets do require cryogenic temperatures to operate, but the cost of this refrigeration is more than compensated by the energy savings. One has only to remember that, in a superconducting machine working with almost zero resistance, the normal losses associated with the flow of electricity in rotor windings of a conventional machine are absent, resulting in a higher efficiency and reduced operating costs. Westinghouse has also developed similar superconducting generators. It is worth pointing out that a big problem in this development at GE was to prevent rotor windings movement under the intense centrifugal and magnetic forces exerted on them. The rotor spins at a speed of 360 rpm. Thus, even an infinitesimally small movement of these components would generate enough heat by friction to quench the superconductors. The GE researchers used a special vacuum epoxy-impregnation process to bond the Nb-Ti superconductors into rock solid modules and strong aluminum supports to hold the windings rigidly.

An important landmark in the development of high-temperature oxide superconductors occurred on March 12, 1997, when Geneva's electric utility, SIG, put into operation an electrical transformer using HTS wires. This transformer was built by ABB; it used the flexible HTS wires made by American Superconductor by the process involving packing of the raw material into hollow silver tubes, drawing into fine filaments, grouping the multifilaments in another metal jacket, further drawing and heat treating to convert the raw material into the oxide superconductor (see Sect. 9.3.3). The wire is then flattened into a ribbon (2.5 × 0.25 mm) that is used to make the transformer coils. The transformer contains 6 km of HTS wires wound into coils, immersed in liquid nitrogen. According to ABB this transformer loses only about one fifth of the ac power losses of the conventional ones. Because HTS wires can carry a higher current density, this new transformer is more compact and lighter than a conventional transformer. It should also be mentioned that liquid nitrogen used as a coolant in the HTS transformer is safer than oils used as insulators in conventional transformers.

Finally, we will describe a very important application of Nb-based sueprconducting composites.

Magnetic Resonance Imaging

The phenomenon of nuclear magnetic resonance (NMR) is exploited in the technique of magnetic resonance imaging (MRI). This technique has a major advantage over X-ray radiography in that it is a noninvasive diagnostic technique, and thus the human body is not exposed to an ionizing radiation. It would be useful to briefly explain the principle behind this important technique. Use is made of the electromagnetic characteristics of the nuclei of elements such as hydrogen, carbon, and phosphorus that are present in the human body. The nuclei of these elements act as bar magnets when placed in a strong magnetic field. The patient is placed in the center of a very powerful magnet. When the magnetic field is turned on, the nuclei of the elements in the patient's body part under examination realign along the magnetic field direction. If we apply a radio frequency field, the nuclei will reorient. And if we repeat this process over and over again, the nuclei will resonate. The resonance frequency is picked up by a sensitive antenna, amplified, and processed by a computer into an image. The MRI images, obtained from the resonance patterns are more detailed and have a higher resolution than traditional techniques of visualization of soft tissues. And best of all, these images are obtained without exposing the patient to a radiation or performing biopsy.

The superconducting solenoid, made from Nb-Ti/Cu composite wire, is immersed in a liquid helium cryogenic Dewar. The liquid helium is consumed at about 4 ml per hour, and each refill of the Dewar lasts about three months. Commercially manufactured NMR (nuclear magnetic resonance) spectrometer systems, also called magnetic resonance imaging systems, for medical diagnostics became available in the 1980s; Figure 9.12 shows one such system. The great advantage of magnetic resonance imaging in clinical diagnostics is that it does not expose the patient to ionizing radiation and its possible harmful side effects. Of course, MRI techniques do not have to use superconducting magnets but certain advantages exist with superconductors, for example, better homogeneity and resolution and higher field strengths than are available with conventional magnets. The disadvantage is that higher fields with superconductors lead to greater shielding problems.

Fig. 9.12. A magnetic resonance imaging system for clinical diagnostics. (Courtesy of Oxford-Airco.)

References

K.K. Chawla (1973a). *Metallography*, **6**, 55.

K.K. Chawla (1973b). *Philos. Mag.*, **28**, 401.

M. Cyrot and D. Pavuna (1992). *Introduction to Superconductivity and High T_c Materials*, World Scientific, Singapore, p. 202.

K. Heine, N. Tenbrink, and M. Thoener (1989). *Appl. Phys. Lett.*, **55**, 2441.

H. Hillmann (1981). In *Superconductor Materials Science*, Plenum, New York, p. 275.

S. Jin (1991). *J. Miner. Met. Mater. Soc.*, **43**, 7.

A.R. Kaufmann and J.J. Pickett (1970). *Bull. Am. Phys. Soc.*, **15**, 833.

J.E. Kunzler, E. Bachler, F.S.L. Hsu, and J.E. Wernick (1961). *Phys. Rev. Lett.*, **6**, 89.

D. Larbalestier (1996). *Science*, **274**, 736.

T. Luhman and M. Suenaga (1976). *Appl. Phys. Lett.*, **29**, 1.

L.J. Masur, E.R. Podtburg, C.A. Craven, A. Otto, Z.L. Wang, and D.M. Kroeger (1994). *J. Miner. Met. Mater. Soc.*, **46**, 28.

T. Matsuda, K. Harada, H. Kasai, O. Kamimura, and A. Tonomura (1996). *Science*, **271**.

E. Nembach and K. Tachikawa (1979). *J. Less Common Met.*, **19**, 1962.

K. Osamura, S.S. Oh, and S. Ochiai (1990). *Superconductor Sci. & Tech.*, **3**, 143.

R. Roberge and S. Foner (1980). In *Filamentary A15 Superconductors*, Plenum Press, New York, p. 241.

R.H. Sandhage, G.N. Riley, Jr., and W.L. Carter (1991). *J. Miner. Met. Mater. Soc.*, **21**, 25.

R.M. Scanlan, W.A. Fietz, and E.F. Koch (1975). *J. Appl. Phys.*, **46**, 2244.

K. Tachikawa (1970). In *Proceedings of the 3rd ICEC*, Illife Science and Technology Publishing, Surrey, UK.

M. Tomsic and A.K. Sarkar (1997). *Superconductor Industry*, **10**, 18.

N. Uno, N. Enomoto, H. Kikuchi, K. Matsumoto, M. Mimura, and M. Nakajima (1991). *Adv. Supercond.* **2**, p. 341.

PART III

CHAPTER 10

Micromechanics of Composites

In this chapter we consider the results of incorporating a reinforcement (fibers, whiskers, particles, etc.) in a matrix to make a composite. It is of great importance to be able to predict the properties of a composite, given the component properties and their geometric arrangement. We examine various micromechanical aspects of composites. A particularly simple case is the *rule-of-mixtures*, a rough tool that considers the composite properties as volume-weighted averages of the component properties. It is important to realize that the rule of mixtures works in only certain simple situations. Composite density is an example where the rule of mixtures is applied readily. In the case of mechanical properties, there are certain restrictions to its applicability. When more precise information is desired, it is better to use more sophisticated approaches based on the theory of elasticity.

10.1 Density

Consider a composite of mass m_c and volume v_c. The total mass of the composite is the sum total of the masses of fiber and matrix, that is,

$$m_c = m_f + m_m \tag{10.1}$$

The subscripts c, f, and m indicate composite, fiber, and matrix, respectively. Note that Eq. (10.1) is valid even in the presence of any voids in the composite. The volume of the composite, however, must include the volume of voids v_v. Thus,

$$v_c = v_f + v_m + v_v \tag{10.2}$$

Dividing Eq. (10.1) by m_c and Eq. (10.2) by v_c and denoting the mass and volume fractions by M_f, M_m and V_f, V_m, V_v, respectively, we can write

$$M_f + M_m = 1 \tag{10.3}$$

and

$$V_f + V_m + V_v = 1 \tag{10.4}$$

The composite density ρ_c $(= m/v)$ is given by

$$\rho_c = \frac{m_c}{v_c} = \frac{m_f + m_m}{v_c} = \frac{\rho_f v_f + \rho_m v_m}{v_c}$$

or

$$\rho_c = \rho_f V_f + \rho_m V_m \tag{10.5}$$

We can also derive an expression for ρ_c in terms of mass fractions. Thus,

$$\rho_c = \frac{m_c}{v_c} = \frac{m_c}{v_f + v_m + v_v} = \frac{m_c}{m_f/\rho_f + m_m/\rho_m + v_v}$$

$$= \frac{1}{M_f/\rho_f + M_m/\rho_m + v_v/m_c}$$

$$= \frac{1}{M_f/\rho_f + M_m/\rho_m + v_v/\rho_c v_c}$$

$$= \frac{1}{M_f/\rho_f + M_m/\rho_m + V_v/\rho_c} \tag{10.6}$$

We can use Eq. (10.6) to indirectly measure the volume fraction of voids in a composite. Rewriting Eq. (10.6), we obtain

$$\rho_c = \frac{\rho_c}{\rho_c[M_f/\rho_f + M_m/\rho_m] + V_v}$$

or

$$V_v = 1 - \rho_c \left(\frac{M_f}{\rho_f} + \frac{M_m}{\rho_m} \right) \tag{10.7}$$

Example 10.1

A thermoplastic matrix contains 40 wt. % glass fiber. If the density of the matrix ρ_m is 1.1 g cm^{-3} while that of glass fiber, ρ_f, is 2.5 g cm^{-3}, what is the density of the composite? Assume that no voids are present.

Solution

Consider 100 g of the composite:

Amount of glass fiber, $m_f = 40$ g

Amount of matrix, $m_m = 60$ g

Volume of the composite, v_c is the sum of the volumes of fiber, v_f and matrix, v_m

$$v_c = v_m + v_f$$
$$= (m_m/\rho_m) + (m_f/\rho_f) = (60/1.1 + 40/2.5) \, \text{cm}^3$$
$$= 54.5 + 16 = 70.5 \, \text{cm}^3$$

The density of the composite, is

$$\rho_c = 100\,\text{g}/70.5\,\text{cm}^3 = 1.42\,\text{g}\,\text{cm}^{-3}$$

10.2 Mechanical Properties

In this section, we first describe some of the methods for predicting elastic constants, thermal properties, and transverse stresses in fibrous composites and then we treat the mechanics of load transfer.

10.2.1 Prediction of Elastic Constants

Consider a unidirectional composite such as the one shown in Figure 10.1. Assume that plane sections of this composite remain plane after deformation. Let us apply a force P_c in the fiber direction. Now, if the two components adhere perfectly and if they have the same Poisson's ratio, then each component will undergo the same longitudinal elongation Δl. Thus, we can write for the strain in each component

$$\varepsilon_f = \varepsilon_m = \varepsilon_{cl} = \frac{\Delta l}{l} \tag{10.8}$$

where ε_{cl} is the strain in the composite in the longitudinal direction. This is called the *isostrain* or *action-in-parallel* situation. It was first treated by Voigt [1910]. If both fiber and matrix are elastic, we can relate the stress σ in the two components to the strain ε_l by Young's modulus E. Thus,

$$\sigma_f = E_f \varepsilon_{cl} \quad \text{and} \quad \sigma_m = E_m \varepsilon_{cl}$$

Let A_c be the cross-sectional area of the composite, A_m, that of the matrix, and A_f, that of all the fibers. Then, from the equilibrium of forces in the fiber

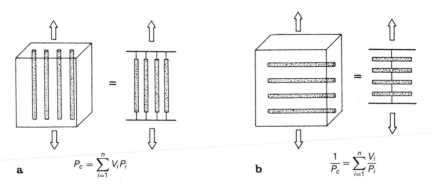

$$P_c = \sum_{i=1}^{n} V_i P_i$$

$$\frac{1}{P_c} = \sum_{i=1}^{n} \frac{V_i}{P_i}$$

a   **b**

Fig. 10.1. Unidirectional composite: **a** isostrain or action in parallel, **b** isostress or action in series.

direction, we can write

$$P_c = P_f + P_m$$

or

$$\sigma_{cl} A_c = \sigma_f A_f + \sigma_m A_m \tag{10.9}$$

From Eqs. (10.8) and (10.9), we get

$$\sigma_{cl} A_c = (E_f A_f + E_m A_m)\varepsilon_{cl}$$

or

$$E_{cl} = \frac{\sigma_{cl}}{\varepsilon_{cl}} = E_f \frac{A_f}{A_c} + E_m \frac{A_m}{A_c}$$

Now, for a given length of a composite, $A_f/A_c = V_f$ and $A_m/A_c = V_m$. Then the preceding expression can be simplified to

$$E_{cl} = E_f V_f + E_m V_m = E_{11} \tag{10.10}$$

Equation (10.10) is called the rule of mixtures for Young's modulus in the fiber direction. A similar expression can be obtained for the composite longitudinal strength from Eq. (10.9), namely,

$$\sigma_{cl} = \sigma_f V_f + \sigma_m V_m \tag{10.11}$$

For properties in the transverse direction, we can represent the simple unidirectional composite by what is called the *action-in-series* or *isostress* situation; see Figure 10.1b. In this case, we group the fibers together as a continuous phase normal to the stress. Thus, we have equal stresses in the two components and the model is equivalent to that treated by Reuss [1929]. For loading transverse to the fiber direction, we have

$$\sigma_{ct} = \sigma_f = \sigma_m$$

while the total displacement of the composite in the thickness direction, t_c, is the sum of displacements of the components, that is,

$$\Delta t_c = \Delta t_m + \Delta t_f$$

Dividing throughout by t_c, the gage length of the composite, we obtain

$$\frac{\Delta t_c}{t_c} = \frac{\Delta t_m}{t_c} + \frac{\Delta t_f}{t_c}$$

Now $\Delta t_c/t_c = \varepsilon_{ct}$, strain in the composite in the transverse direction, while Δt_m and Δt_f equal the strains in the matrix and fiber times their respective gage lengths; that is, $\Delta t_m = \varepsilon_m t_m$ and $\Delta t_f = \varepsilon_f t_f$. Then

$$\varepsilon_{ct} = \frac{\Delta t_c}{t_c} = \frac{\Delta t_m}{t_m} \frac{t_m}{t_c} + \frac{\Delta t_f}{t_f} \frac{t_f}{t_c}$$

or

$$\varepsilon_{ct} = \varepsilon_m \frac{t_m}{t_c} + \varepsilon_f \frac{t_f}{t_c} \qquad (10.12)$$

For a given cross-sectional area of the composite under the applied load, the volume fractions of fiber and matrix can be written as

$$V_m = \frac{t_m}{t_c} \quad \text{and} \quad V_f = \frac{t_f}{t_c}$$

This simplifies Eq. (10.12) to

$$\varepsilon_{ct} = \varepsilon_m V_m + \varepsilon_f V_f \qquad (10.13)$$

Considering both components to be elastic and remembering that $\sigma_{ct} = \sigma_f = \sigma_m$ in this case, we can write Eq. (10.13) as

$$\frac{\sigma_{ct}}{E_{ct}} = \frac{\sigma_{ct}}{E_m} V_m + \frac{\sigma_{ct}}{E_f} V_f$$

or

$$\frac{1}{E_{ct}} = \frac{V_m}{E_m} + \frac{V_f}{E_f} = \frac{1}{E_{22}} \qquad (10.14)$$

The relationships given by Eqs. (10.5), (10.10), (10.11), (10.13), and (10.14) are commonly referred to as rules-of-mixtures. Figure 10.2 shows the plots of Eqs. (10.10) and (10.14). The reader should appreciate that these relationships and their variants are but rules of thumb obtained from a simple strength of materials approach. More comprehensive micromechanical models, based on the theory of elasticity, can and should be used to obtain the elastic constants of fibrous composites. We describe below, albeit very briefly, some of these.

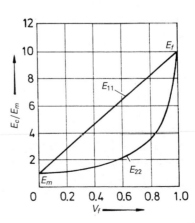

Fig. 10.2. Variation of longitudinal modulus (E_{11}) and transverse modulus (E_{22}) with fiber volume fraction (V_f).

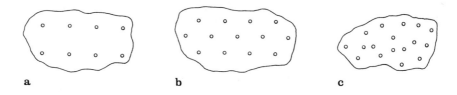

Fig. 10.3. Various fiber arrays in a matrix: **a** square, **b** hexagonal, and **c** random.

10.2.2   Micromechanical Approach

The states of stress and strain can each be described by six components. An anisotropic body with no symmetry elements present requires 21 independent elastic constants to relate stress and strain (Nye, 1969). This is the most general case of elastic anisotropy. An elastically isotropic body, on the other hand, is the simplest case; it needs only two independent elastic constants. In such a body, when a tensile stress σ_z is applied in the z direction, a tensile strain ε_z results in that direction. In addition to this, because of the Poisson's ratio effect, two equal compressive strains ($\varepsilon_x = \varepsilon_y$) result in the x and y directions. In a generally anisotropic body, the two transverse strain components are not equal. In fact, as we shall see in Chapter 11, in such a body, tensile loading can result in tensile and shear strains. The large number of independent elastic constants (21 in the most general case, i.e., no symmetry elements) represents the degree of elastic complexity in a given system. Any symmetry elements present will reduce the number of independent elastic constants (Nye, 1969).

A composite containing uniaxially aligned fibers will have a plane of symmetry perpendicular to the fiber direction (i.e., material on one side of the plane will be the mirror image of the material on the other side). Figure 10.3 shows square, hexagonal, and random fiber arrays in a matrix. A square array of fibers, for example, will have symmetry planes parallel to the fibers as well as perpendicular to them. Such a material is an orthotropic material (three mutually perpendicular planes of symmetry) and possesses nine independent elastic constants (Nye, 1969). Hexagonal and random arrays of aligned fibers are transversely isotropic and have five independent elastic constants. These five constants as well as the stress-strain relationships, as derived by Hashin and Rosen (1964) and Rosen (1973), are given in Table 10.1. There are two Poisson's ratios: one gives the transverse strain caused by an axially applied stress, and the other gives the axial strain caused by a transversely applied stress. The two are not independent but are related (see Sect. 11.3). Thus, the number of independent elastic constants for a transversely isotropic composite is five. Note that the total number of independent elastic constants in Table 10.1 is five (count the number of Cs).

A summary of the elastic constants for a transversely isotropic composite in terms of the elastic constants of the two components is given in Table 10.2

Table 10.1. Elastic moduli of a transversely isotropic fibrous composite

$E = C_{11} - \dfrac{2C_{12}^2}{C_{22} + C_{23}}$	$K_{23} = \frac{1}{2}(C_{22} + C_{23})$
$G = G_{12} = G_{13} = G_{44}$	$G_{23} = \frac{1}{2}(C_{22} - C_{23})$
$v = v_{13} = v_{31} = \frac{1}{2}\left(\dfrac{C_{11} - E}{K_{23}}\right)^{1/2}$	$v_{23} = \dfrac{K_{23} - \phi G_{23}}{K_{23} + \phi G_{23}}$
$E_2 = E_3 = \dfrac{4G_{23}K_{23}}{K_{23} + \phi G_{23}}$	$\phi = 1 + \dfrac{4K_{23}v^2}{E}$

Stress-Strain Relationships

$\varepsilon_{11} = \dfrac{1}{E_1}[\sigma_{11} - v(\sigma_{22} + \sigma_{33})]$	$\varepsilon_{22} = \varepsilon_{33} = \dfrac{1}{E_2}(\sigma_{22} - v\sigma_{33}) - \dfrac{v}{E}\sigma_{11}$
$\gamma_{12} = \gamma_{13} = \dfrac{1}{G}\sigma_{12}$	$\gamma_{23} = \dfrac{2(1 + v_{23})}{E_2}\sigma_{23}$

Source: Adapted with permission from Hashin and Rosen (1964)

(Chamis, 1983). Because the plane 2–3 is isotropic in Figure 10.4, the properties in directions 2 and 3 are identical. The matrix is treated as an isotropic material while the fiber is treated as an anisotropic material. Thus, E_m and v_m are the two constants required for the matrix while five constants (E_{f1}, E_{f2}, G_{f12}, G_{f23}, and v_{f12}) are required for the fiber. The expressions for the five independent constants (E_{11}, E_{22}, G_{12}, G_{23}, and v_{12}) are given in Table 10.2.

Frequently, composite structures are fabricated by stacking thin sheets of unidirectional composites called *plies* in an appropriate orientation sequence

Table 10.2. Elastic constants of a transversely isotropic composite in terms of component constants (matrix isotropic, fiber anisotropic)

Longitudinal modulus	$E_{11} = E_{f1}V_f + E_m V_m$
Transverse modulus	$E_{22} = E_{33} = \dfrac{E_m}{1 - \sqrt{V_f}(1 - E_m/E_{f2})}$
Shear modulus	$G_{12} = G_{13} = \dfrac{G_m}{1 - \sqrt{V_f}(1 - G_m/G_{f12})}$
Shear modulus	$G_{23} = \dfrac{G_m}{1 - \sqrt{V_f}(1 - G_m/G_{f23})}$
Poisson's ratio	$v_{12} = v_{13} = v_{f12}V_f + v_m V_m$
Poisson's ratio	$v_{23} = \dfrac{E_{22}}{2G_{23}} - 1$

Source: Adapted with permission from Chamis (1983).

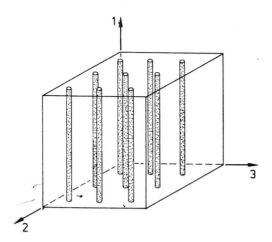

Fig. 10.4. A transversely isotropic fiber composite: plane transverse to fibers (2–3 plane) is isotropic.

dictated by elasticity theory (see Chapter 11). It is of interest to know properties, such as the elastic constants and the strength characteristics of a ply. In particular, it is of value if we are able to predict the lamina characteristics starting from the individual component characteristics. Later in the macromechanical analysis (Chapter 11), we treat a ply as a homogeneous but thin orthotropic material. Elastic constants in the thickness direction can be ignored in such a ply, leaving four independent elastic constants, namely, E_{11}, E_{22}, v_{12}, and G_{12}, i.e., one less than the number for a thick but transversely isotropic material. The missing constant is G_{23}, the transverse shear modulus in the 2–3 plane normal to the fiber axis.

A brief description of the various micromechanical techniques used to predict the elastic constants follows and then we give an account of a set of empirical equations, called *Halpin-Tsai equations*, that can be used under certain conditions to predict the elastic constants of a fiber composite.

In the so-called self-consistent field methods (Chamis and Sendeckyj, 1968), approximations of phase geometries are made and a simple representation of the response field is obtained. The phase geometry is represented by one single fiber embedded in a matrix cylinder. This outer cylinder is embedded in an unbounded homogeneous material whose properties are taken to be equivalent to those of average composite properties. The matrix under a uniform load at infinity introduces a uniform strain field in the fiber. Elastic constants are obtained from this strain field. The results obtained are independent of fiber arrangements in the matrix and are reliable at low-fiber-volume fractions (V_f), reasonable at intermediate V_f, and unreliable at high V_f (Hill, 1964). Exact methods deal with specific geometries, for example, fibers arranged in a hexagonal, square, or rectangular array in a matrix. The elasticity problem is then solved by a series development, a complex variable technique, or a numerical solution. The approach of Eshelby (1957, 1959) considers an infinite matrix containing an ellipsoidal inclusion. Modifications of the Eshelby method have been made by Mori and Tanaka (1973).

The variational or bounding methods focus on the upper and lower bounds on elastic constants. When the upper and lower bounds coincide, the property is determined exactly. Frequently, the upper and lower bounds are well separated. When these bounds are close enough, we can safely use them as indicators of the material behavior. It turns out that this is the case for longitudinal properties of a unidirectional lamina. Hill (1965) derived bounds for the ply elastic constants that are analogous to those derived by Hashin and Rosen (1964) and Rosen (1973). In particular, Hill put rigorous bounds on the longitudinal Young's modulus, E_{11}, in terms of the bulk modulus in plane strain (k_p), Poisson's ratio (v), and the shear modulus (G) of the two phases. No restrictions were made on the fiber form or packing geometry. The term k_p is the modulus for lateral dilation with zero longitudinal strain and is given by

$$k_p = \frac{E}{2(1 - 2v)(1 + v)}$$

The bounds on the longitudinal modulus, E_{11}, are

$$\frac{4V_f V_m (v_f - v_m)^2}{(V_f/k_{pm}) + (V_m/k_{pf}) + 1/G_m} \leq E_{11} - E_f V_f - E_m V_m$$
$$\leq \frac{4V_f V_m (v_f - v_m)^2}{(V_f/k_{pm}) + (V_m/k_{pf}) + 1/G_f} \tag{10.15}$$

Equation (10.15) shows that the deviations from the rule of mixtures [Eq. (10.10)] are quite small ($<2\%$). We may verify this by substituting some values of practical composites such as carbon or boron fibers in an epoxy matrix or a metal matrix composite such as tungsten in a copper matrix. Note that the deviation from the rule of mixtures value comes from the $(v_m - v_f)^2$ factor. For $v_f = v_m$, we have E_{11} given exactly by the rule of mixtures.

Hill (1965) also showed that for a unidirectionally aligned fiber composite

$$v_{12} \gtrless v_f V_f + v_m V_m \quad \text{accordingly as} \quad (v_f - v_m)(k_{pf} - k_{pm}) \gtrless 0 \tag{10.16}$$

Generally, $v_f < v_m$ and $E_f \gg E_m$. Then, v_{12} will be less than that predicted by the rule of mixtures ($= v_f V_f + v_m V_m$). It is easy to see that the bounds on v_{12} are not as close as the ones on E_{11}. This is because $v_f - v_m$ appears in the case of v_{12} [Eq. (10.16)] while $(v_f - v_m)^2$ appears in the case of E_{11} [Eq. (10.15)]. In the case where $v_f - v_m$ is very small, the bounds are close enough to allow us to write

$$v_{12} \simeq v_f V_f + v_m V_m \tag{10.17}$$

10.2.3 Halpin-Tsai Equations

Halpin, Tsai, and Kardos (Halpin and Tsai, 1967; Halpin and Kardos, 1976; Kardos, 1971) empirically developed some generalized equations that readily give satisfactory results compared to the complicated expressions. They are also useful in determining the properties of composites that contain discontinuous fibers oriented in the loading direction. One writes a single equation of the form

$$\frac{p}{p_m} = \frac{1 + \xi \eta V_f}{1 - \eta V_f} \tag{10.18}$$

$$\eta = \frac{p_f/p_m - 1}{p_f/p_m + \xi} \tag{10.19}$$

where p represents composite moduli, for example, E_{11}, E_{22}, G_{12}, or G_{23}; p_m and p_f are the corresponding matrix and fiber moduli, respectively; V_f is the fiber volume fraction; and ξ is a measure of the reinforcement that depends on boundary conditions (fiber geometry, fiber distribution, and loading conditions). The term ξ, is an empirical factor that is used to make Eq. (10.18) conform to the experimental data.

The function η in Eq. (10.19) is constructed in such a way that when $V_f = 0$, $p = p_m$ and when $V_f = 1$, $p = p_f$. Furthermore, the form of η is such that

$$\frac{1}{p} = \frac{V_m}{p_m} + \frac{V_f}{p_f} \quad \text{for } \xi \to 0$$

and

$$p = p_f V_f + p_m V_m \quad \text{for } \xi \to \infty$$

These two extremes (not necessarily tight) bound the composite properties. Thus, values of ξ between 0 and ∞ will give an expression for p between these extremes. Some typical values of ξ are given in Table 10.3. Thus, we

Table 10.3. Values of ξ for some uniaxial composites

Modulus	ξ
E_{11}	$2(l/d)$
E_{22}	0.5
G_{12}	1.0
G_{21}	0.5
K	0

Source: Adapted from Nielsen (1974), courtesy of Marcel Dekker, Inc.

can cast the Halpin-Tsai equations for the transverse modulus as

$$\frac{E_{22}}{E_m} = \frac{1 + \xi\eta V_f}{1 - \eta V_f} \quad \text{and} \quad \eta = \frac{E_f/E_m - 1}{E_f/E_m + \xi} \tag{10.20}$$

Comparing these expressions with exact elasticity solutions, one can obtain the value of ξ. Whitney [1973] suggests $\xi = 1$ or 2 for E_{22}, depending on whether a hexagonal or square array of fibers is used.

Nielsen [1974] modified the Halpin-Tsai equations to include the maximum packing fraction ϕ_{max} of the reinforcement. His equations are

$$\frac{p}{p_m} = \frac{1 + \xi\eta V_f}{1 - \eta\Psi V_f}$$

$$\eta = \frac{p_f/p_m - 1}{p_f/p_m + \xi} \tag{10.21}$$

$$\Psi \simeq 1 + \left(\frac{1 - \phi_{max}}{\phi_{max}^2}\right) V_f$$

where ϕ_{max} is the maximum packing factor. It allows one to take into account the maximum packing fraction. For a square array of fibers, $\phi_{max} = 0.785$, while for a hexagonal arrangement of fibers, $\phi_{max} = 0.907$. In general, ϕ_{max} is between these two extremes and near the random packing, $\phi_{max} = 0.82$.

Example 10.2

Consider a unidirectionally reinforced glass fiber/epoxy composite. The fibers are continuous and 60% by volume. The tensile strength of glass fibers is 1 GPa and the Young's modulus is 70 MPa. The tensile strength of the epoxy matrix is 60 MPa and its Young's modulus is 3 GPa. Compute the Young's modulus and the tensile strength of the composite in the longitudinal direction.

Solution

Young's modulus of the composite in the longitudinal direction is given by

$$E_{cl} = 70 \times 0.6 + 3 \times 0.4 = 42 + 1.2 = 43.2\,\text{GPa}$$

To calculate the tensile strength of the composite in the longitudinal direction, we need to determine which component, fiber or matrix, has the lower failure strain. The failure strain of the fiber is

$$\varepsilon = \sigma_f/E_f = 1/70 = 0.014$$

while that of the matrix is

$$\varepsilon_m = (60 \times 10^{-3})/3 = 0.020$$

Thus, $\varepsilon_f < \varepsilon_m$, i.e., fibers fail first. At that strain, assuming a linear stress-strain curve for the epoxy matrix, the matrix strength is $\sigma'_m = E_m\varepsilon_f =$

$3 \times 0.014 = 0.042\,\text{GPa} = 42\,\text{MPa}$. Then, we get the composite tensile strength as

$$\sigma_c = 0.6 \times 1 + 0.4 \times 0.042$$

$$= 0.6 + 0.0168 = 617\,\text{MPa}$$

10.2.4 Transverse Stresses

When a fibrous composite consisting of components with different elastic moduli is uniaxially loaded, stresses in transverse directions arise because of the Poisson's ratio differences between the matrix and fiber, that is, because the two components have different contractile tendencies. Here we follow Kelly's [1970] treatment of this important but, unfortunately, not well appreciated subject.

Consider a unit fiber reinforced composite consisting of a single fiber (radius a) surrounded by its shell of matrix (outer radius b) as shown in Figure 10.5. The composite as a whole is thought to be built of an assembly of such unit composites, a reasonably valid assumption at moderate fiber volume fractions. We axially load the composite in direction z. Owing to the obvious cylindrical symmetry, we treat the problem in polar coordinates, r, θ, and z. It follows from the axial symmetry that the stress and strain are independent of angle θ and are functions only of r, which simplifies the problem. We can write Hooke's law for this situation as

$$\begin{bmatrix} e_r & 0 & 0 \\ 0 & e_\theta & 0 \\ 0 & 0 & e_z \end{bmatrix} = \frac{1+v}{E} \begin{bmatrix} \sigma_r & 0 & 0 \\ 0 & \sigma_\theta & 0 \\ 0 & 0 & \sigma_z \end{bmatrix} - \frac{v}{E}(\sigma_r + \sigma_\theta + \sigma_z) \begin{bmatrix} 1 & 0 & 0 \\ 0 & 1 & 0 \\ 0 & 0 & 1 \end{bmatrix}$$

$$(10.22)$$

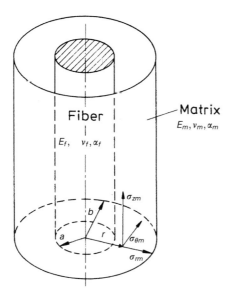

Fig. 10.5. A single fiber surrounded by its matrix shell.

where e is the strain, σ is the stress, v is the Poisson's ratio, E is Young's modulus in the longitudinal direction, and the subscripts r, θ, and z refer to the radial, circumferential, and axial directions, respectively. The only equilibrium equation for this problem is

$$\frac{d\sigma_r}{dr} + \frac{\sigma_r - \sigma_\theta}{r} = 0 \tag{10.23}$$

Also, for the plane strain condition, we can write for the strain components, in terms of displacements,

$$e_r = \frac{du_r}{dr} \quad e_\theta = \frac{u_r}{r} \quad e_z = \text{const} \tag{10.24}$$

where u_r is the radial displacement.

From Eq. (10.22) we have, after some algebraic manipulation,

$$\frac{\sigma_\theta}{K} = (1 - v)e_\theta + v(e_r + e_z)$$
$$\frac{\sigma_r}{K} = (1 - v)e_r + v(e_\theta + e_z) \tag{10.25}$$

where

$$K = \frac{E}{(1 + v)(1 - 2v)}$$

From Eqs. (10.24) and (10.25), we get

$$\frac{\sigma_\theta}{K} = v\frac{du_r}{dr} + (1 - v)\frac{u_r}{r} + ve_z$$
$$\frac{\sigma_r}{K} = (1 - v)\frac{du_r}{dr} + v\frac{u_r}{r} + ve_z \tag{10.26}$$

Substituting Eq. (10.26) in Eq. (10.23), we obtain the following differential equation in terms of the radial displacement u_r:

$$\frac{d^2u_r}{dr^2} + \frac{1}{r}\frac{du_r}{dr} - \frac{u_r}{r^2} = 0 \tag{10.27}$$

Equation (10.27) is a common differential equation in elasticity problems with rotational symmetry (Love, 1952), and its solution is

$$u_r = Cr + \frac{C'}{r} \tag{10.28}$$

where C and C' are constants of integration to be determined by using boundary conditions. Now, Eq. (10.28) is valid for displacements in both components, that is, fiber and matrix. Let us designate the central component by subscript 1 and the sleeve by subscript 2. Thus, we can write the displacements in the two components as

$$u_{r1} = C_1 r + \frac{C_2}{r}$$

$$u_{r2} = C_3 r + \frac{C_4}{r}$$

$$(10.29)$$

The boundary conditions can be expressed as follows:

1. At the free surface, the stress is zero, that is, $\sigma_{r2} = 0$ at $r = b$.
2. At the interface, the continuity condition requires that at $r = a$, $u_{r1} = u_{r2}$ and $\sigma_{r1} = \sigma_{r2}$.
3. The radial displacement must vanish along the symmetry axis, that is, at $r = 0$, $u_{r1} = 0$.

The last boundary condition immediately gives $C_2 = 0$, because otherwise u_{r1} will become infinite at $r = 0$. Applying the other boundary conditions to Eqs. (10.26) and (10.29), we obtain three equations with three unknowns. Knowing these integration constants, we obtain u and thus the stresses in the two components. It is convenient to develop an expression for radial pressure p at the interface. At the interface $r = a$, if we equate σ_{r2} to $-p$, then after some tedius manipulations it can be shown that

$$p = \frac{2e_z(v_2 - v_1)V_2}{V_1/k_{p2} + V_2/k_{p1} + 1/G_2} \qquad (10.30)$$

where k_p is the plane strain bulk modulus equal to $E/2(1 + v)(1 - 2v)$. The expressions for the stresses in the components involving p are

Component 1:

$$\sigma_{r1} = \sigma_{\theta 1} = -p$$

$$\sigma_{z1} = E_1 e_z - 2v_1 p$$

$$(10.31)$$

Component 2:

$$\sigma_{r2} = p\left(\frac{a^2}{b^2 - a^2}\right)\left(1 - \frac{b^2}{r^2}\right)$$

$$\sigma_{\theta 2} = p\left(\frac{a^2}{b^2 - a^2}\right)\left(1 + \frac{b^2}{r^2}\right)$$

$$(10.32)$$

$$\sigma_{z2} = E_2 e_z + 2v_2 p\left(\frac{a^2}{b^2 - a^2}\right)$$

Note that p is positive when the central component 1 is under compression, that is, when $v_1 < v_2$.

Figure 10.6 shows the stress distribution schematically in a fiber composite (1 = fiber, 2 = matrix). We can draw some inferences from Eqs. (10.31) and (10.32) and Figure 10.6.

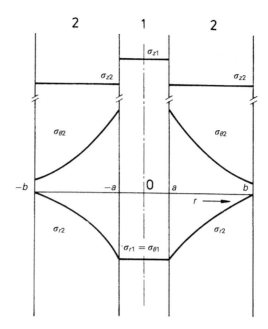

Fig. 10.6. Three-dimensional stress distribution (schematic) in the unit composite shown in Figure 10.5. Transverse stresses (σ_r and σ_θ) result from the differences in the Poisson's ratios of the fiber and matrix.

1. Axial stress is uniform in components 1 and 2, although its magnitude is different in the two and depends on the respective elastic constants.
2. In the central component 1, σ_{r1} and $\sigma_{\theta1}$ are equal in magnitude and sense. In the sleeve 2, σ_{r2} and $\sigma_{\theta2}$ vary as $1 - b^2/r^2$ and $1 + b^2/r^2$, respectively.
3. When the Poisson's ratio difference $(v_2 - v_1)$ goes to zero, σ_r and σ_θ go to zero; that is, the rheological interaction will vanish.
4. Because of the relatively small difference in the Poisson's ratios of the components of a composite, the transverse stresses that develop in the elastic regime will be relatively small. If one of the components deforms plastically $(v \to 0.5)$, then Δv can become significant and so will the transverse stresses.

10.3 Thermal Properties

Thermal energy is responsible for the atomic or molecular vibration about a mean position in any material. As the temperature of the materials is increased, the amplitude of thermal energy-induced vibrations is increased and the interatomic or intermolecular spacing increases, i.e., an expansion of the body occurs. Most materials show such an expansion with increasing temperature. In general, the thermal expansion of a material is greater in the liquid state than in the crystalline state, with the transition occurring at the melting point. In the case of a glassy material, such a transition occurs at the glass transition temperature. Over a certain range of temperature, one can relate the temperature interval and thermal strain by a coefficient, called the *coefficient of thermal expansion*. In the case of a linear strain, the linear thermal expansion coefficient, α, is a second-rank symmetric tensor, and is related to the strain tensor, ε, by the following relationship:

$$\varepsilon_{ij} = \alpha_{ij}\Delta T \tag{10.33}$$

where ΔT is the temperature change. The thermal expansion coefficient, α, generally does not have a constant value over a very large range of temperature. Thus, we can define α_{ij} in a more general way by taking into account this variation with temperature as follows:

$$\alpha_{ij} = \delta\varepsilon_{ij}/\delta T$$

If we consider a volumetric strain, the volumetric coefficient of thermal expansion, β, is given by

$$\beta_{ij} = \frac{1}{V}\left(\frac{\delta V}{\delta T}\right)$$

where V is the volume and T is the temperature. For small strains, it can be easily shown that

$$\beta_{ij} = 3\alpha_{ij}$$

The volumetric expansion coefficient, β, is equal to the sum of the diagonal terms of the strain tensor, i.e.,

$$\beta = \varepsilon_{11} + \varepsilon_{22} + \varepsilon_{33} = 3\alpha \tag{10.34a}$$

or

$$\alpha = \frac{1}{3}[\varepsilon_{11} + \varepsilon_{22} + \varepsilon_{33}] \tag{10.34b}$$

As we said earlier, only over some specified range of temperature can the coefficient of thermal expansion be treated as a constant. Consider a temperature range ΔT over which α is a constant. We can write Eq. (10.33) in an extended form as

$$\begin{vmatrix} \varepsilon_{11} & \varepsilon_{12} & \varepsilon_{13} \\ & \varepsilon_{22} & \varepsilon_{23} \\ & & \varepsilon_{33} \end{vmatrix} = \begin{vmatrix} \alpha_{11} & \alpha_{12} & \alpha_{13} \\ & \alpha_{22} & \alpha_{23} \\ & & \alpha_{33} \end{vmatrix}\Delta T \tag{10.35a}$$

Or, using the contracted notation (see Sect. 11.1), we can write

$$\begin{vmatrix} \varepsilon_1 \\ \varepsilon_2 \\ \varepsilon_3 \\ \varepsilon_4 \\ \varepsilon_5 \\ \varepsilon_6 \end{vmatrix} = \begin{vmatrix} \alpha_1 \\ \alpha_2 \\ \alpha_3 \\ \alpha_4 \\ \alpha_5 \\ \alpha_6 \end{vmatrix}\Delta T \tag{10.35b}$$

If an arbitrary direction $[hkl]$ has direction cosines n_1, n_2, and n_3, then we can write for the linear thermal expansion coefficient, α_{hkl}, in that direction

$$\alpha_{hkl} = n_1^2 \alpha_1 + n_2^2 \alpha_2 + n_3^2 \alpha_3 \qquad (10.36)$$

In a transversely isotropic fibrous composite (i.e., hexagonal symmetry), we have $\alpha_1 = \alpha_2 = \alpha_\perp$, perpendicular to the fiber axis and $\alpha_3 = \alpha_\parallel$, parallel to the fiber axis. Then, remembering that $n_1^2 + n_2^2 + n_3^2 = 1$, Eq. (10.36) becomes

$$\alpha_{hkl} = (n_1^2 + n_2^2)\alpha_1 + n_3^2 \alpha_3$$

$$\alpha_{hkl} = \alpha_\perp \sin^2 \theta + \alpha_\parallel \cos^2 \theta \qquad (10.37a)$$

$$\alpha_{hkl} = \alpha_\perp + (\alpha_\parallel - \alpha_\perp) \cos^2 \theta \qquad (10.37b)$$

where θ is the angle between direction $[hkl]$ and the fiber axis.

10.3.1 Expressions for Thermal Expansion Coefficients of Composites

Various equations have been proposed for obtaining the thermal expansion coefficients of a composite, knowing the material constants of the components and their geometric arrangements. Different equations predict very different values of expansion coefficients for a given composite. Almost all expressions, however, predict expansion coefficient values that are different from those given by a simple rule of mixtures $(= \alpha_f V_f + \alpha_m V_m)$. This is because these equations take into account the important fact that the presence of a reinforcement, with an expansion coefficient less than that of the matrix, introduces a mechanical constraint on the matrix. A fiber will cause a greater constraint on the matrix than a particle. Let us consider the expansion coefficients of a particulate and a fibrous composite.

One can regard a particulate composite as a homogeneous material in a statistical sense, i.e., assuming a uniform distribution of the particles in the matrix. Let us denote the volume fractions of the two phases making a particulate composite by V_1 and V_2 $(= 1 - V_1)$. Various researchers have derived bounds and given expressions for the expansion coefficients and other transport properties such as thermal conductivity. Kerner (1956) developed the following expression for the volumetric expansion coefficient of a composite consisting of spherical particles dispersed in a matrix:

$$\beta_c = \beta_m V_m + \beta_p V_p - (\beta_m - \beta_p) V_p \left[\frac{1/K_m - 1/K_p}{V_m/K_p + V_p/K_m + 0.75 G_m} \right]$$

where subscripts c, m, and p denote the composite, matrix, and reinforcement, respectively; β is the volumetric expansion coefficient, and K denotes the bulk modulus. Kerner's expression does not differ significantly from the rule of mixtures because the particle reinforcement constrains the matrix much less than fibers. The coefficient of linear thermal expansion according to Turner (1946) is given by:

$$\alpha_c = \frac{\alpha_m V_m K_m + \alpha_p V_p K_p}{V_p K_p + V_m K_m}$$

where the symbols have the significance given earlier. Turner's expression, generally, gives an expansion coefficient much lower than the rule-of-mixtures value.

Unidirectionally aligned fibrous composites have two (or sometimes three) thermal expansion coefficients: α_{cl} in the longitudinal direction and α_{ct} in the transverse direction. Fibers generally have a lower expansion coefficient than that of the matrix, and thus the former mechanically constrain the latter. This results in an α_{cl} smaller than α_{ct} for the composite. At low fiber volume fractions, it is not unusual to find the transverse expansion of a fibrous composite, α_{ct}, greater than that of the matrix in isolation. The long stiff fibers prevent the matrix from expanding in the longitudinal direction, and as a result the matrix is forced to expand more than usual in the transverse direction. It should be pointed out that in the case of some CMCs, this situation can be reversed. For example, in the case of alumina fibers ($\alpha = 8 \times 10^{-6} \, K^{-1}$) in a low-expansion glass or ceramic matrix, it is the matrix that will constrain the fibers, i.e., the situation in this case is the reverse of the one commonly encountered.

We give below some expressions for the coefficients of thermal expansion of unidirectionally reinforced fiber composites. All of these analyses involve the following assumptions:

1. The bonding between the fiber and matrix is perfect and mechanical in nature, i.e., no chemical interaction is allowed.
2. The fibers are continuous and perfectly aligned.
3. The properties of the constituents do not change with temperature.

Schapery (1969) used energy methods to derive the following expressions for expansion coefficient of a fibrous composite, assuming Poisson's ratios of the components are not very different. The longitudinal expansion coefficient for the composite is

$$\alpha_{cl} = \frac{\alpha_m E_m V_m + \alpha_f E_f V_f}{E_m V_m + E_f V_f} \tag{10.38}$$

and the transverse expansion coefficient is

$$\alpha_{ct} \simeq (1 + v_m)\alpha_m V_m + (1 + v_f)\alpha_f V_f - \alpha_{cl}\bar{v} \tag{10.39a}$$

$$\bar{v} = v_f V_f + v_m V_m$$

For high fiber volume fractions, $V_f > 0.2$ or 0.3, α_{ct} can be approximated by

$$\alpha_{ct} \simeq (1 + v_m)\alpha_m V_m + \alpha_f V_f \tag{10.39b}$$

The longitudinal and transverse coefficients of thermal expansion for

Fig. 10.7. Longitudinal and transverse linear thermal expansion coefficients versus fiber volume fraction for alumina fiber in an aluminum matrix.

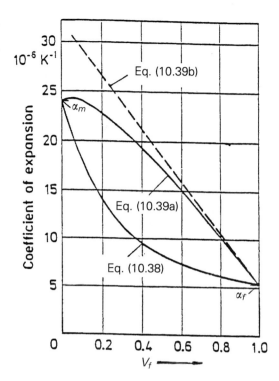

alumina fibers in an aluminum matrix are plotted in Figure 10.7. Note the marked anisotropy in the expansion for aligned fibrous composites.

The anisotropy in expansion can be reduced somewhat if the composite contains randomly oriented short fibers or whiskers in three dimensions. We can write its isotropic thermal expansion coefficient

$$\alpha \simeq \frac{\alpha_{cl} + 2\alpha_{ct}}{3} \tag{10.40}$$

where α_{cl} and α_{ct} are given by Eqs. (10.38) and (10.39).

It should be pointed out the length of fiber as well as fiber orientation have an effect on the coefficients of thermal expansion (Marom and Weinberg, 1975; Vaidya et al., 1994). It would appear that the fiber length has a more sensitive effect on the expansion characteristics because the constraint on the matrix is highest when the fibers are continuous. An extreme case of a low constraint is one of a matrix containing spherical particles, i.e., aspect ratio = 1. Continuous fibers with an aspect ratio of infinity represent the other extreme in constraint.

The effect of the fiber length on the thermal expansion can be incorporated into Schapery's equation as (Marom and Weinberg, 1975; Vaidya and

Chawla, 1994)

$$\alpha_{11} = \frac{k\alpha_f V_f E_f + \alpha_m V_m E_m + \alpha_i V_i E_i}{k V_f E_f + V_m E_m + V_i E_i}$$

where the value of k is given by

$$k = \frac{l}{2l_c} \quad (l < l_c, 0 < k < 0.5)$$

$$k = 1 - l_c/2l \quad (l > l_c, 0 < k < 1)$$

where l is the length and l_c is the critical length of the fiber (see Sect. 10.4).

Chamis (1983) used a simple force balance to arrive at the following expressions for α_1 and α_2:

$$\alpha_1 = \left| \frac{E_{f1} \alpha_{f1} V_f + E_m \alpha_m V_m}{E_{f1} V_f + E_m V_m} \right|$$

$$\alpha_2 = \alpha_{f2} \sqrt{V_f} + (1 - \sqrt{V_f}) \left(1 + V_f v_m \frac{E_{f1}}{E_c} \right) \alpha_m$$

Rosen and Hashin (1970) used a plane strain model to derive expressions for α_1 and α_2:

$$\alpha_1 = \bar{\alpha} + \left[\frac{\alpha_f - \alpha_m}{1/K_f - 1/K_m} \right] \left[\frac{3(1 - 2v_c)}{E_c} - \frac{1}{K_c} \right]$$

$$\alpha_2 = \bar{\alpha} + \left| \frac{\alpha_f - \alpha_m}{1/K_f - 1/K_m} \right| \left| \frac{3}{2k_c} - \frac{3(1 - 2v_c)v_c}{E_c} - \frac{1}{K_c} \right|$$

where $\bar{\alpha} = \alpha_f V_f + \alpha_m V_m$, K_c is the composite bulk modulus, E_c is the composite elastic modulus, and k_c is the composite transverse bulk modulus.

Many applications of composites require controlled thermal expansion characteristics. All the expressions given earlier predict thermal expansion coefficients of composites based on continuum mechanics. That is, no account is taken of the effect of fiber or particle size on the coefficient of thermal expansion (CTE). Xu et al. (1994) examined the effect of particle size on the CTE of TiC/Al XD^{TM} composite, with two TiC particle sizes, 0.7 and 4 μm. The results of this work showed the effect of particle size on the thermal expansion coefficient of particle reinforced aluminum composites in terms of the degree of constraint on the matrix expansion. Careful TEM work showed lattice distortion at the interfacial zone. The effect of particle size on the CTE of composite was related to the volume fraction of interfacial zone through which the constraint occurs. A phenomenological approach that takes into account the different degree of constraint on the matrix expansion based on the TEM work allows us to compute the CTE of this composite with different particle sizes.

10.3.2   Expressions for Thermal Conductivity of Composites

The heat flow in a material is proportional to the temperature gradient, and the constant of proportionality is called the *thermal conductivity*. Thus, in the most general form, using indicial notation, we can write

$$q_i = -k_{ij}\, dT/dx_j$$

where q_i is the heat flux along the x_i axis, dT/dx_j is the temperature gradient across a surface perpendicular to the x_j axis, and k_{ij} is the thermal conductivity. As should be evident from the two indices, thermal conductivity is also a second-rank tensor. Although k_{ij} is not a symmetric tensor in the most general case, it is a symmetric tensor for most crystal systems. For an isotropic material, k_{ij} reduces to a scalar number, k. For an orthotropic material, we shall have three constants along the three principal axes, viz., k_{11}, k_{22}, and k_{33}. For a transversely isotropic material such as a unidirectionally reinforced fibrous composite, there will be two constants: thermal conductivity in the axial direction, k_{cl}, and that in the transverse direction, k_{ct}. The thermal conductivity in the axial direction, k_{cl}, can be predicted by a rule-of-mixtures-type expression (Behrens, 1968)

$$k_1 = k_{cl} = k_{f1}V_f + k_mV_m \tag{10.41}$$

where k_{f1} is the thermal conductivity of the fiber in the axial direction, k_m is that of the isotropic matrix, and V_f and V_m are the volume fractions of the fiber and matrix, respectively.

In a transverse direction, the thermal conductivity of a unidirectionally aligned fiber composite (i.e., transversely isotropic) can be approximated by the action-in-series model discussed earlier. This would give

$$k_{ct} = k_2 = k_{f2}k_m/(k_{f2}V_f + k_mV_m)$$

More complicated expressions have been derived for the transverse thermal conductivity. In real composites, it is likely that the thermal contact at the fiber/matrix interface will be less than perfect because of thermal mismatch, etc. An expression that takes into account the fact that the interface region may have a different thermal conductance than the matrix or the fiber is as follows (Hasselman and Johnson, 1987):

$$
\begin{aligned}
k_{ct} = k_2 = k_m[&(k_f/k_m - 1 - k_f/ah_i)V_f \\
&+ (1 + k_f/k_m + k_f/ah_i)]/[(1 - k_f/k_m + k_f/ah_i)V_f \\
&+ (1 + k_f/k_m + k_f/ah_i)]
\end{aligned} \tag{10.42}
$$

where k_m is the matrix thermal conductivity, k_f is the transverse thermal conductivity of the fibers, V_f is the fiber volume fraction, a is the fiber radius, and h_i is the thermal conductance of the interface region. It can be seen that the effect of interfacial conductivity is governed by the magnitude of nondimensional parameter, k_f/ah_i. h_i will have a value of infinity for perfect thermal contact and it will be equal to zero for a pore. Bhatt et al. (1992) studied the important role of interface in controlling the effective thermal conductivity of composites. In particular, one can use the measurement of thermal conductivity as a nondestructive tool to determine the integrity of the fiber/matrix interface.

We can also use Halpin-Tsai-Kardos equations to obtain the following expression for the transverse thermal conductivity of a composite containing unidirectionally aligned fibers:

$$k_{c2} = k_{c3} = k_{ct} = (1 + \eta V_f)/(1 - \eta V_f)k_m$$

and

$$\eta = [(k_{f2}/k_m) - 1]/[(k_{f2}/k_m) + 1]$$

where we have taken ζ equal to 1 in the Halpin-Tsai-Kardos equation.

If we know the thermal conductivity in directions 1 and 2, we can find the thermal conductivity k_x and k_y in any arbitrary directions, x and y, respectively, by the following equations:

$$k_x = k_1 \cos^2 \theta + k_2 \sin^2 \theta$$

$$k_y = k_1 \sin^2 \theta + k_2 \cos^2 \theta$$

$$k_{xy} = (k_2 - k_1) \sin \theta \cos \theta$$

where θ is the angle measured from the x axis to the 1 axis and k_{xy} can be considered to be a thermal coupling coefficient.

A summary of the general expressions for the thermal properties of transversely isotropic fiber composites is given in Table 10.4.

Table 10.4. Thermal properties of a transversely isotropic composite (matrix isotropic, fiber anisotropic)

Heat capacity	$C = \dfrac{1}{\rho}(V_f \rho_f C_f + V_m \rho_m C_m)$
Longitudinal conductivity	$k_{11} = V_f k_{f1} + V_m k_m$
Transverse conductivity	$k_{22} = k_{33} = (1 - \sqrt{V_f})k_m + \dfrac{k_m \sqrt{V_f}}{1 - \sqrt{V_f}(1 - k_m/k_{f2})}$
Longitudinal thermal expansion coefficient	$\alpha_{11} = \dfrac{V_f E_{f1} \alpha_{f1} + V_m E_m \alpha_m}{E_{f1} V_f + E_m V_m}$
Transverse thermal expansion coefficient	$\alpha_{22} = \alpha_{33} = \alpha_{f2}\sqrt{V_f} + \alpha_m(1 - \sqrt{V_f})\left(1 + \dfrac{V_f v_m E_{f1}}{E_{f1}V_f + E_m V_m}\right)$

Source: Adapted with permission form Chamis (1983).

Example 10.3

A unidirectional glass fiber reinforced epoxy has 50 vol. % of fiber. Estimate its thermal conductivity parallel to the fibers. Given $k_{glass} = 0.9$ W/mK, $k_{epoxy} = 0.15$ W/mK.

Solution

The longitudinal thermal conductivity is given by

$$k_{cl} = V_m k_m + V_f k_f$$
$$= 0.5 \times 0.15 + 0.5 \times 0.9$$
$$= 0.075 + 0.45 = 0.525 \, \text{W/mK}$$

Example 10.4

Derive an expression for the heat capacity or specific heat, C_p, of a composite.

Solution

The total quantity of heat in the composite, Q_c, is the sum of heats in the fiber (Q_f) and matrix (Q_m). Thus,

$$Q_c = Q_f + Q_m$$

The quantity of heat is heat capacity times the mass, i.e.,

$$Q = mC_p = v\rho C_p$$

where m is the mass, v is the volume, ρ is the density, and C_p is the heat capacity or specific heat. Using the subscripts c, f, and m for composite, fiber, and matrix, respectively, we can write

$$v_c \rho_c C_{pc} = v_f \rho_f C_{pf} + v_m \rho_m C_{pm}$$

Remembering that $v_f/v_c = V_f$, the fiber volume fraction, and $v_m/v_c = V_m$, the matrix volume fraction, we can write

$$\rho_c C_{pc} = V_f \rho_f C_{pf} + V_m \rho_m C_{pm}$$

or

$$C_{pc} = 1/\rho_c [V_f \rho_f C_{pf} + V_m \rho_m C_{pm}]$$

10.3.3   Hygral and Thermal Stresses

Hygroscopy, meaning water absorption, can be a problem with polymeric resins and natural fibers because it can lead to swelling. Swelling can lead to

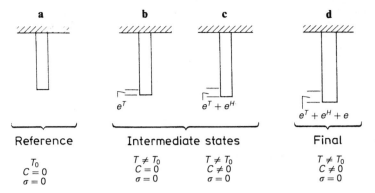

Fig. 10.8. a Strain-free reference state; **b** thermal strain (e^T); **c** hygral (e^H) and thermal (e^T) strains; and **d** final state: hygral (e^H), thermal (e^T), and mechanical strain.

stresses if, as is likely in a composite, the material is not allowed to expand freely because of the presence of the matrix. Such stresses resulting from moisture absorption or hygroscopy are called *hygral stresses*. Thermal stresses will result in a material when the material is not allowed to expand or contract freely because of a constraint. When both hygral and thermal stresses are present, we call them *hygrothermal stresses*, and mathematically we can treat them together.

Consider the passage of a composite from a reference state, where the body is stress-free and relaxed at a temperature T, concentration of moisture $C = 0$, and external stress $\sigma = 0$, to a final state where the body has hygrothermal as well as external stresses. We can consider that the final state of the composite with hygrothermal stresses and external loading is attained via two intermediate stages shown in Figure 10.8. The final strain in the body can then be written

$$e_i = \underbrace{e_i^T \mid e_i^H}_{\text{nonmechaical strains}} + \underbrace{S_{ij}\sigma_j}_{\text{mechanical strain}} \qquad (10.43)$$

If we regard the composite as transversely isotropic, we can write

$$e_{xy}^T = e_{xy}^H = 0$$

$$e_y^T = e_z^T \quad \text{and} \quad e_y^H = e_z^H$$

$$e_i^T = \alpha_i(T - T_0)$$

$$e_i^H = \beta_i C$$

$$(10.44)$$

where the α_i are thermal expansion coefficients (K^{-1}), the β_i are nondimensional swelling coefficients, T is the temperature, T_0 is the equilibrium temperature, C is the concentration of water vapor, and the superscripts H and T indicate hygral and thermal strain components, respectively.

Total volumetric hygrothermal strain can be expressed as the sum of the diagonal terms of the strain matrix. It is important to note that the thermal and hygral effects are dilational only, i.e., they cause only expansion or contraction but do not affect the shear components. Thus,

$$\frac{\Delta V}{V} = e_x^T + e_y^T + e_z^T + e_x^H + e_y^H + e_z^H$$

$$= e_x^T + 2e_y^T + e_x^H + 2e_y^H \tag{10.45}$$

Hygrothermal stresses are very important in polymer matrix composites (see also Chapters 3 and 4).

We first consider the constitutive equations for an isotropic material and then for an anisotropic material. Consider an isotropic material, in plane stress, subjected to a temperature change, ΔT, and a change in moisture content, ΔC. We can write the constitutive equations for this case as follows

$$\varepsilon_1 = \sigma_1/E - v\sigma_2/E + \beta\,\Delta C + \alpha\,\Delta T$$

$$\varepsilon_2 = v\sigma_1/E - \sigma_2/E + \beta\,\Delta C + \alpha\,\Delta T$$

$$\varepsilon_6 = \sigma_6/G$$

Note that, as pointed out earlier, ΔC and ΔT do not produce shear strains. For a specially orthotropic, unidirectionally reinforced fiber reinforced lamina, we can write

$$\begin{vmatrix} \varepsilon_1 \\ \varepsilon_2 \\ \varepsilon_6 \end{vmatrix} = \begin{vmatrix} S_{11} & S_{12} & 0 \\ & S_{22} & 0 \\ & & S_{66} \end{vmatrix} \begin{vmatrix} \sigma_1 \\ \sigma_2 \\ \sigma_6 \end{vmatrix} + \begin{bmatrix} \alpha_1 \\ \alpha_2 \\ 0 \end{bmatrix} \Delta C + \begin{bmatrix} \beta_1 \\ \beta_2 \\ 0 \end{bmatrix} \Delta T$$

In a PMC, the matrix is likely to absorb more moisture than the fiber, which will lead to $\beta_2 > \beta_1$. In an off-axis, generally orthotropic, unidirectionally reinforced fiber reinforced lamina, we shall have the following expression

$$\begin{bmatrix} \varepsilon_x \\ \varepsilon_y \\ \varepsilon_s \end{bmatrix} = \begin{bmatrix} \overline{S_{11}} & \overline{S_{12}} & \overline{S_{16}} \\ & \overline{S_{22}} & \overline{S_{26}} \\ & & \overline{S_{66}} \end{bmatrix} \begin{bmatrix} \sigma_x \\ \sigma_y \\ \sigma_s \end{bmatrix} + \begin{bmatrix} \alpha_x \\ \alpha_y \\ \alpha_s \end{bmatrix} \Delta C + \begin{bmatrix} \beta_x \\ \beta_y \\ \beta_s \end{bmatrix} \Delta T$$

Thermal Stresses

During curing or solidification of the matrix around fibers, a high magnitude of shrinkage stresses can result. The interfacial pressure developed during curing is akin to that obtained upon embedding a cylinder of radius $r + \delta r$ in a cylindrical hole of radius r. Specifically, the thermal stresses generated depend on the fiber volume fraction, fiber geometry, thermal mismatch ($\Delta\alpha$), and the modulus ratio, E_f/E_m. Generally, $\alpha_m > \alpha_f$; that is, on cooling from T_0 to T ($T_0 > T$), the matrix would tend to contract more than the fibers, causing the fibers to experience axial compression. In extreme cases, fiber buckling can also lead to the generation of interfacial shear stresses. This problem of thermal stresses in composite materials is a most serious and important problem. It is worth repeating that thermal stresses are internal stresses that arise when there exists a constraint on the free dimensional change of a body (Chawla, 1973a). In the absence of this constraint, the body can experience free thermal strains without any accompanying thermal stresses. The constraint can have its origin in (1) a temperature gradient, (2) crystal structure anisotropy (e.g., noncubic structure), (3) phase transformations resulting in a volume change, and (4) a composite material made of dissimilar materials (i.e., materials having different α's). The thermal gradient problem is a serious one in ceramic materials in general. A thermal gradient ΔT is inversely related to the thermal diffusivity a of a material. Thus,

$$\Delta T = \phi\left(\frac{1}{a}\right) = \phi\left(\frac{C_p\rho}{k}\right) \tag{10.46}$$

where C_p is the specific heat, ρ is the density, and k is the thermal conductivity. Metals generally have high thermal diffusivity and any thermal gradients that might develop are dissipated rather quickly. It should be emphasized that in composite materials even a uniform temperature change (i.e., no temperature gradient) will result in thermal stresses owing to the ever-present thermal mismatch (Chawla, 1973a). Thermal stresses resulting from a thermal mismatch will generally have an expression of the form

$$\sigma = f(E\Delta\alpha\Delta T) \tag{10.47}$$

We describe below the three-dimensional thermal stress state in a composite consisting of a central fiber surrounded by its shell of matrix; see Figure 10.5.

The elasticity problem is basically the same as the same discussed for transverse stresses. We use polar coordinates, r, θ, and z (see Fig. 10.5). Axial symmetry makes shear stresses go to zero and the principal stresses are independent of θ. At low volume fractions, the outer cylinder is the matrix and the inner cylinder is the fiber. The expression for strain has an $\alpha\Delta T$ component. Thus, for component 2,

$$e_{r2} = \frac{\sigma_{r2}}{E_2} - \frac{v_2}{E_2}(\sigma_{\theta2} + \sigma_{z2}) + \alpha_2\Delta T$$

$$e_{\theta2} = \frac{\sigma_{\theta2}}{E_2} - \frac{v_2}{E_2}(\sigma_{r2} + \sigma_{z2}) + \alpha_2\Delta T$$

$$e_{z2} = \frac{\sigma_{z2}}{E_2} - \frac{v_2}{E_2}(\sigma_{r2} + \sigma_{\theta2}) + \alpha_2\Delta T$$

The resultant stresses in 1 and 2 will have the form

Component 1   *Component 2*

$$\sigma_{r1} = A_1 \qquad\qquad \sigma_{r2} = A_2 - \frac{B_2}{r^2}$$

$$\sigma_{\theta 1} = A_1 \qquad\qquad \sigma_{\theta 2} = A_2 + \frac{B_2}{r^2} \qquad\qquad (10.48)$$

$$\sigma_{z1} = C_1 \qquad\qquad \sigma_{z2} = C_2$$

The following boundary conditions exist for our problem:

1. At the interface $r = a$, $\sigma_{r1} = \sigma_{r2}$ for stress continuity.
2. At the free surface $r = b$, $\sigma_{r2} = 0$.
3. The resultant of axial stress σ_z on a section $z =$ constant is zero.
4. Radial displacements in the two components are equal at the interface; that is, at $r = a$, $u_{r1} = u_{r2}$.
5. The radial displacement in component 1 must vanish at the symmetry axis; that is, at $r = 0$, $u_{r1} = 0$.

Using these boundary conditions, it is possible to determine the constants given in Eq. (10.48). The final equations for the matrix sleeve are given here (Poritsky, 1934; Chawla, 1973b):

$$\sigma_r = A\left(1 - \frac{b^{2\,\prime}}{r^2}\right) \qquad \sigma_\theta = A\left(1 + \frac{b^{2\,\prime}}{r^2}\right) \qquad \sigma_z = B \qquad (10.49)$$

where

$$A = -\left[\frac{E_m(\alpha_m - \alpha_f)\Delta T (a/b)^2}{1 + (a/b)^2(1 - 2v)[(b/a)^2 - 1]E_m/E_f}\right]$$

$$B = \frac{A}{(a/b)^2}$$

$$\times \left[2v\left(\frac{a}{b}\right)^2 + \frac{1 + (a/b)^2(1 - 2v) + (a/b)^2(1 - 2v)[(b/a)^2 - 1]E_m/E_f}{1 + [(b/a)^2 - 1]E_m/E_f}\right]$$

$$v_m = v_f = v$$

A plot of σ_r, σ_θ, and σ_z against r/a, where r is the distance in the radial direction and a is the fiber radius, is shown in Figure 10.9 for the system W/Cu for two fiber volume fractions (Chawla and Metzger, 1972). Note the change in stress level of σ_z with V_f. This thermoelastic solution can provide information about the magnitude of the elastic stresses involved and whether or not the elastic state will be exceeded. Also, if the matrix deforms plastically in response to these thermal stresses, the equations can tell where the

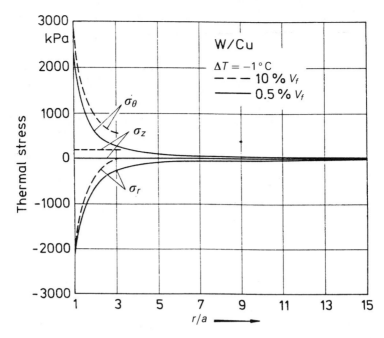

Fig. 10.9. Three-dimensional thermal stress state in a tungsten fiber/copper matrix composite for two different volume fractions. Note the change in σ_z level with V_f. [From Chawla and Metzger (1972), used with permission.]

plastic deformation will begin. Chawla and Metzger (1972) and Chawla (1973a; 1973b; 1974; 1976a; 1976b), in a series of studies with metal matrix composites, showed that the magnitude of the thermal stresses generated is large enough to deform the soft metallic matrix plastically. Depending on the temperatures involved, the plastic deformation could involve slip, cavitation, grain boundary sliding and/or migration. They measured the dislocation densities in the copper matrix of tungsten filament/copper single-crystal composites by the etch-pitting technique and showed that the dislocation densities were higher near the fiber/matrix interface than away from the interface, indicating that the plastic deformation, in response to thermal stresses, initiated at the interface. Figure 10.10 shows the variation of dislocation density ($\simeq$ etch pit density) versus distance from the interface. The increase in the dislocation density in the plateau region with V_f (in Fig. 10.10) is due to a higher σ_z with higher V_f value (see Fig. 10.9). Tresca or von Mises yield criteria applied to the stress situation existing in Figure 10.9 will show that the matrix plastic flow starts at the interface. Dislocation generation due to thermal mismatch between reinforcement and a metallic matrix has been observed in TEM by many researchers (e.g., Aresenault and Fisher, 1983; Vogelsang et al., 1986).

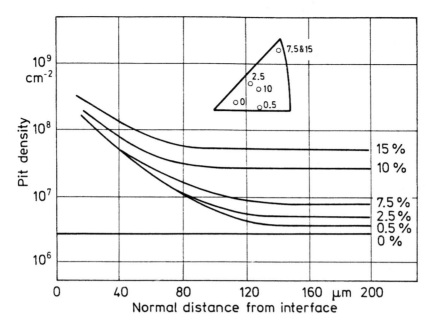

Fig. 10.10. Variation of dislocation density ($\simeq$ pit density) with distance from the interface. The higher dislocation density in the plateau region with high V_f is due to a higher σ_z with high V_f (Fig. 10.9). [From Chawla and Metzger (1972), used with permission.]

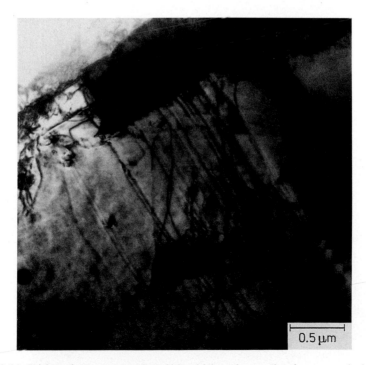

Fig. 10.11. Dislocations generated at SiC whiskers in an aluminum matrix in an in situ thermal cycling experiment done in a high-voltage electron microscope. [From Vogelsang et al. (1986), used with permission.]

Thermal Stresses in Particulate Composites

The particulate form of reinforcement can result in a considerably reduced degree of anisotropy. However, so long as a thermal mismatch exists between the particle and the matrix, thermal stresses will be present in such composites as well. Consider a particulate composite consisting of small particles distributed in a matrix. If we regard this composite as an assembly of elastic spheres of uniform size embedded in an infinite elastic continuum, then it can be shown from the theory of elasticity (see, for example, Timoshenko and Goodier, 1951) that an axially symmetrical stress distribution will result around each particle. Figure 10.12 shows a schematic of such a particle reinforced composite. The particle radius is a while the surrounding matrix sphere has a radius b. This elasticity problem has spherical symmetry, therefore the use of spherical coordinates, r, θ, and ϕ as indicated in Figure 10.13a makes for a simple analysis. We have the following stress, strain, and displacement components:

$$\sigma_r, \sigma_\theta = \sigma_\phi$$

$$\varepsilon_r, \varepsilon_\theta = \varepsilon_\phi$$

$$u_r = u, \quad \text{independent of } \theta \text{ or } \phi$$

The equilibrium equation is

$$\frac{d\sigma_r}{dr} + \frac{2}{r}(\sigma_r - \sigma_\theta) = 0 \tag{10.50}$$

while the strain-displacement relationships are

$$\varepsilon_r = \frac{du}{dr} \quad \varepsilon_\theta = \frac{u}{r} \tag{10.51}$$

Substituting Eq. (10.51) in Eq. (10.50), we get the governing differential equation for our problem:

$$\frac{d^2u}{dr^2} + \frac{2}{r}\frac{du}{dr} - \frac{2}{r^2}u = 0 \tag{10.52}$$

The solution to this differential equation is

$$u = Ar + \frac{C}{r^2}$$

We apply the following boundary conditions:

1. Stress vanishes at the free surface (i.e., at $r = b$).

Fig. 10.12. A particle reinforce composite.

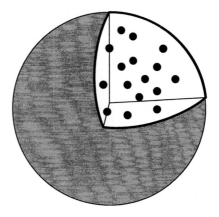

2. The radial stress at the interface $(r = a)$ is the interfacial pressure, P, i.e.,

$$\sigma_r(a) = -P$$

The particle has a hydrostatic state of stress with the stress components

$$\sigma_{rp} = P = \text{constant} = \sigma_{\theta p} \qquad (10.53)$$

while the stresses in the matrix are

$$\sigma_{rm} = \frac{P}{1 - V_p}\left[\frac{a^3}{r^3} - V_p\right] \qquad (10.54)$$

$$\sigma_{\theta m} = -\frac{P}{1 - V_p}\left[\frac{1}{2}\frac{a^3}{r^3} + V_p\right] \qquad (10.55)$$

$$P = \frac{(\alpha_m - \alpha_p)\Delta T}{\left[\dfrac{0.5(1 + v_m) + (1 - 2v_m)V_p}{E_m(1 - V_p)} + \dfrac{1 - 2v_p}{E_p}\right]} \qquad (10.56)$$

$$V_p = \left(\frac{a}{b}\right)^3$$

where V_p is the particle volume fraction, a is the particle radius, b is the matrix radius, and other symbols have the significance given earlier. Figure 10.13a shows the three-dimensional stress distribution in a particulate composite. Note the different stress distribution in a particulate composite from in a fibrous composite. The particle is under a uniform pressure, P, while the matrix has different radial and tangential stress components. The radial and tangential components in the matrix vary with distance, as shown in Figure 10.13b. The radial component goes to zero at the free surface, $r = b$, as per our boundary conditions. The tangential component has a nonzero value at the free surface.

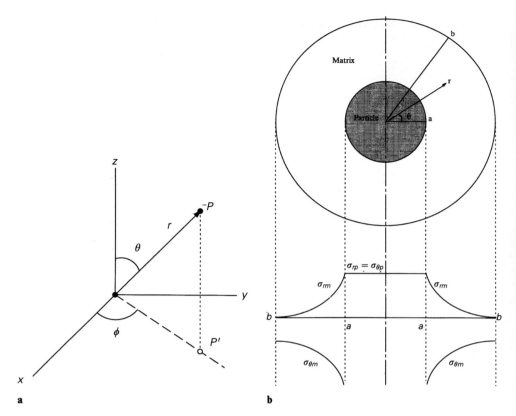

Fig. 10.13. a Spherical coordinate system. **b** Stress distribution in a particulate composite.

10.4   Mechanics of Load Transfer from Matrix to Fiber

The topic of load transfer from the matrix to the fiber has been treated by a number of researchers (Cox, 1952 ; Dow, 1963; Schuster and Scala, 1964; Kelly, 1973). The matrix holds the fibers together and transmits the applied load to the fibers, the real load-bearing component in most cases. Let us focus our attention on a high-modulus fiber embedded in a low-modulus matrix. Figure 10.14a shows the situation prior to the application of an external load. We assume that the fiber and matrix are perfectly bonded and that the Poisson's ratios of the two are the same. Imagine vertical lines running through the fiber/matrix interface in a continuous manner in the unstressed state, as shown in Figure 10.14a. Now let us load this composite axially as shown in Figure 10.14b. We assume that no direct loading of the fibers occurs. Then the fiber and the matrix experience locally different axial displacements because of the different elastic moduli of the components. Different axial displacements in the fiber and the matrix mean that shear

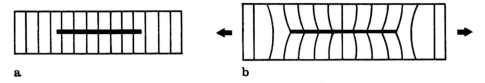

Fig. 10.14. A high-modulus fiber embedded in a low-modulus matrix: **a** before deformation, **b** after deformation.

strains are being produced in the matrix on planes parallel to the fiber axis and in a direction parallel to the fiber axis. Under such circumstances, our imaginary vertical lines of the unstressed state will become distorted, as shown in Figure 10.14b. Transfer of the applied load to the fiber thus occurs by means of these shear strains in the matrix. Termonia (1987) used a finite difference numerical technique to model the elastic strain field perturbance in composite consisting of an embedded fiber in a matrix when a uniform far field strain imposed on the matrix. An originally uniform orthogonal mesh around the fiber became distorted when an axial tensile stress was applied.

It is instructive to examine the stress distribution along the fiber/matrix interface. There are two important cases: (1) both the matrix and fiber are elastic, and (2) the matrix is plastic and the fiber is elastic. Fibers such as boron, carbon, and ceramic fibers are essentially elastic right up to fracture. Metallic matrices show elastic and plastic deformation before fracture, while polymeric and ceramic matrices can be treated, for all practical purposes, as elastic up to fracture.

10.4.1 Fiber Elastic–Matrix Elastic

Consider a fiber of length l embedded in a matrix subjected to a strain; see Figure 10.15. We assume that (1) there exists a perfect bonding between the fiber and matrix (i.e., there is no sliding between them) and (2) the Poisson's ratios of the fiber and matrix are equal, which implies an absence of transverse stresses when the load is applied along the fiber direction. Let the displacement of a point at a distance x from one extremity of the fiber be u in the presence of a fiber and v in the absence of a fiber. Then we can write for the transfer of load from the matrix to the fiber

$$\frac{dP_f}{dx} = B(u - v) \qquad (10.57)$$

where P_f is the load on the fiber and B is a constant that depends on the geometric arrangement of fibers, the matrix type, and moduli of the fiber and matrix. Differentiating Eq. (10.57), we get

$$\frac{d^2P_f}{dx^2} = B\left(\frac{du}{dx} - \frac{dv}{dx}\right) \qquad (10.58)$$

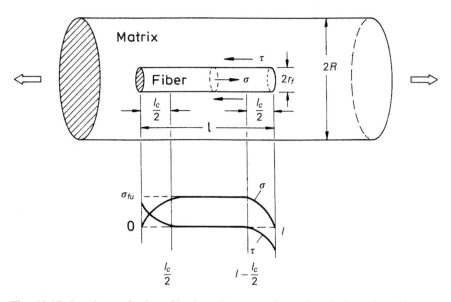

Fig. 10.15. Load transfer in a fiber/matrix composite and variation of tensile stress (σ_f) in the fiber and interfacial shear stress (τ) with distance along the interface.

We have

$$\frac{du}{dx} = \text{strain in fiber} = \frac{P_f}{E_f A_f}$$

$$\frac{dv}{dx} = \text{strain in the matrix away from the fiber}$$

$$= \text{imposed strain, } e.$$

Thus, Eq. (10.58) can be rewritten

$$\frac{d^2 P_f}{dx^2} = B\left(\frac{P_f}{A_f E_f} - e\right) \tag{10.59}$$

A solution of this differential equation is

$$P_f = E_f A_f e + S \sinh \beta x + T \cosh \beta x \tag{10.60}$$

where

$$\beta = \left(\frac{B}{A_f E_f}\right)^{1/2} \tag{10.61}$$

We use the following boundary condition to evaluate the constants S and T:

$$P_f = 0 \quad \text{at } x = 0 \quad \text{and} \quad x = l$$

Putting in these values and using the half-angle trigonometric formulas, we get the following result:

$$P_f = E_f A_f e \left[1 - \frac{\cosh \beta(l/2 - x)}{\cosh(\beta l/2)} \right] \quad \text{for } 0 < x < l/2 \qquad (10.62)$$

or

$$\sigma_f = \frac{P_f}{A_f} = E_f e \left[1 - \frac{\cosh \beta(l/2 - x)}{\cosh(\beta l/2)} \right] \quad \text{for } 0 < x < l/2 \qquad (10.63)$$

The maximum possible value of strain in the fiber is the imposed strain e, and thus the maximum stress is eE_f. Therefore, if we have a long enough fiber, the stress in the fiber will increase from the two ends to a maximum value, $\sigma_{fu} = E_f e$. It can be shown readily that the average stress in the fiber is

$$\bar{\sigma}_f = \frac{E_f e}{l} \int_0^l \left[1 - \frac{\cosh \beta(l/2 - x)}{\cosh(\beta l/2)} \right] dx = E_f e \left[1 - \frac{\tanh(\beta l/2)}{\beta l/2} \right] \qquad (10.64)$$

We can obtain the variation of shear stress τ along the fiber/matrix interface by considering the equilibrium of forces acting over an element of fiber (radius r_f). Thus, we can write from Figure 10.15

$$\frac{dP_f}{dx} dx = 2\pi r_f \, dx \, \tau \qquad (10.65)$$

Now P_f, the tensile load on the fiber, is equal to $\pi r_f^2 \sigma_f$. Substituting this in Eq. (10.65), we get

$$\tau = \frac{1}{2\pi r_f} \frac{dP_f}{dx} = \frac{r_f}{2} \frac{d\sigma_f}{dx} \qquad (10.66)$$

From Eqs. (10.63) and (10.66), we obtain

$$\tau = \frac{E_f r_f e \beta}{2} \frac{\sinh \beta(l/2 - x)}{\cosh(\beta l/2)} \qquad (10.67)$$

Figure 10.15 shows the variation of τ and σ_f with distance x. The maximum shear stress, in Eq. (10.67), will be the smaller of the following two shear stresses: (1) the shear yield stress of the matrix or (2) the shear strength of the fiber/matrix interface. Whichever of these two shear stresses is attained first will control the load transfer phenomenon and should be used in Eq. (10.67).

Now we can determine the constant B. The value of B depends on the fiber packing geometry. Consider Figure 10.15 again and let the fiber length l be much greater than the fiber radius r_f. Let $2R$ be the average fiber spacing (center to center). Let us also denote the shear stress in the fiber direction at a distance r from the axis by $\tau(r)$. Then, at the fiber surface ($r = r_f$), we can write

$$\frac{dP_f}{dx} = -2\pi r_f \tau(r_f) = B(u - v)$$

Thus,

$$B = -\frac{2\pi r_f \tau(r_f)}{u - v} \tag{10.68}$$

Let w be the real displacement in the matrix. Then at the fiber/matrix interface, no sliding being permitted, $w = u$. At a distance R from the center of a fiber, the matrix displacement is unaffected by the fiber presence and $w = v$. Considering the equilibrium of forces acting on the matrix volume between r_f and R, we can write

$$2\pi r \tau(r) = \text{constant} = 2\pi r_f \tau(r_f)$$

or

$$\tau(r) = \frac{\tau(r_f) r_f}{r} \tag{10.69}$$

The shear strain γ in the matrix is given by $\tau(r) = G_m \gamma$, where G_m is the matrix shear modulus. Then

$$\gamma = \frac{dw}{dr} = \frac{\tau(r)}{G_m} = \frac{\tau(r_f) r_f}{G_m r} \tag{10.70}$$

Integrating from r_f to R, we obtain

$$\int_{r_f}^{R} dw = \Delta w = \frac{\tau(r_f) r_f}{G_m} \int_{r_f}^{R} \frac{1}{r} dr = \frac{\tau(r_f) r_f}{G_m} \ln\left(\frac{R}{r_f}\right) \tag{10.71}$$

But, by definition,

$$\Delta w = v - u = -(u - v) \tag{10.72}$$

From Eqs. (10.71) and (10.72) we get

$$\frac{\tau(r_f) r_f}{u - v} = -\frac{G_m}{\ln(R/r_f)} \tag{10.73}$$

From Eqs. (10.68) and (10.73), one obtains

$$B = \frac{2\pi G_m}{\ln(R/r_f)} \tag{10.74}$$

and from Eq. (10.61), one can obtain an expression for the load transfer parameter β:

$$\beta = \left(\frac{B}{E_f A_f}\right)^{1/2} = \left[\frac{2\pi G_m}{E_f A_f \ln(R/r_f)}\right]^{1/2} \tag{10.75}$$

The value of R/r_f is a function of fiber packing. For a square array of fibers $\ln(R/r_f) = \frac{1}{2} \ln(\pi/V_f)$, while for a hexagonal packing $\ln(R/r_f) = \frac{1}{2} \ln(2\pi/\sqrt{3} V_f)$. We can define $\ln(R/r_f) = \frac{1}{2} \ln(\phi_{max}/V_f)$, where ϕ_{max} is the maximum packing factor. Substituting this in Eq. (10.75), we get

$$\beta = \left[\frac{4\pi G_m}{E_f A_f \ln(\phi_{max}/V_f)} \right]^{1/2}$$

Note that the greater the value of the ratio G_m/E_f, the greater is the value of β and the more rapid is the stress increase in the fiber from either end.

More rigorous analyses give results similar to the one above and differ only in the value of β. In all analyses, β is proportional to $(G_m/E_f)^{1/2}$, and differences occur only in the term involving fiber volume fraction, $\ln(R/r_f)$, in the preceding equation.

Termonia (1987) used the finite difference method to show the high shear strains in the matrix near the fiber extremeties. Micro-Raman spectroscopy has been used to study the deformation behavior of organic and inorganic fiber reinforced composites (Galiotis et al., 1985; Day et al., 1987, 1989; Schadler and Galiotis, 1996; Yang et al., 1992; Young et al., 1990). Characteristic Raman spectra can be obtained from these fibers in the undeformed and the deformed states. Under tension, the peaks of the Raman bands shift to lower frequencies. The magnitude of frequency shift is a function of the material, Raman band under consideration, and the Young's modulus of the material. The shift in Raman bands results from changes in force constants due to changes in molecular or atomic bond lengths and bond angles. Micro-Raman spectroscopy is a very powerful technique that allows us to obtain point-to-point variation in strain along the fiber length embedded in a transparent matrix. An example of this for Kevlar aramid 149/epoxy composite is shown in Figure 10.16 (Young, 1994). In Figures 10.16a and b, we see that up to 1% strain, the strain in the fiber builds from the two ends as predicted by the shear lag analysis described earlier, while in the middle of the fiber length the strain in the fiber and matrix are equal. In Figures 10.16c and d, we see that as fiber fractures at different sites along its length, the strain in fiber drops at those sites.

Let us reexamine Figure 10.15. As per our boundary condition, the normal stress is zero at the two extremities of the fiber. The normal stress, σ, rises from the two ends to a maximum value along most of the fiber length, provided we have a long enough fiber. This gives rise to the concept of a critical fiber length for load transfer (see the next section). The shear stress is maximum at the fiber ends. Matrix yielding or interfacial failure would be expected to start at the fiber ends.

10.4.2 Fiber Elastic–Matrix Plastic

It should be clear from the preceding discussion that to load high-strength fibers in a ductile matrix to their maximum strength, the matrix shear strength must be large. A metallic matrix will flow plastically in response to the high shear stresses developed. Of course, if the fiber/matrix interface is

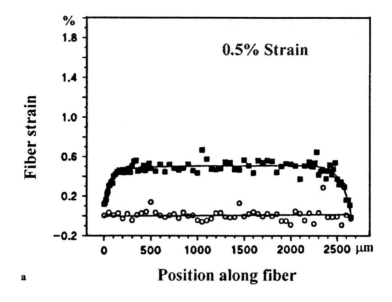

a

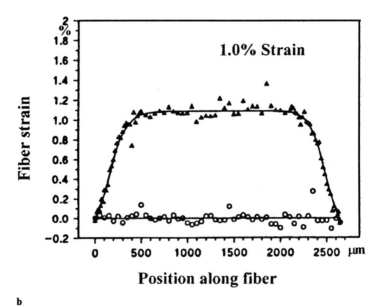

b

Fig. 10.16. Micro-Raman spectra of Kevlar aramid 149 when the Kevlar/epoxy composite is loaded axially. (a) and (b) strain build up in the fiber from the two ends. (c) and (d) fiber fracture leads to a drop in strain at those sites. [After Young (1994).]

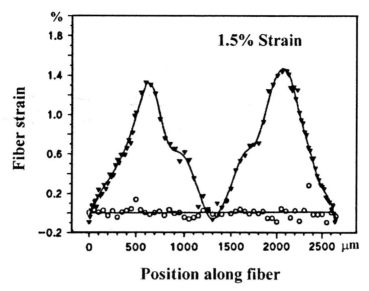

c

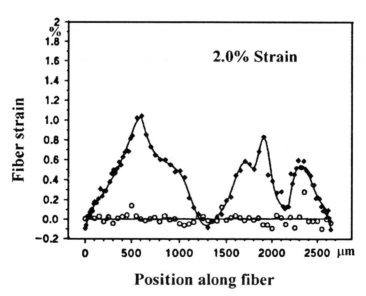

d

Fig. 10.16 (*continued*)

weaker, it will fail first. In MMCs, assuming that the plastically deforming matrix does not work-harden, the shear stress at the fiber surface, $\tau(r_f)$, will have an upper limit of τ_y, the matrix shear yield strength. In PMCs and CMCs, frictional slip at the interface is more likely than plastic flow of the matrix. In the case of PMCs and CMCs, therefore, the limiting shear stress will be the interface strength in shear, τ_i. The term τ_i should replace τ_y in what follows for PMCs and CMCs. If the polymer shrinkage during curing results in a radial pressure p on the fibers, then τ_y should be replaced by μp as $\tau_i = \mu p$, where μ is the coefficient of sliding friction between the fiber and matrix (Kelly, 1973). The equilibrium of forces, then, over a fiber length of $l/2$ gives

$$\sigma_f \frac{\pi d^2}{4} l = \tau_y \pi d \frac{l}{2}$$

or

$$\frac{l}{d} = \frac{\sigma_f}{2\tau_y}$$

We consider $l/2$ and not l because the fiber is being loaded from both ends. Given a sufficiently long fiber, it should be possible to load it to its breaking stress σ_{fu} by means of load transfer through the matrix flowing plastically around it. Let $(l/d)_c$ be the maximum fiber length-to-diameter ratio necessary to accomplish this. We call this ratio (l/d) the aspect ratio of a fiber and $(l/d)_c$ is the critical aspect ratio necessary to attain the breaking stress of the fiber, σ_{fu}. Then we can write

$$\left(\frac{l}{d}\right)_c = \frac{\sigma_{fu}}{2\tau_y} \tag{10.76}$$

For a given fiber diameter d, we can think of a critical fiber length l_c. Thus,

$$\frac{l_c}{d} = \frac{\sigma_{fu}}{2\tau_y} \tag{10.77}$$

Over a length l_c, the load in the fiber builds up from both ends. Strain builds in a likewise manner. Beyond l_c (i.e., in the middle portion of the fiber), the local displacements in the matrix and fiber are the same, and the fiber carries the major load while the matrix carries only a minor portion of the applied load. Equation (10.77) tells us that the fiber length l must be equal to or greater than l_c for the fiber to be loaded to its maximum stress, σ_{fu}. For $l < l_c$, the matrix will flow plastically around the fiber and will load it to a stress in its central portion given by

$$\sigma_f = 2\tau \frac{l}{d} < \sigma_{fu} \tag{10.78}$$

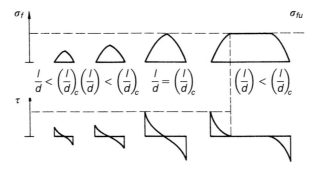

Fig. 10.17. Schematic of variation of tensile stress in a fiber (σ_f) and interface shear stress (τ) with different fiber aspect ratios (l/d).

This is shown in Figure 10.17. An examination of this figure shows that even for $l/d > (l/d)_c$ the average stress in the fiber will be less than the maximum stress to which it is loaded in its central region. In fact, one can write for the average fiber stress

$$\bar{\sigma}_f = \frac{1}{l}\int_0^l \sigma_f \, dx = \frac{1}{l}[\sigma_f(l - l_c) + \beta\sigma_f l_c] = \frac{1}{l}[\sigma_f l - l_c(\sigma_f - \beta\sigma_f)]$$

or

$$\bar{\sigma}_f = \sigma_f\left(1 - \frac{1 - \beta}{l/l_c}\right) \tag{10.79}$$

where $\beta\sigma_f$ is the average stress in the fiber over a portion $l_c/2$ of its length at both ends. One can thus regard β as a load transfer function. Its value will be precisely 0.5 for an ideally plastic material; that is, the increase in stress in the fiber over the portion $l_c/2$ will be linear. The longitudinal strength of a composite containing short fibers but well aligned will always be less than that of a composite containing unidirectionally aligned continuous fibers. For the strength of short fiber composite, per rule of mixtures, we can write:

$$\sigma_c = \bar{\sigma}_f V_f + \sigma'_m V_m$$

$$\sigma_c = \sigma_f V_f\left(1 - \frac{1 - \beta}{l/l_c}\right) + \sigma'_m(1 - V_f) \tag{10.80}$$

If $\beta = 0.5$,

$$\sigma_c = \sigma_f V_f\left(1 - \frac{l_c}{2l}\right) + \sigma'_m(1 - V_f)$$

where σ'_m is the in situ matrix stress at the strain under consideration. Suppose that in a whisker-reinforced metal the whiskers have an $l/l_c = 10$; then it can easily be shown that the strength of such a composite containing discontinuous but aligned fibers will be 95% of that of a composite containing continuous fibers. Thus, as long as the fibers are reasonably long compared to the load transfer length, there is not much loss of strength owing to their discontinuous nature. The stress concentration effect at the ends of the discontinuous fibers has been neglected in this simple analysis.

10.5 Load transfer in Particulate Composites

The shear lag model described in the previous section is suitable for explaining the load transfer from the matrix to a high aspect ratio reinforcement via shear along the interface parallel to the loading direction. No direct tensile loading of the reinforcement occurs. Such a model will not be expected to work for a particulate composite. A modified shear lag model (Fukuda and Chou, 1981; Nardone and Prewo, 1986) takes into account tensile loading at the particle ends. According to this modified shear lag model, the yield strength of a particulate composite is given by

$$\sigma_{yc} = \sigma_{ym}[1 + (L + t)/4L]V_p + \sigma_{ym}(1 - V_p)$$

where σ_{ym} is the yield stress of the unreinforced matrix, V_p is the particle volume fraction, L is the length of the particle perpendicular to the applied load and t is the length of the particle parallel to the loading direction.

For the case of a composite with equiaxed particles (aspect ratio $= 1$), the above expression reduces to

$$\sigma_{yc} = \sigma_{ym}(1 + 0.5\,V_p)$$

Note that this expression predicts a linear but modest increase in the strength of the composite with particle volume fraction. However, no account is taken of the particle size or other microstructural parameters.

References

R.J. Arsenault and R.M. Fisher (1983). *Scripta Met.*, **17**, 67.

E. Behrens (1968). *J. Composite Mater.*, **2**, 2.

H. Bhatt, K.Y. Donaldson, D.P.H. Hasselman, and R.T. Bhatt (1992). *J. Mater. Sci.*, **27**, 6653.

C.C. Chamis (1983). *NASA Tech. Memo. 83320*, presented at the 38th Annual Conference of the Society of Plastics Industry (SPI), Houston, TX.

C.C. Chamis and G.P. Sendecky (1968). *J. Composite Mater.*, **2**, 332.

K.K. Chawla (1973a). *Metallography*, **6**, 155.

K.K. Chawla (1973b). *Philos. Mag.*, **28**, 401.

K.K. Chawla (1974). In *Grain Boundaries in Engineering Materials*, Claitor's Publishing Division, Baton Rouge, LA, p. 435.

K.K. Chawla (1976a). *J. Mater. Sci.*, **11**, 1567.

K.K. Chawla (1976b). In *Proceedings of the International Conference on Composite Materials/1975*, TMS-AIME, New York, p. 535.

K.K. Chawla and M. Metzger (1972). *J. Mater. Sci.*, **7**, 34.

H.L. Cox (1952). *Brit. J. App. Phys.*, **3**, 122.

R.J. Day, I.M. Robinson, M. Zakikhani and R.J. Young (1987). *Polymer*, **28**, 1833.

R.J. Day, V. Piddock, R. Taylor, R.J. Young, and M. Zakikhani (1989). *J. Mater. Sci.*, **24**, 2898.

N.F. Dow (1963). *General Electric Report No. R63-SD-61*.

J.D. Eshelby (1957). *Proc. Roy. Soc.*, **A241**, 376.

J.D. Eshelby (1959). *Proc. Roy. Soc.*, **A252**, 561.

H. Fukuda and T.W. Chou (1981). *J. Comp. Mater.*, **15**, 79.

C. Galiotis, I.M. Robinson, R.J. Young, B.J.E. Smith, and D.N. Batchelder (1985). *Polym. Commun.*, **26**, 354.

J.C. Halpin and J.L. Kardos (1976). *Polym. Eng. Sci.*, **16**, 344.

J.C. Halpin and S.W. Tsai (1967). "Environmental Factors Estimation in Composite Materials Design," *AFML TR 67-423*.

Z. Hashin and B.W. Rosen (1964). *J. Appl. Mech.*, **31**, 233.

D.P.H. Hasselman and L.F. Johnson (1987). *J. Composite Mater.*, **27**, 508.

R. Hill (1964). *J. Mech. Phys. Solids*, **12**, 199.

R. Hill (1965). *J. Mech. Phys. Solids*, **13**, 189.

J.L. Kardos (1971). *CRC Crit. Rev. Solid State Sci.*, **3**, 419.

A. Kelly (1970). In *Chemical and Mechanical Behavior of Inorganic Materials*, Wiley-Interscience, New York, p. 523.

A. Kelly (1973). *Strong Solids*, second ed., Clarendon Press, Oxford, p. 157.

A. Kelly and H. Lilbolt (1971). *Philos. Mag.*, **20**, 175.

E.H. Kerner (1956). *Proc. Phys. Soc. London*, **B69**, 808.

A.E.H. Love (1952). *A Treatise on the Mathematical Theory of Elasticity*, 4th ed., Dover, New York, p. 144.

G.D. Marom and A. Weinberg (1975). *J. Mater. Sci.*, **10**, 1005.

T. mori and K. Tanaka (1973). *Acta Met.*, **21**, 571.

V.C. Nardone and K.M. Prewo (1986). *Scripta Met.*, **20**, 43.

L.E. Nielsen (1974). *Mechanical Properties of Polymers and Composites*, Vol. 2, Marcel Dekker, New York.

J.F. Nye (1969). *Physical Properties of Crystals*, Oxford University Press, London, p. 131.

H. Poritsky (1934). *Physics*, **5**, 406.

A. Reuss (1929). *Z. Angew. Math. Mech.*, **9**, 49.

B.W. Rosen (1973). *Composites*, **4**, 16.

B.W. Rosen and Z. Hashin (1970). *Int. J. Eng. Sci.*, **8**, 157.

L.S. Schadler and C. Galiotis (1995). *International Materials Reviews* **40**, 116.

R.A. Schapery (1969). *J. Composite Mater.*, **2**, 311.

D.M. Schuster and E. Scala (1964). *Trans. Met. Soc.-AIME*, **230**, 1635.

Y. Termonia (1987). *J. Mater. Sci*, **22**, 504.

S. Timoshenko and J.N. Goodier (1951). *Theory of Elasticity*, McGraw-Hill, New York, p. 416.

P.S. Turner (1946). *J. Res. Natl. Bur. Stand.*, **37**, 239.

M. Vogelsang, R.J. Arsenault, and R.M. Fisher (1986). *Met. Trans. A*, **7A**, 379.

W. Voigt (1910). *Lehrbuch der Kristallphysik*, Teubner, Leipzig.

R.U. Vaidya and K.K. Chawla (1994). *Composites Science and Technology*, **50**, 13.

R.U. Vaidya, R. Venkatesh, and K.K. Chawla (1994). *Composites*, **25**, 308.

J.M. Whitney (1973). *J. Struct. Div.*, 113.

Z.R. Xu, K.K. Chawla, R. Mitra, and M.E. Fine (1994). *Scripta Met. et Mater.*, **31**, 1525.

X. Yang, X. Hu, R.J. Day, and R.J. Young (1992). *J. Mater. Sci.*, **27**, 1409.

R.J. Young, R.J. Day, and M. Zakikhani (1990). *J. Mater. Sci.*, **25**, 127.

R.J. Young (1994). In *High-Performance Composites: Commonalty of Phenomena*, K.K. Chawla, P.K. Liaw, and S.G. Fishman (Eds.), TMS, Warrendale, PA, p. 263.

Suggested Reading

B.D. Agarwal and L.J. Broutman (1980). *Analysis and Performance of Fiber Composites*, John Wiley & Sons, New York.

A. Kelly (1973). *Strong Solids*, 2nd ed., Clarendon Press, Oxford.

S. Nemat-Nasser and M. Hori (1993). *Micromechanics: Overall Properties of Heterogenous Materials*, North-Holland, Amsterdam.

M.R. Piggott (1980). *Load-Bearing Fibre Composites*, Pergamon Press, Oxford.

V.K. Tewary (1978). *Mechanics of Fibre Composites*, Halsted Press, New York.

CHAPTER 11

Macromechanics of Composites

Laminated fibrous composites are made by bonding together two or more laminae. The individual unidirectional laminae or plies are oriented in such a manner that the resulting structural component has the desired mechanical and/or physical characteristics in different directions. Thus, one exploits the inherent anisotropy of fibrous composites to design a composite material with appropriate properties.

In Chapter 10 we treated the micromechanics of fibrous composites, that is, how to obtain the composite properties when the properties of the matrix and the fiber and their geometric arrangements are known. While micromechanics is very useful in analyzing the composite behavior, we use the information obtained from a micromechanical analysis of a thin unidirectional lamina (or in the case of a lack of such analytical information, we must determine experimentally the properties of a lamina) as input for a macromechanical analysis of a laminated composite. Figure 11.1 shows this concept schematically (McCullough, 1971). Once we have determined, analytically or otherwise, the characteristics of a fibrous lamina, we ignore its detailed microstructural nature and simply treat it as a homogeneous, orthotropic sheet. A laminated composite is made by stacking a number of such orthotropic sheets at specific orientations to get composite materials with the desired characteristics. We then use the existing theory of laminated plates or shells to analyze macromechanically such laminated composites.

To appreciate the significance of such a macromechanical analysis, we first review the basic ideas of the elastic constants of a bulk isotropic material and a lamina, a lamina as an orthotropic sheet, and finally the use of classical laminated plate theory to analyze macromechanically the laminated composites. The reader is referred to some standard texts on elasticity (Love, 1952; Timoshenko and Goodier, 1951; Nye, 1969) for a detailed review of elasticity.

11.1 Elastic Constants of an Isotropic Material

Stress is defined as force per unit area of a body. We can represent the stress acting at a point in a solid by the stress components acting on the surfaces of

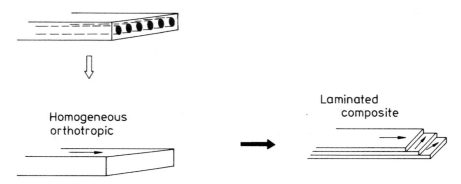

Fig. 11.1. Macromechanical analysis of laminate composites. A unidirectional ply is treated as a homogeneous, orthotropic material. Many such plies are stacked in an appropriate order (following laminated plate or shell theory) to make the composite. [Reprinted from McCullough (1971), courtesy of Marcel Dekker, Inc.].

an elemental cube at that point. There are nine stress components (three normal and six shear) acting on the front faces of an elemental cube. The component σ_{ij} represents the force per unit area in the i direction on a face whose normal is the j direction. Rotational equilibrium requires that $\sigma_{ij} = \sigma_{ji}$. Thus, we are left with six stress components; $i = j$ gives the normal stresses while $i \neq j$ gives three shear stresses.

The displacement of a point in a deformed body with respect to its original position in the undeformed state can be represented by a vector **u** with components u_1, u_2, and u_3; these components are the projections of **u** on the x_i, x_2, and x_3 axes. *Strain* is defined as the ratio of change in length to original length. We can define the strain components in terms of the first derivatives of the displacement components as follows:

$$\varepsilon_{ij} = \frac{1}{2} \left(\frac{\partial u_i}{\partial x_j} + \frac{\partial u_j}{\partial x_i} \right).$$

Similar to stress components, $i = j$ are the normal strains while $i \neq j$ gives the shear strains. It should be noted that ε_{ij} for $i \neq j$ gives half of the engineering shear strain, γ_{ij}; that is,

$$\gamma_{ij} = 2\varepsilon_{ij} = \frac{\partial u_i}{\partial x_j} + \frac{\partial u_j}{\partial x_i}$$

The relationship between stress and strain in linear elasticity is described by Hooke's law. Hooke's law states that, for small strains, stress is linearly proportional to strain. For the simple case of a stress applied unidirectionally to an isotropic solid, we can write Hooke's law as

$$\sigma = E\varepsilon \tag{11.1}$$

where σ is the unidirectional stress, ε is the strain in the applied stress direction, and E is Young's modulus. We have omitted the indices in this simple unidirectional case.

In its most generalized form, Hooke's law can be written in the indicial or tensorial notation as

$$\sigma_{ij} = C_{ijkl}\varepsilon_{kl} \tag{11.2}$$

where C_{ijkl} are the elastic constants or stiffnesses. Equation (11.2), when written out in an expanded form, will have 81 elastic constants. It is general practice to use a contracted matrix notation for writing stresses, strains, and elastic constants. The contracted notation is especially useful for matrix algebra operations.

We use C_{mn} for C_{ijkl}, σ_m for σ_{ij}, and ε_n for ε_{kl} as per the following procedure:

ij or kl	11	22	33	23	31	12
m or n	1	2	3	4	5	6

Then Eq. (11.2) can be rewritten

$$\sigma_m = C_{mn}\varepsilon_n \tag{11.3}$$

It can be shown from symmetry considerations that $C_{mn} = C_{nm}$.

Conversely, we can write

$$\varepsilon_m = S_{mn}\sigma_n \tag{11.4}$$

where S_{mn}, the compliance matrix, is the inverse of the stiffness matrix C_{mn}.

In the expanded form, we have

$$
\begin{bmatrix}
\sigma_1 \\
\sigma_2 \\
\sigma_3 \\
\sigma_4 \\
\sigma_5 \\
\sigma_6
\end{bmatrix}
=
\begin{bmatrix}
C_{11} & C_{12} & C_{13} & C_{14} & C_{15} & C_{16} \\
 & C_{22} & C_{23} & C_{24} & C_{25} & C_{26} \\
 & & C_{33} & C_{34} & C_{35} & C_{36} \\
 & & & C_{44} & C_{45} & C_{46} \\
 & & & & C_{55} & C_{56} \\
 & & & & & C_{66}
\end{bmatrix}
\begin{bmatrix}
\varepsilon_1 \\
\varepsilon_2 \\
\varepsilon_3 \\
\varepsilon_4 \\
\varepsilon_5 \\
\varepsilon_6
\end{bmatrix}
\tag{11.5}
$$

Note that σ_4, σ_5, and σ_6 now represent the shear stresses while ε_4, ε_5, and ε_6 represent the engineering shear strains. The dashed line along the diagonal indicates that the matrix is symmetrical. Equation (11.5) thus gives 21 independent elastic constants in the most general case, i.e., with no symmetry elements present.

For most materials, the number of independent elastic constants is further reduced because of the various symmetry elements present. For example, only three elastic constants are independent for cubic systems. For isotropic materials where elastic properties are independent of direction, only two constants are independent. For isotropic materials, Eq. (11.5) reduces to

$$
\begin{bmatrix} \sigma_1 \\ \sigma_2 \\ \sigma_3 \\ \sigma_4 \\ \sigma_5 \\ \sigma_6 \end{bmatrix} = \begin{bmatrix} C_{11} & C_{12} & C_{12} & 0 & 0 & 0 \\ & C_{11} & C_{12} & 0 & 0 & 0 \\ & & C_{11} & 0 & 0 & 0 \\ & & & \dfrac{C_{11}-C_{12}}{2} & 0 & 0 \\ & & & & \dfrac{C_{11}-C_{12}}{2} & 0 \\ & & & & & \dfrac{C_{11}-C_{12}}{2} \end{bmatrix} \begin{bmatrix} \varepsilon_1 \\ \varepsilon_2 \\ \varepsilon_3 \\ \varepsilon_4 \\ \varepsilon_5 \\ \varepsilon_6 \end{bmatrix}
$$

$$(11.6)$$

In terms of the compliance matrix, for isotropic materials, we can write

$$
\begin{bmatrix} \varepsilon_1 \\ \varepsilon_2 \\ \varepsilon_3 \\ \varepsilon_4 \\ \varepsilon_5 \\ \varepsilon_6 \end{bmatrix} = \begin{bmatrix} S_{11} & S_{12} & S_{12} & 0 & 0 & 0 \\ & S_{11} & S_{12} & 0 & 0 & 0 \\ & & S_{11} & 0 & 0 & 0 \\ & & & 2(S_{11}-S_{12}) & 0 & 0 \\ & & & & 2(S_{11}-S_{12}) & 0 \\ & & & & & 2(S_{11}-S_{12}) \end{bmatrix} \begin{bmatrix} \sigma_1 \\ \sigma_2 \\ \sigma_3 \\ \sigma_4 \\ \sigma_5 \\ \sigma_6 \end{bmatrix}
$$

$$(11.7)$$

Only C_{11} and C_{12} (or S_{11} and S_{12}) are the independent constants. Engineers frequently use elastic constants such as Young's modulus E, Poisson's ratio v, shear modulus G, and bulk modulus K. Only two of these are independent because E, G, v, and K are interrelated:

$$ E = 2G(1+v) \quad \text{and} \quad K = \frac{E}{3(1-2v)} $$

The relationships between these engineering constants and compliances are as follows:

$$ E = \frac{1}{S_{11}} \quad v = -\frac{S_{12}}{S_{11}} \quad G = \frac{1}{2(S_{11}-S_{12})} $$

and the compliances are related to the stiffnesses as follows:

$$ S_{11} = \frac{C_{11}+C_{12}}{(C_{11}-C_{12})(C_{11}+2C_{12})} $$

$$ S_{12} = -\frac{C_{12}}{(C_{11}-C_{12})(C_{11}+2C_{12})} $$

11.2 Elastic Constants of a Lamina

We can make a laminated composite by stacking a sufficiently large number of laminae. A *lamina*, the unit building block of a composite, can be considered to be in a state of generalized plane stress. This implies that the through thickness stress components are zero. Thus $\sigma_3 = \sigma_4 = \sigma_5 = 0$. Then Eqs. (11.6) and (11.7) are reduced, for an isotropic lamina, to

$$
\begin{bmatrix} \sigma_1 \\ \sigma_2 \\ \sigma_6 \end{bmatrix} = \begin{bmatrix} C_{11} & C_{12} & 0 \\ C_{12} & C_{11} & 0 \\ 0 & 0 & \dfrac{C_{11} - C_{12}}{2} \end{bmatrix} \begin{bmatrix} \varepsilon_1 \\ \varepsilon_2 \\ \varepsilon_6 \end{bmatrix} \tag{11.8}
$$

$$
\begin{bmatrix} \varepsilon_1 \\ \varepsilon_2 \\ \varepsilon_6 \end{bmatrix} = \begin{bmatrix} S_{11} & S_{12} & 0 \\ S_{12} & S_{11} & 0 \\ 0 & 0 & 2(S_{11} - S_{12}) \end{bmatrix} \begin{bmatrix} \sigma_1 \\ \sigma_2 \\ \sigma_6 \end{bmatrix} \tag{11.9}
$$

Equations (11.8) and (11.9) describe the stress-strain relationships for an isotropic lamina, for example, an aluminum sheet. A fiber reinforced lamina, however, is not an isotropic material. It is an orthotropic material; that is, it has three mutually perpendicular axes of symmetry. Relationships become slightly more complicated when we have orthotropy rather than isotropy.

Recall that a fiber reinforced lamina or ply is a thin sheet (~ 0.1 mm) containing oriented fibers. Generally, the fibers are oriented unidirectionally as in a prepreg but fibers in the form of a woven roving may also be used. Several such thin laminae are stacked in a specific order of fiber orientation, cured, and bonded into a laminated composite. Because the behavior of a laminated composite depends on the characteristics of individual laminae, and with due regard to their directionality, we now discuss the elastic behavior of an orthotropic lamina.

For the case of an orthotropic material with the coordinate axes parallel to the symmetry axes of the material, the array of elastic constants is given by

$$
[S_{ij}] = \begin{bmatrix} S_{11} & S_{12} & S_{13} & 0 & 0 & 0 \\ & S_{22} & S_{23} & 0 & 0 & 0 \\ & & S_{33} & 0 & 0 & 0 \\ & & & S_{44} & 0 & 0 \\ & & & & S_{55} & 0 \\ & & & & & S_{66} \end{bmatrix} \tag{11.10}
$$

A similar expression can be written for C_{ij}. Taking into account the fact that a lamina is a thin orthotropic material, that is, through thickness components are zero, we can write the compliance matrix for an orthotropic lamina by eliminating the terms involving the z axis:

$$[S_{ij}] = \begin{bmatrix} S_{11} & S_{12} & 0 \\ & S_{22} & 0 \\ & & S_{66} \end{bmatrix} \tag{11.11}$$

We can rewrite in full form Hooke's law for a thin orthotropic lamina, with natural and geometric axes coinciding, as follows:

$$\begin{bmatrix} \varepsilon_1 \\ \varepsilon_2 \\ \varepsilon_6 \end{bmatrix} = \begin{bmatrix} S_{11} & S_{12} & 0 \\ S_{12} & S_{22} & 0 \\ 0 & 0 & S_{66} \end{bmatrix} \begin{bmatrix} \sigma_1 \\ \sigma_2 \\ \sigma_6 \end{bmatrix} \tag{11.12}$$

Conversely,

$$\begin{bmatrix} \sigma_1 \\ \sigma_2 \\ \sigma_6 \end{bmatrix} = \begin{bmatrix} Q_{11} & Q_{12} & 0 \\ Q_{12} & Q_{22} & 0 \\ 0 & 0 & Q_{66} \end{bmatrix} \begin{bmatrix} \varepsilon_1 \\ \varepsilon_2 \\ \varepsilon_6 \end{bmatrix} \tag{11.13}$$

It is customary to use the symbol Q_{ij} rather than C_{ij} for thin material. The Q_{ij} are called *reduced stiffnesses*. The relationships between Q_{ij} and S_{ij} can easily be shown to be

$$Q_{11} = \frac{S_{22}}{S_{11}S_{22} - S_{12}^2}$$

$$Q_{12} = -\frac{S_{12}}{S_{11}S_{22} - S_{12}^2}$$

$$Q_{22} = \frac{S_{11}}{S_{11}S_{22} - S_{12}^2} \tag{11.14}$$

$$Q_{66} = \frac{1}{S_{66}}$$

Also,

$$Q_{ij} = C_{ij} - \frac{C_{i3}C_{j3}}{C_{33}} \quad (i,j = 1,2,6)$$

Note that three-dimensional orthotropy requires nine independent elastic constants [Eq. (11.10)], while bidimensional orthotropy requires only four [Eq. (11.11)]. For an isotropic material (two- or three-dimensional), one just needs two independent elastic constants [Eqs. (11.6) through (11.9)].

It is worth emphasizing that Eqs. (11.12) and (11.13), showing terms with indices 16 and 26 to be zero, represent a special case of orthotropy when the principal material axes of symmetry (the fiber direction [1] and the direction transverse to it [2]) coincide with the geometric directions. If this is not so, that is, if the material symmetry axes and the geometric axes do not coincide,

Fig. 11.2. An off-axis unidirectional lamina.

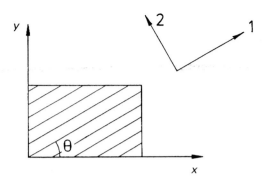

which is a more general case of orthotropy, then we shall have a fully populated elastic constant matrix and the stress-strain relationships become

$$
\begin{bmatrix} \sigma_x \\ \sigma_y \\ \sigma_s \end{bmatrix} = \begin{bmatrix} \bar{Q}_{11} & \bar{Q}_{12} & \bar{Q}_{16} \\ \bar{Q}_{12} & \bar{Q}_{22} & \bar{Q}_{26} \\ \bar{Q}_{16} & \bar{Q}_{26} & \bar{Q}_{66} \end{bmatrix} \begin{bmatrix} \varepsilon_x \\ \varepsilon_y \\ \varepsilon_s \end{bmatrix} \tag{11.15}
$$

where the $\bar{Q}_{ij}$ matrix is called the transformed reduced stiffness matrix because it is obtained by transforming Q_{ij} (specially orthotropic) to $\bar{Q}_{ij}$ (generally orthotropic). We can perform the transformation of axes as will be shown and obtain $\bar{Q}_{ij}$ from Q_{ij}.

Figure 11.2 shows the situation for a unidirectional composite lamina where the two sets of axes do not coincide. The properties in the 1–2 system are known, and we wish to determine them in the x-y system or vice versa. Both stress and strain are second-rank tensors. A second-rank tensor, T_{ij}, transforms as

$$
T_{ij} = a_{ik} a_{jl} T_{kl}
$$

where a_{ik} and a_{jl} are the direction cosines. Table 11.1 gives the direction cosines for the transformation of axes shown in Figure 11.2. Angle θ is positive when the x-y axes are rotated counterclockwise with respect to the 1–2 axes. This transformation of axes is carried out easily in the matrix form (see Appendix A). For stresses we can write

$$
\begin{bmatrix} \sigma_1 \\ \sigma_2 \\ \sigma_6 \end{bmatrix} = [T]_\sigma \begin{bmatrix} \sigma_x \\ \sigma_y \\ \sigma_s \end{bmatrix} \tag{11.16}
$$

while for strains,

$$
\begin{bmatrix} \varepsilon_1 \\ \varepsilon_2 \\ \varepsilon_6 \end{bmatrix} = [T]_\varepsilon \begin{bmatrix} \varepsilon_x \\ \varepsilon_y \\ \varepsilon_s \end{bmatrix} \tag{11.17}
$$

Table 11.1. Direction cosines

Direction	x	y	
1	$a_{11} = m$	$a_{12} = n$	$m = \cos\theta$
2	$a_{21} = -n$	$a_{22} = m$	$m = \sin\theta$

where $[T]_\sigma$ and $[T]_\varepsilon$ are the transformation matrices for stress and strain transformations, respectively, and are given by

$$[T]_\sigma = \begin{bmatrix} m^2 & n^2 & 2mn \\ n^2 & m^2 & -2mn \\ -mn & mn & m^2 - n^2 \end{bmatrix} \tag{11.18}$$

$$[T]_\varepsilon = \begin{bmatrix} m^2 & n^2 & mn \\ n^2 & m^2 & -mn \\ -2mn & 2mn & m^2 - n^2 \end{bmatrix} \tag{11.19}$$

where $m = \cos\theta$ and $n = \sin\theta$. This method of using different transformation matrices for stress and stain transformations avoids the need of putting the factor $\frac{1}{2}$ before the engineering shear strains to convert them to tensorial strain components suitable for transformation. Multiplying both sides of Eq. (11.16) by $[T]_\sigma^{-1}$ and remembering that $[T]_\sigma[T]_\sigma^{-1} = [I]$, the identity matrix, we get

$$\begin{bmatrix} \sigma_x \\ \sigma_y \\ \sigma_s \end{bmatrix} = [T]_\sigma^{-1} \begin{bmatrix} \sigma_1 \\ \sigma_2 \\ \sigma_6 \end{bmatrix} \tag{11.20}$$

$[T]_\sigma^{-1}$ can be obtained from $[T]_\sigma$ by simply substituting $-\theta$ for θ. Appendix A gives the procedure for obtaining the inverse of a given matrix. In this particular case, substituting $-\theta$ for θ in Eq. (11.18) results in

$$[T]_\sigma^{-1} = \begin{bmatrix} m^2 & n^2 & -2mn \\ n^2 & m^2 & 2mn \\ mn & -mn & m^2 - n^2 \end{bmatrix} \tag{11.21}$$

where $m = \cos\theta$ and $n = \sin\theta$. Substituting Eq. (11.13). in Eq. (11.20), we obtain

$$\begin{bmatrix} \sigma_x \\ \sigma_y \\ \sigma_s \end{bmatrix} = [T]_\sigma^{-1}[Q] \begin{bmatrix} \varepsilon_1 \\ \varepsilon_2 \\ \varepsilon_6 \end{bmatrix} \tag{11.22}$$

If we now substitute Eq. (11.17) in Eq. (11.22), we arrive at

$$\begin{bmatrix} \sigma_x \\ \sigma_y \\ \sigma_s \end{bmatrix} = [T]_\sigma^{-1}[Q][T]_\varepsilon \begin{bmatrix} \varepsilon_x \\ \varepsilon_y \\ \varepsilon_s \end{bmatrix} = [\bar{Q}] \begin{bmatrix} \varepsilon_x \\ \varepsilon_y \\ \varepsilon_s \end{bmatrix} \tag{11.23}$$

where

$$[\bar{Q}] = [T]_\sigma^{-1}[Q][T]_\varepsilon \tag{11.24}$$

$[\bar{Q}]$ is the stiffness matrix for a generally orthotropic lamina whose components in expanded form are written as follows ($m = \cos\theta$, $n = \sin\theta$):

$$\bar{Q}_{11} = Q_{11}m^4 + 2(Q_{12} + 2Q_{66})m^2n^2 + Q_{22}n^4$$

$$\bar{Q}_{12} = (Q_{11} + Q_{22} - 4Q_{66})m^2n^2 + Q_{12}(m^4 + n^4)$$

$$\bar{Q}_{22} = Q_{11}n^4 + 2(Q_{12} + 2Q_{66})m^2n^2 + Q_{22}m^4$$

$$\bar{Q}_{16} = (Q_{11} - Q_{12} - 2Q_{66})m^3n + (Q_{12} - Q_{22} + 2Q_{66})mn^3 \tag{11.25}$$

$$\bar{Q}_{26} = (Q_{11} - Q_{12} - 2Q_{66})mn^3 + (Q_{12} - Q_{22} + 2Q_{66})m^3n$$

$$\bar{Q}_{66} = (Q_{11} + Q_{22} - 2Q_{12} - 2Q_{66})m^2n^2 + Q_{66}(m^4 + n^4)$$

Note that although $\bar{Q}_{ij}$ is a completely filled matrix, only four of its components are independent: $\bar{Q}_{16}$ and $\bar{Q}_{26}$ are linear combinations of the other four.

A corresponding stress-strain relationship in terms of compliances of a generally orthotropic lamina can be obtained:

$$\begin{bmatrix} \varepsilon_x \\ \varepsilon_y \\ \varepsilon_s \end{bmatrix} = \begin{bmatrix} S_{11} & S_{12} & S_{16} \\ S_{12} & S_{22} & S_{26} \\ S_{16} & S_{26} & S_{66} \end{bmatrix} \begin{bmatrix} \sigma_x \\ \sigma_y \\ \sigma_s \end{bmatrix} \tag{11.26}$$

In a generally orthotropic lamina wherein we have nonzero 16 and 26 terms, a unidirectional normal stress σ_x has both normal as well as shear strains as responses and vice versa; that is, there is a coupling between the normal and shear effects. In the case of a specially orthotropic lamina where the 16 and 26 terms are zero, we have normal stresses producing normal strains and shear stresses producing shear strains and vice versa. In this there is no coupling between the normal and shear components. We present more information about such coupling effects in Section 11.5.

11.3 Relationships between Engineering Constants and Reduced Stiffnesses and Compliances

Consider the thin lamina shown in Figure 11.3 with the natural axes coinciding with the geometric axes. The conventional engineering constants in this case are Young's moduli in direction 1 (E_1) and direction 2 (E_2), the principal shear modulus G_6, and the principal Poisson's ratio ν_1. The Poisson's ratio ν_1, when the lamina is strained in direction 1, is equal to $-\varepsilon_2/\varepsilon_1$. The Poisson's ratio ν_2, when the lamina is strained in direction 2, equals $-\varepsilon_1/\varepsilon_2$.

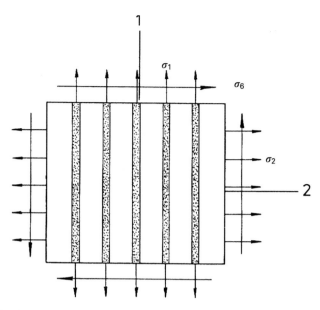

Fig. 11.3. A thin lamina with natural (or material) axes coinciding with the geometric axes.

We wish to relate these five conventional engineering constants to the four independent elastic constants, the reduced stiffnesses Q_{ij}. Let us consider that σ_1 is the only nonzero component in Eq (11.13). Then we can write

$$\sigma_1 = Q_{11}\varepsilon_1 + Q_{12}\varepsilon_2$$

$$\sigma_2 = Q_{12}\varepsilon_1 + Q_{22}\varepsilon_2 = 0$$

Solving for ε_1 and ε_2, we get

$$\varepsilon_1 = \frac{Q_{22}}{Q_{11}Q_{22} - Q_{12}^2}\,\sigma_1$$

and

$$\varepsilon_2 = -\frac{Q_{12}}{Q_{11}Q_{22} - Q_{12}^2}\,\sigma_1$$

By definition, we have $E_1 = \sigma_1/\varepsilon_1$. Thus,

$$E_1 = \frac{Q_{11} - Q_{22} - Q_{12}^2}{Q_{22}} \tag{11.27}$$

and

$$\nu_1 = -\frac{\varepsilon_2}{\varepsilon_1} = \frac{Q_{12}}{Q_{22}} \tag{11.28}$$

If we repeat this procedure with σ_2 as the only nonzero stress component in Eq. (11.13), we obtain

$$E_2 = \frac{\sigma_2}{\varepsilon_2} = \frac{Q_{11}Q_{22} - Q_{12}^2}{Q_{11}} \tag{11.29}$$

and

$$\nu_2 = -\frac{\varepsilon_1}{\varepsilon_2} = \frac{Q_{12}}{Q_{11}} \tag{11.30}$$

If we consider that σ_6 is the only nonzero component, we can get

$$G_6 = \frac{\sigma_6}{\varepsilon_6} = Q_{66} \tag{11.31}$$

Note that only four of the five constants are independent because

$$\nu_1 E_2 = \nu_2 E_1 \tag{11.32}$$

or

$$\frac{E_1}{E_2} = \frac{\nu_1}{\nu_2} \tag{11.33}$$

We can solve Eqs. (11.27)–(11.30) for the Q_{ij} to give

$$Q_{11} = \frac{E_1}{1 - \nu_1\nu_2} \quad Q_{22} = \frac{E_2}{1 - \nu_1\nu_2} \quad Q_{12} = \frac{\nu_1 E_2}{1 - \nu_1\nu_2} = \frac{\nu_2 E_1}{1 - \nu_1\nu_2}$$

and $Q_{66} = Q_6$ is given by Eq. (11.31). Similarly, we can show that the relationships between compliances and engineering constants are as follows:

$$S_{11} = \frac{1}{E_1} \quad S_{22} = \frac{1}{E_2} \quad S_{12} = -\frac{\nu_1}{E_1} = -\frac{\nu_2}{E_2} \quad S_{66} = \frac{1}{G_6}$$

11.4   Variation of Lamina Properties with Orientation

In Section 11.2 we obtained the relationships between Q_{ij} and $\bar{Q}_{ij}$. It is of interest to obtain similar relationships for conventional engineering constants referred to geometric axes x-y (E_x, E_y, G_s, and ν_x) in terms of engineering constants referred to material symmetry axes 1-2 (E_1, E_2, G_6, and ν_1). Consider Eqs. (11.16) and (11.18) and let σ_x be the only nonzero stress component. Then

$$\sigma_1 = \sigma_x m^2 \tag{11.34a}$$

$$\sigma_2 = \sigma_x n^2 \tag{11.34b}$$

$$\sigma_6 = -\sigma_x mn \tag{11.34c}$$

By Hooke's law, we can write for the strains in a lamina

$$\varepsilon_1 = \frac{1}{E_1}(\sigma_1 - \nu_1 \sigma_2) \tag{11.35a}$$

$$\varepsilon_2 = \frac{1}{E_2}(\sigma_2 - \nu_2 \sigma_1) \tag{11.35b}$$

$$\varepsilon_6 = \frac{\sigma_6}{G_6} \tag{11.35c}$$

From Eqs. (11.34)–(11.35), we get

$$\varepsilon_1 = \sigma_x\left(\frac{m^2}{E_1} - \nu_1 \frac{n^2}{E_1}\right) = \sigma_x\left(\frac{m^2}{E_1} - \frac{\nu_2}{E_2}n^2\right) \tag{11.36a}$$

$$\varepsilon_2 = \sigma_x\left(\frac{n^2}{E_2} - \nu_2 \frac{m^2}{E_2}\right) = \sigma_x\left(\frac{n^2}{E_2} - \frac{\nu_1}{E_1}m^2\right) \tag{11.36b}$$

$$\varepsilon_6 = -\frac{\sigma_x mn}{G_6} \tag{11.36c}$$

Because we have the strain transformation given by Eq. (11.17), we can write the inverse of Eq. (11.17) as

$$\begin{bmatrix} \varepsilon_x \\ \varepsilon_y \\ \varepsilon_s \end{bmatrix} = [T]_\varepsilon^{-1} \begin{bmatrix} \varepsilon_1 \\ \varepsilon_2 \\ \varepsilon_6 \end{bmatrix}$$

where $[T]_\varepsilon^{-1}$ can be obtained by substituting $-\theta$ for θ in Eq. (11.19). In the expanded form, we have

$$\varepsilon_x = m^2\varepsilon_1 + n^2\varepsilon_2 - mn\varepsilon_6 \tag{11.37a}$$

$$\varepsilon_y = n^2\varepsilon_1 + m^2\varepsilon_2 + mn\varepsilon_6 \tag{11.37b}$$

$$\varepsilon_s = 2(\varepsilon_1 - \varepsilon_2)mn + \varepsilon_6(m^2 - n^2) \tag{11.37c}$$

Substituting Eq. (11.36) in Eq. (11.37), we obtain

$$\varepsilon_x = \sigma_x\left[\frac{m^4}{E_1} + \frac{n^4}{E_2} + \left(\frac{1}{G_6} - \frac{2\nu_1}{E_1}\right)m^2n^2\right] \tag{11.38a}$$

$$\varepsilon_y = -\sigma_x\left[\frac{\nu_1}{E_1} - \left(\frac{1}{E_1} + \frac{2\nu_1}{E_1} + \frac{1}{E_2} - \frac{1}{G_6}\right)m^2n^2\right] \tag{11.38b}$$

$$\varepsilon_s = -\sigma_x(2mn)\left[\frac{\nu_1}{E_1} + \frac{1}{E_2} - \frac{1}{2G_6} - m^2\left(\frac{1}{E_1} + \frac{2\nu_1}{E_1} + \frac{1}{E_2} - \frac{1}{G_6}\right)\right] \tag{11.38c}$$

Now $E_x = \sigma_x/\varepsilon_x$ by definition. Combining this with Eq. (11.38a), we obtain

$$\frac{1}{E_x} = \frac{m^4}{E_1} + \frac{n^4}{E_2} + \left(\frac{1}{G_6} - \frac{2\nu_1}{E_1}\right)m^2n^2 \tag{11.39}$$

E_y can be obtained from E_x by substituting $\theta + 90°$ for θ in Eq. (11.39):

$$\frac{1}{E_y} = \frac{n^4}{E_1} + \frac{m^4}{E_2} + \left(\frac{1}{G_6} - \frac{2v_1}{E_1}\right)m^2n^2 \qquad (11.40)$$

where $v_x = -\varepsilon_y/\varepsilon_x$ when σ_x is the applied stress. Then from Eqs. (11.38b) and (11.39), we obtain

$$\frac{v_x}{E_2} = -\frac{\varepsilon_y}{E_x\varepsilon_x} = -\frac{\varepsilon_y}{\sigma_x} = \frac{v_1}{E_1} - \left(\frac{1}{E_1} + \frac{2v_1}{E_1} + \frac{1}{E_2} - \frac{1}{G_6}\right)m^2n^2$$

or

$$v_x = E_x\left[\frac{v_1}{E_1} - \left(\frac{1}{E_1} + \frac{2v_1}{E_1} + \frac{1}{E_2} - \frac{1}{G_6}\right)m^2n^2\right] \qquad (11.41)$$

Similarly, it can be shown that

$$v_y = E_y\left[\frac{v_2}{E_2} - \left(\frac{1}{E_1} + \frac{1}{E_2} + \frac{2v_1}{E_1} - \frac{1}{G_6}\right)m^2n^2\right] \qquad (11.42)$$

Taking σ_s to be the only nonzero stress component, noting that $\varepsilon_s = \sigma_s/G_s$, and applying Hooke's law, we obtain

$$\frac{1}{G_s} = \frac{1}{G_6} + 4m^2n^2\left(\frac{1+v_1}{E_1} + \frac{1+v_2}{E_2} - \frac{1}{G_6}\right) \qquad (11.43)$$

Figure 11.4 shows the variations of E_x, E_y, G_s, v_x, and v_y with fiber orientation θ for a 50% V_f carbon/epoxy composite. The other relevant data used in Figure 11.4 are $E_1 = 240$ GPa, $E_2 = 8$ GPa, $G_6 = 6$ GPa, and $v_1 = 0.26$.

11.5 Analysis of Laminated Composites

Now that we have discussed the analysis of an individual lamina, we proceed to discuss the macroscopic analysis of laminated composites (Jones, 1975; Christensen, 1979; Tsai and Hahn, 1980; Halpin, 1984; Daniel and Ishai, 1994). In this analysis, the individual identities of fiber and matrix are ignored. Each individual lamina is treated as a homogeneous, orthotropic sheet, and the laminated composite is analyzed using the classical theory of laminated plates.

It would be in order at this point to describe how a multidirectional laminate can be defined by using a code to designate the stacking sequence of laminae. Figure 11.5 shows an example of a stacking sequence. Such a stacking sequence can be described by the following code:

$$[0_2/90_2/-45_3/45_3]_s$$

This code says that starting from the bottom of the laminate, that is, at $z = -h/2$, we have a group of two plies at $0°$ orientation; then two plies at

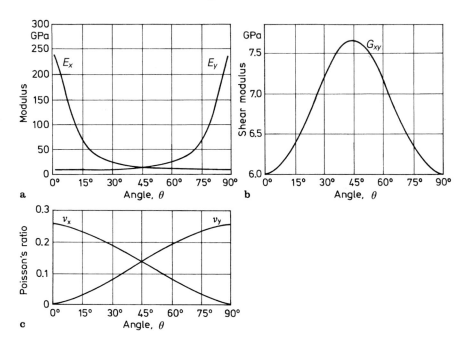

Fig. 11.4. Variation of elastic constants with fiber orientation angle for a 50 v/o carbon/epoxy composite: **a** longitudinal and transverse Young's moduli (E_x and E_y), **b** shear modulus (G_{xy}) and **c** Poisson's ratio (v_x and v_y).

90° orientation; followed by a group of three plies at −45° orientation; and lastly, a group of three plies at +45° orientation. The subscript s indicates that the laminate is symmetrical with respect to the midplane ($z = 0$); that is, the top half of the laminate is a mirror image of the bottom half. It is not necessary that a laminate composite by symmetrical. If the top half has the sequence opposite to that of the bottom half, then we shall have an asymmetrical laminate. In any event, we may represent the total stacking sequence of the laminate shown in Figure 11.5 in the following way:

$$[0_2/90_2/-45_3/45_6/-45_3/90_2/0_2]_T$$

where the subscript T indicates that the code represents the whole of the laminate thickness. Note that we have merged the two middle groups of the same ply orientation into one group. If the laminate composite consists of an odd number of laminae, the midplane will lie in the central ply.

11.5.1 Basic Assumptions

We assume that the laminate thickness is small compared to its lateral dimensions. Therefore, stresses acting on the interlaminar planes in the interior of the laminate, that is, away from the free edges, are negligibly small

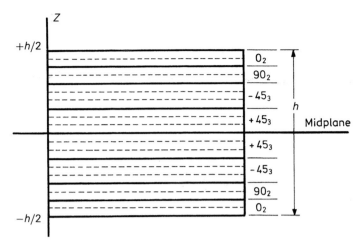

Fig. 11.5. A laminate composite with the stacking sequence given by $[0_1^\circ/90_2^\circ/-45_3^\circ/45_3^\circ]_s$.

(we shall see later that the situation is different at the free edges). We also assume that there exists a perfect bond between any two laminae. That being so, the laminae cannot slide over each other, and we have continuous displacements across the bond. We make yet another important assumption: namely, a line originally straight and perpendicular to the laminate midplane remains so after deformation. Actually, this follows from the perfect bond assumption, which does not allow sliding between the laminae.

Finally, we have the so-called Kirchhoff assumption, which states that in-plane displacements are linear functions of the thickness, and therefore the interlaminar shear strains, ε_{xz} and ε_{yz}, are negligible. With these assumptions we can reduce the laminate behavior to a two-dimensional analysis of the laminate midplane.

We have the following strain-displacement relationships:

$$\varepsilon_x = \frac{\partial u}{\partial x} \quad \varepsilon_{xy} = \frac{\partial u}{\partial y} + \frac{\partial v}{\partial x}$$

$$\varepsilon_y = \frac{\partial v}{\partial y} \quad \varepsilon_{xz} = \frac{\partial u}{\partial z} + \frac{\partial w}{\partial x} \qquad (11.44)$$

$$\varepsilon_z = \frac{\partial w}{\partial z} \quad \varepsilon_{yz} = \frac{\partial v}{\partial z} + \frac{\partial w}{\partial y}$$

Here, u, v, and w are the displacements in the x, y, and z directions, respectively. For $i \neq j$, the ε_{ij} represent engineering shear strain components equal to twice the tensorial shear components. As per Kirchhoff's assumption, the in-plane displacements are linear functions of the thickness coordinate z. Then

$$u = u_0(x, y) + zF_1(x, y) \quad v = v_0(x, y) + zF_2(x, y) \qquad (11.45)$$

where u_0 and v_0 are displacements of the midplane. It also follows from Kirchhoff's assumptions that interlaminar shear strains ε_{xz} and ε_{yz} are zero. Therefore, from Eqs. (11.44) and (11.45) we obtain

$$\varepsilon_{xz} = F_1(x, y) + \frac{\partial w}{\partial x} = 0$$

$$\varepsilon_{yz} = F_2(x, y) + \frac{\partial w}{\partial y} = 0$$

It therefore follows that

$$F_1(x, y) = -\frac{\partial w}{\partial x} \quad \text{and} \quad F_2(x, y) = -\frac{\partial w}{\partial y} \tag{11.46}$$

The normal strain in the thickness direction, ε_z, is negligible; thus we can write

$$w = w(x, y)$$

That is, the vertical displacement of any point does not change in the thickness direction.

Substituting Eq. (11.46) into Eq. (11.45), we obtain

$$\varepsilon_x = \frac{\partial u}{\partial x} = \frac{\partial u_0}{\partial x} - z \frac{\partial^2 w}{\partial x^2} = \varepsilon_x^0 + z K_x \tag{11.47a}$$

$$\varepsilon_y = \frac{\partial v}{\partial y} = \frac{\partial v_0}{\partial y} - z \frac{\partial^2 w}{\partial y^2} = \varepsilon_y^0 + z K_y \tag{11.47b}$$

$$\varepsilon_{xy} = \frac{\partial u}{\partial y} + \frac{\partial v}{\partial x} = \frac{\partial u_0}{\partial y} + \frac{\partial v_0}{\partial x} - 2z \frac{\partial^2 w}{\partial x \partial y} = \varepsilon_{xy}^0 + z K_{xy} \tag{11.47c}$$

Denoting ε_{xy} by ε_s and K_{xy} by K_s, as per our notation, we can rewrite the expression for ε_{xy} as

$$\varepsilon_s = \varepsilon_s^0 + z K_s \tag{11.47d}$$

Here, ε_x^0, ε_y^0, and ε_s^0 are the midplane strains, while K_x, K_y, and K_s are the plate curvatures. We can represent these quantities in a compact form as follows:

$$\begin{bmatrix} \varepsilon_x^0 \\ \varepsilon_y^0 \\ \varepsilon_s^0 \end{bmatrix} = \begin{bmatrix} \partial u_0 / \partial x \\ \partial u_0 / \partial y \\ \partial u_0 / \partial y + \partial v_0 / \partial x \end{bmatrix} \tag{11.48}$$

and

$$\begin{bmatrix} K_x \\ K_y \\ K_s \end{bmatrix} = - \begin{bmatrix} \partial^2 w / \partial x^2 \\ \partial^2 w / \partial y^2 \\ 2 \partial^2 w / \partial x \partial y \end{bmatrix} \tag{11.49}$$

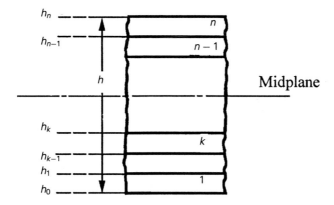

Fig. 11.6. A laminated composite made up of n stacked plies.

Equation (11.47) can be put into the following form:

$$\begin{bmatrix} \varepsilon_x \\ \varepsilon_y \\ \varepsilon_s \end{bmatrix} = \begin{bmatrix} \varepsilon_x^0 \\ \varepsilon_y^0 \\ \varepsilon_s^0 \end{bmatrix} + z \begin{bmatrix} K_x \\ K_y \\ K_s \end{bmatrix} \tag{11.50}$$

11.5.2   Constitutive Relationships for Laminated Composites

Consider a composite made of n stacked layers or plies; see Figure 11.6. Let h be the thickness of the laminated composite. Then we can write, for the kth layer, the following constitutive relationship:

$$[\sigma]_k = [\overline{Q}]_k [\varepsilon]_k \tag{11.51}$$

From the theory of laminated plates, we have the, strain-displacement relationships given by Eq. (11.50). We can rewrite Eq. (11.48) as

$$[\varepsilon] = [\varepsilon^0] + z[K] \tag{11.52}$$

Substituting Eq. (11.52) in Eq. (11.51), for the kth ply we get

$$[\sigma]_k = [\overline{Q}]_k [\varepsilon^0] + z[\overline{Q}]_k [K] \tag{11.53}$$

Because the stresses in a laminated composite vary from ply to ply, it is convenient to define laminate force and moment resultants as shown in Figure 11.7. These resultants of stresses and moments acting on a laminate cross section, defined as follows, provide us with a statically equivalent system of forces and moments acting at the midplane of the laminated composite. In the most general case, such a composite will have σ_x, σ_y, σ_z, σ_{xy}, σ_{yz}, and σ_{zx} as the stress components. Our laminated composite, however, is in a state of plane stress. Thus, we shall have only σ_x, σ_y, and σ_{xy} ($= \sigma_s$). We

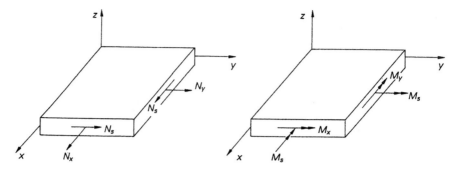

Fig. 11.7. Force (N) and moment (M) resultants in a laminated composite.

define the three corresponding stress resultants as

$$N_x = \int_{-h/2}^{h/2} \sigma_x \, dz$$

$$N_y = \int_{-h/2}^{h/2} \sigma_y \, dz \tag{11.54}$$

$$N_s = \int_{-h/2}^{h/2} \sigma_s \, dz$$

These stress resultants have the dimensions of force per unit length and are positive in the same direction as the corresponding stress components. These resultants give the total force per unit length acting at the midplane. Additionally, moments are applied at the midplane which are equivalent to the moments produced by the stresses with respect to the midplane. We define the moment resultants as

$$M_x = \int_{-h/2}^{h/2} \sigma_x z \, dz$$

$$M_y = \int_{-h/2}^{h/2} \sigma_y z \, dz \tag{11.55}$$

$$M_{xy} = M_s = \int_{-h/2}^{h/2} \sigma_s z \, dz$$

This system of three stress resultants [Eq. (11.54)] and three moment resultants [Eq. (11.55)] is statically equivalent to actual stress distribution through the thickness of the composite laminate.

From Eqs. (11.52) and (11.53), we can write for the stress resultants a summation over the n plies:

$$
\begin{bmatrix} N_x \\ N_y \\ N_s \end{bmatrix} = \sum_{k=1}^{n} \int_{h_{k-1}}^{h_k} \begin{bmatrix} \sigma_x \\ \sigma_y \\ \sigma_s \end{bmatrix}_k dz
$$

$$
= \sum_{k=1}^{n} \left(\int_{h_{k-1}}^{h_k} \begin{bmatrix} \bar{Q}_{11} & \bar{Q}_{12} & \bar{Q}_{16} \\ \bar{Q}_{12} & \bar{Q}_{22} & \bar{Q}_{26} \\ \bar{Q}_{16} & \bar{Q}_{26} & \bar{Q}_{66} \end{bmatrix} \begin{bmatrix} \varepsilon_x^0 \\ \varepsilon_y^0 \\ \varepsilon_s^0 \end{bmatrix} dz \right.
$$

$$
\left. + \int_{h_{k-1}}^{h_k} \begin{bmatrix} \bar{Q}_{11} & \bar{Q}_{12} & \bar{Q}_{16} \\ \bar{Q}_{12} & \bar{Q}_{22} & \bar{Q}_{26} \\ \bar{Q}_{16} & \bar{Q}_{26} & \bar{Q}_{66} \end{bmatrix} \begin{bmatrix} K_x \\ K_y \\ K_s \end{bmatrix} z \, dz \right) \tag{11.56}
$$

Note that $[\varepsilon^0]$ and $[K]$ are not functions of z and in a given ply $[\bar{Q}]$ is not a function of z. Thus, we can simplify the preceding expression to

$$
\begin{bmatrix} N_x \\ N_y \\ N_s \end{bmatrix} = \sum_{k=1}^{n} \left(\begin{bmatrix} \bar{Q}_{11} & \bar{Q}_{12} & \bar{Q}_{16} \\ \bar{Q}_{12} & \bar{Q}_{22} & \bar{Q}_{26} \\ \bar{Q}_{16} & \bar{Q}_{26} & \bar{Q}_{66} \end{bmatrix} \begin{bmatrix} \varepsilon_x^0 \\ \varepsilon_y^0 \\ \varepsilon_s^0 \end{bmatrix} \int_{h_{k-1}}^{h_k} dz \right.
$$

$$
\left. + \begin{bmatrix} \bar{Q}_{11} & \bar{Q}_{12} & \bar{Q}_{16} \\ \bar{Q}_{12} & \bar{Q}_{22} & \bar{Q}_{26} \\ \bar{Q}_{16} & \bar{Q}_{26} & \bar{Q}_{66} \end{bmatrix} \begin{bmatrix} K_x \\ K_y \\ K_s \end{bmatrix} \int_{h_{k-1}}^{h_k} z \, dz \right) \tag{11.57}
$$

We can rewrite Eq. (11.57) as

$$
\begin{bmatrix} N_x \\ N_y \\ N_s \end{bmatrix} = \begin{bmatrix} A_{11} & A_{12} & A_{16} \\ A_{12} & A_{22} & A_{26} \\ A_{16} & A_{26} & A_{66} \end{bmatrix} \begin{bmatrix} \varepsilon_x^0 \\ \varepsilon_y^0 \\ \varepsilon_s^0 \end{bmatrix} + \begin{bmatrix} B_{11} & B_{12} & B_{16} \\ & B_{22} & B_{26} \\ B_{16} & B_{26} & B_{66} \end{bmatrix} \begin{bmatrix} K_x \\ K_y \\ K_s \end{bmatrix} \tag{11.58}
$$

or

$$
[N] = [A][\varepsilon^0] + [B][K] \tag{11.59}
$$

where

$$
A_{ij} = \sum_{k=1}^{n} (\bar{Q}_{ij})_k (h_k - h_{k-1}) \tag{11.60}
$$

and

$$
B_{ij} = \frac{1}{2} \sum_{k=1}^{n} (\bar{Q}_{ij})_k (h_k^2 - h_{k-1}^2) \tag{11.61}
$$

Similarly, from Eqs. (11.52) and (11.55), we can write for the moment resultants

$$
\begin{bmatrix} M_x \\ M_y \\ M_s \end{bmatrix} = \begin{bmatrix} B_{11} & B_{12} & B_{16} \\ B_{12} & B_{22} & B_{26} \\ B_{16} & B_{26} & B_{66} \end{bmatrix} \begin{bmatrix} \varepsilon_x^0 \\ \varepsilon_y^0 \\ \varepsilon_s^0 \end{bmatrix} + \begin{bmatrix} D_{11} & D_{12} & D_{16} \\ D_{12} & D_{22} & D_{26} \\ D_{16} & D_{26} & D_{66} \end{bmatrix} \begin{bmatrix} K_x \\ K_y \\ K_s \end{bmatrix} \quad (11.62)
$$

or

$$
[M] = [B][\varepsilon^0] + [D][K] \quad (11.63)
$$

where

$$
D_{ij} = \frac{1}{3} \sum_{k=1}^{n} (\bar{Q}_{ij})_k (h_k^3 - h_{k-1}^3) \quad (11.64)
$$

and the B_{ij} are given by Eq. (11.61).

We may combine Eqs. (11.59) and (11.63) and write the constitutive equations for the laminate composite in a more compact form. Thus,

$$
\left[\frac{N}{M} \right] = \left[\begin{array}{c|c} A & B \\ \hline B & D \end{array} \right] \left[\frac{\varepsilon^0}{K} \right] \quad (11.65)
$$

To appreciate the significance of the preceding expressions, let us examine the expression for N_x:

$$
N_x = A_{11}\varepsilon_x^0 + A_{12}\varepsilon_y^0 + A_{16}\varepsilon_s^0 + B_{11}K_x + B_{12}K_y + B_{16}K_s
$$

We note that the stress resultant is a function of the midplane tensile strains (ε_x^0 and ε_y^0), the midplane shear strain (ε_s^0), the bending curvatures (K_x and K_y), and the twisting (K_s). This is a much more complex situation than that observed in a homogeneous plate where tensile loads result in only tensile strains. In a laminated plate we have coupling between tensile and shear, tensile and bending, and tensile and twisting effects. Specifically, the terms A_{16} and A_{26} bring in the tension-shear coupling, while the terms B_{16} and B_{26} represent the tension-twisting coupling. The D_{16} and D_{26} terms in a similar expression for M_x represent flexure-twisting coupling.

Under certain conditions, the stress and moment resultants become uncoupled. It is instructive to examine the conditions under which some of these simplifications can result. The A_{ij} terms are the sum of ply $\bar{Q}_{ij}$ times the ply thickness [Eq. (11.60)]. Thus, the A_{ij} will be zero if the positive contributions of some laminae are nullified by the negative contributions of others. Now the Q_{ij} terms of a ply are derived from orthotropic stiffnesses and, because of the form of transformation equations (11.25), $\bar{Q}_{11}$, $\bar{Q}_{12}$, $\bar{Q}_{22}$, and $\bar{Q}_{66}$ are always positive. This means that A_{11}, A_{12}, A_{22}, and A_{66} are always positive. $\bar{Q}_{16}$ and $\bar{Q}_{26}$, however, are zero for 0° and 90° orientations and can be positive or negative for θ between 0° and 90°. In fact, $\bar{Q}_{16}$ and $\bar{Q}_{26}$ are odd functions of θ; that is, for equal positive and negative orientations, they will be equal in magnitude but opposite in sign. In particular, $\bar{Q}_{16}$ and $\bar{Q}_{26}$ for a

$+\theta$ orientation are equal to but opposite in sign to $\bar{Q}_{16}$ and $\bar{Q}_{26}$ values for a $-\theta$ orientation. Thus, if for each $+\theta$ ply, we have another identical ply of the same thickness at $-\theta$, then we shall have what is called a *specially orthotropic laminate* with respect to in-plane stresses and strains; that is, $A_{16} = A_{26} = 0$. The relative position of such plies in the stacking sequence does not matter.

The B_{ij} terms are sums of terms involving $\bar{Q}_{ij}$ and differences of the square of z terms for the top (h_k) and bottom (h_{k-1}) of each ply. Thus, the B_{ij} terms are even functions of h_k, which means that they are zero if the laminate composite is symmetrical with respect to thickness. In other words, the B_{ij} are zero if we have for each ply above the midplane a ply identical in properties and orientation and at an equal distance below the midplane. Such a laminate is called a *symmetric laminate* and will have B_{ij} identically zero. This simplifies the constitutive equations, and symmetric laminates are considerably easier to analyze. Additionally, because of the absence of bending-stretching coupling in symmetric laminates, they do not have the problem of warping encountered in nonsymmetric laminates and caused by in-plane forces induced by thermal contractions occurring during the curing of the resin matrix. Symmetric laminates will only experience tensile strains at the midplane but no flexure. The reader should realize that the origin of the $[B]$ matrix lies not in the intrinsic orthotropy of the laminae, but in the heterogeneous (nonsymmetric) stacking sequence of the plies. Thus, a two-ply composite consisting of isotropic materials such as aluminum and steel will show a nonzero $[B]$.

The bending matrix D_{ij} terms are defined in terms of $\bar{Q}_{ij}$ and the difference between h_k^3 and h_{k-1}^3. The geometrical contribution $(h_k^3 - h_{k-1}^3)$ is always positive. Thus, as explained above for A_{ij}, D_{11}, D_{12}, D_{22}, and D_{66} are always positive. Recall that $\bar{Q}_{16}$ and $\bar{Q}_{26}$ are odd functions of θ. D_{16} and D_{26} are therefore zero for all plies oriented at $0°$ or $90°$ because these plies have $\bar{Q}_{16} = \bar{Q}_{26} = 0$. D_{16} and D_{26} can also be made zero if, for each ply oriented at $+\theta$ and at a given distance above the midplane, we have an identical ply at an equal distance below the midplane but oriented at $-\theta$. This follows from the property of the odd function of θ; that is, $\bar{Q}_{16}(+\theta) = -\bar{Q}_{16}(-\theta)$, $\bar{Q}_{26}(+\theta) = -\bar{Q}_{26}(-\theta)$, while $(h_k^3 - h_{k-1}^3)$ is the same for both plies. Note, however, that such a laminated composite does not have a midplane of symmetry; that is, $B_{ij} \neq 0$. In fact, D_{16} and D_{26} are not zero for any midplane symmetric laminate except for unidirectional laminates $(0°$ or $90°)$ and crossplied laminates $(0°/90°)$. We can make D_{16} and D_{26} arbitrarily small, however, by using a large enough number of plies stacked at $\pm\theta$. This is because the contributions of $+\theta$ plies to D_{16} and D_{26} are opposite in sign to those of $-\theta$ plies, and although their locations are different distances from the midplane, they tend to cancel each other.

Yet another simple stacking sequence is the quasi-isotropic sequence. Such a laminated composite can be made by having plies of identical properties oriented in such a way that the angle between any two adjacent layers is

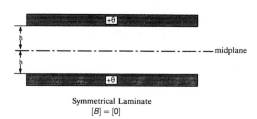

Fig. 11.8. Some special laminates.

Symmetrical Laminate
$[B] = [0]$

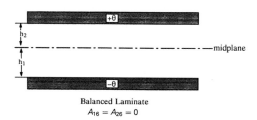

Balanced Laminate
$A_{16} = A_{26} = 0$

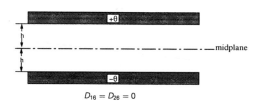

$D_{16} = D_{26} = 0$

$2\pi/n$, where n is the number of plies. Such a laminate has $[A]$ independent of orientation in the plane. We call such a stacking sequence *quasi-isotropic*, because $[B]$ and $[D]$ are not necessarily isotropic. Figure 11.8 summarizes some special laminates for which some coupling coefficients go to zero.

The important results of this section can be summarized as follows:

$$[\sigma]_k = [\bar{Q}]_k [\varepsilon]_k$$

where $1 \le k \le n$ and $i, j = 1, 2, 6$.

$$\varepsilon_i = \varepsilon_i^0 + zK_i$$

$$N_i = \int_{-h/2}^{h/2} \sigma_i \, dz$$

$$M_i = \int_{-h/2}^{h/2} \sigma_i \, dz$$

$$N_i = A_{ij}\varepsilon_j^0 + B_{ij}K_j$$

$$M_i = B_{ij}\varepsilon_j^0 + D_{ij}K_j$$

$$A_{ij} = \sum_{k=1}^{n}(\bar{Q}_{ij})_k(h_k - h_{k-1})$$

$$B_{ij} = \frac{1}{2}\sum_{k=1}^{n}(\bar{Q}_{ij})_k(h_k^2 - h_{k-1}^2)$$

$$D_{ij} = \frac{1}{3}\sum_{k=1}^{n}(\bar{Q}_{ij})_k(h_k^3 - h_{k-1}^3)$$

Symmetric laminates

$$\bar{Q}(z) = \bar{Q}(-z)$$

Anti-symmetric laminates

$$\bar{Q}(z) = -\bar{Q}(-z)$$

Example 11.1

A laminate is made up by stacking $0°$ and $45°$ plies as shown below:

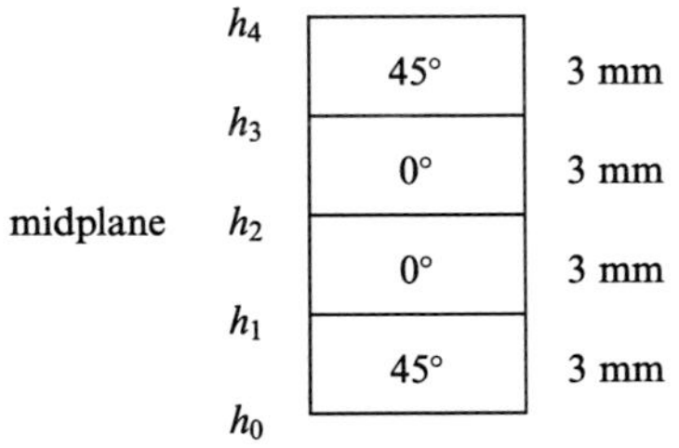

h_4		
	$45°$	3 mm
h_3		
	$0°$	3 mm
midplane h_2		
	$0°$	3 mm
h_1		
	$45°$	3 mm
h_0		

The $[Q_{ij}]_{0°}$ and $[\bar{Q}_{ij}]_{45°}$ matrices are:

$$[Q_{ij}]_{0°} = \begin{bmatrix} 140 & 5 & 0 \\ & 5 & 0 \\ & & 5 \end{bmatrix} \text{GPa}$$

$$[\bar{Q}_{ij}]_{45°} = \begin{bmatrix} 50 & 35 & 30 \\ & 50 & 30 \\ & & 35 \end{bmatrix} \text{GPa}$$

Compute the $[A]$, $[B]$, and $[D]$ matrices for this laminate.

Solution

$$(\bar{Q}_{ij})_{45°} = \begin{bmatrix} 50 & 35 & 30 \\ & 50 & 30 \\ & & 35 \end{bmatrix}$$

$$(\bar{Q}_{ij})_{0°} = (Q_{ij})_{0°} = \begin{bmatrix} 140 & 5 & 0 \\ & 5 & 0 \\ & & 5 \end{bmatrix}$$

Let us now compute the submatrices, $[A]$, $[B]$, and $[D]$.

$$A_{ij} = \sum_{k=1}^{4} (\bar{Q}_{ij})_k (h_k - h_{k-1})$$

$$= (\bar{Q}_{ij})_1 (h_1 - h_0) + (\bar{Q}_{ij})_2 (h_2 - h_1) + (\bar{Q}_{ij})_3 (h_3 - h_2)$$
$$+ (\bar{Q}_{ij})_4 (h_4 - h_3)$$

$$A_{ij} = (\bar{Q}_{ij})_1 [(-3) - (-6)] + (\bar{Q}_{ij})_2 [0 - (-3)] + (\bar{Q}_{ij})_3 [3 - 0]$$
$$+ (\bar{Q}_{ij})_4 [6 - 3]$$

$$= 3[(\bar{Q}_{ij})_1 + (\bar{Q}_{ij})_2 + (\bar{Q}_{ij})_3 + (\bar{Q}_{ij})_4]$$

$$(\bar{Q}_{ij})_1 = (\bar{Q}_{ij})_4 = (\bar{Q}_{ij})_{45°}$$

$$(\bar{Q}_{ij})_2 = (\bar{Q}_{ij})_3 = (\bar{Q}_{ij})_{0°}$$

$$A = 6[(\bar{Q}_{ij})_{45°} + (\bar{Q}_{ij})_{0°}]$$

$$= 6 \left\{ \begin{bmatrix} 50 & 35 & 30 \\ & 50 & 30 \\ & & 35 \end{bmatrix} + \begin{bmatrix} 140 & 5 & 0 \\ & 5 & 0 \\ & & 5 \end{bmatrix} \right\}$$

$$= \begin{bmatrix} 1140 & 240 & 180 \\ & 330 & 180 \\ & & 240 \end{bmatrix} \times 10^6 \, \text{N/m}$$

We have a symmetrical laminate, therefore

$$[B] = [0]$$

$$3D = \sum_{k=1}^{4} (\bar{Q}_{ij})_k (h_k^3 - h_{k-1}^3)$$

$$= (\bar{Q}_{ij})_1 (h_1^3 - h_0^3) + (\bar{Q}_{ij})_2 (h_2^3 - h_1^3) + (\bar{Q}_{ij})_3 (h_3^3 - h_2^3)$$
$$+ (\bar{Q}_{ij})_4 (h_4^3 - h_3^3)$$

$$= (\bar{Q}_{ij})_1[(-3)^3 - (-6)^3] + (\bar{Q}_{ij})_2[0 - (-3)^3] + (\bar{Q}_{ij})_3[3^3 - 0]$$
$$+ (\bar{Q}_{ij})_4[6^3 - 3^3]$$
$$= (\bar{Q}_{ij})_1(-27 + 216) + (\bar{Q}_{ij})_2(0 + 27) + (\bar{Q}_{ij})_3(27 - 0)$$
$$+ (\bar{Q}_{ij})_4(216 - 27)$$
$$3D = 189(\bar{Q}_{ij})_1 + 27(\bar{Q}_{ij})_2 + 27(\bar{Q}_{ij})_3 + 189(\bar{Q}_{ij})_4$$
$$(\bar{Q}_{ij})_1 = (\bar{Q}_{ij})_4 = (\bar{Q}_{ij})_{45°}$$
$$(\bar{Q}_{ij})_2 = (\bar{Q}_{ij})_3 = (\bar{Q}_{ij})_{0°}$$
$$D = 126(\bar{Q}_{ij})_{45°} + 18(\bar{Q}_{ij})_{0°}$$

$$D = \left\{ 126 \begin{bmatrix} 50 & 35 & 30 \\ & 50 & 30 \\ & & 35 \end{bmatrix} + 18 \begin{bmatrix} 140 & 5 & 0 \\ & 5 & 0 \\ & & 5 \end{bmatrix} \right\} \times 10^9 \, \frac{N}{m^2} \times 10^{-9} \, m^3$$

$$D = \begin{bmatrix} 8820 & 4500 & 3780 \\ & 6390 & 3780 \\ & & 4500 \end{bmatrix} N \cdot m$$

11.6 Stresses and Strains in Laminate Composites

We saw in Section 11.5 that strains produced in a lamina under load depend on the midplane strains, plate curvatures, and distances from the midplane. Midplane strains and plate curvatures can be expressed as functions of an applied load system, that is, in terms of stress and moment resultants. We derived the general constitutive equation (11.65) for laminate composites. We can invert Eq. (11.65) partially or fully and obtain explicit expressions for $[\varepsilon^0]$ and $[K]$. We use Eqs. (11.59) and (11.63) for this purpose. Solving Eq. (11.59) for midplane strains, we obtain

$$[\varepsilon^0] = [A]^{-1}[N] - [A]^{-1}[B][K] \tag{11.66}$$

Substituting Eq. (11.66) in Eq. (11.63), we obtain

$$[M] = [B][A]^{-1}[N] - ([B][A]^{-1}[B] - [D])[K] \tag{11.67}$$

Combining Eqs. (11.66) and (11.67), we obtain a partially inverted form of the constitutive equation:

$$\begin{bmatrix} \varepsilon^0 \\ \overline{M} \end{bmatrix} = \begin{bmatrix} A^* & | & B^* \\ \overline{C^*} & | & \overline{D^*} \end{bmatrix} \begin{bmatrix} N \\ \overline{K} \end{bmatrix} \tag{11.68}$$

where

$$[A^*] = [A]^{-1}$$

$$[B^*] = -[A]^{-1}[B]$$

$$[C^*] = [B][A]^{-1} = -[B^*]^T \tag{11.69}$$

$$[D^*] = [D] - [B][A]^{-1}[B]$$

From Eqs. (11.66) and (11.69), we can write

$$[\varepsilon^0] = [A^*][N] + [B^*][K] \tag{11.70}$$

$$[M] = [C^*][N] + [D^*][K] \tag{11.71}$$

From Eq. (11.71), we solve for $[K]$ and obtain

$$[K] = [D^*]^{-1}[M] - [D^*]^{-1}[C^*][N] \tag{11.72}$$

Substituting this value of $[K]$ [Eq. (11.72)] in Eq. (11.70), we obtain

$$[\varepsilon^0] = [A^*][N] + [B^*]([D^*]^{-1}[M] - [D^*]^{-1}[C^*][N])$$

$$= ([A^*] - [B^*][D^*]^{-1}[C^*])[N] + [B^*][D^*]^{-1}[M] \tag{11.73}$$

We can combine Eqs. (11.72) and (11.70) to obtain the fully inverted form:

$$\left[\frac{\varepsilon^0}{K}\right] = \left[\begin{array}{c|c} A' & B' \\ \hline C' & D' \end{array}\right]\left[\frac{N}{M}\right] = \left[\begin{array}{c|c} A' & B' \\ \hline B' & D' \end{array}\right]\left[\frac{N}{M}\right] \tag{11.74}$$

where

$$[A'] = [A^*] - [B^*][D^*]^{-1}[C^*]$$

$$= [A^*] + [B^*][D^*]^{-1}[B^*]^T \quad ([C^*] = -[B^*]^T)$$

$$[B'] = [B^*][D^*]^{-1}$$

$$[C'] = -[D^*]^{-1}[C^*] = [D^*]^{-1}[B^*]^T = [B']^T = [B']$$

$$[D'] = [D^*]^{-1}$$

Equations (11.65), (11.68), and (11.74) are useful forms of the laminate constitutive relationships. We note that each form involves using the elastic properties of the lamina (from the $\bar{Q}_{ij}$ values for each lamina) and the ply stacking sequence (z coordinate).

11.7   Interlaminar Stresses and Edge Effects

The classical lamination theory used in Section 11.5 to describe the laminate composite behavior is rigorously correct for an infinite laminate composite.

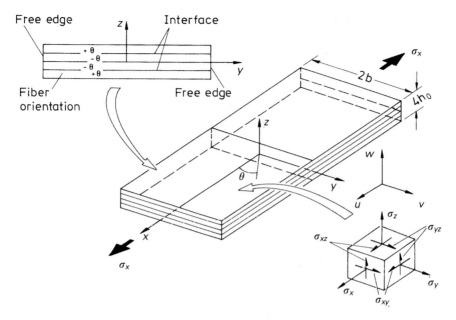

Fig. 11.9. A four-ply laminate ($\pm\theta$, thickness $4h_0$) under a uniform axial strain. [From Pipes and Pagano (1970), used with permission.]

It turns out that the assumption of a generalized plane stress state is quite valid in the interior of the laminate, that is, away from the free edges. At and near the free edges (extending a distance approximately equal to the laminate thickness) there exists, in fact, a three-dimensional state of stress. Under certain circumstances, there can be rather large interlaminar stresses present at the free edges, which can lead to delamination of plies or matrix cracking at the free edges and thereby cause failure. Researchers studied these aspects quite extensively and have clarified a number of issues. Pipes and Pagano [1970] considered a four-ply laminate, $\pm\theta$ and thickness $4h_0$, under a uniform axial strain as shown in Figure 11.9. They used a finite difference method to obtain the numerical results for a carbon/epoxy composite system having plies at $\pm 45°$. The classical lamination theory states that in each ply there exists a state of plane stress with σ_x as the axial component and σ_{xy} ($=\sigma_s$) as the in-plane shear stress component. As per the lamination theory, the stress components vary from layer to layer, but they are constant within each layer. This is correct for an infinitely wide laminate. It is incorrect for a finite-width laminate because the in-plane shear stress must vanish at the free edge surface. Figure 11.9 shows the stress distribution at the interface $z = h_0$ as a function of y/b, where $2b$ is the laminate width. The in-plane shear stress $\sigma_{xy}(=\sigma_s)$ converges to the value predicted by the lamination theory for $y/b < 0.5$, that is, away from the free edge. The axial stress component σ_x is also in accord with the lamination theory prediction for $y/b < 0.5$. The

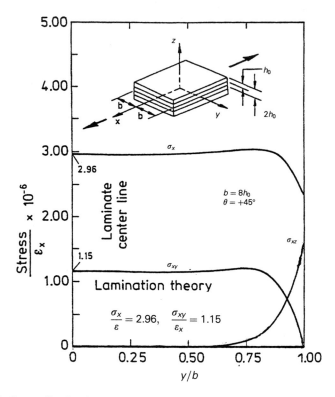

Fig. 11.10. Stress distribution at the interface $z = h_0$ as a function of y/b, where $2b$ is the laminate width. Note the high value of σ_{xy} at the free edge; σ_{xy} falls approximately to zero at $y/b = 0.5$. [From Pipes and Pagano (1970), used with permission.]

stress components σ_y, σ_z, and σ_{yz} increase near the free edge but they are quite small. The interlaminar shear stress σ_{xz} however, has a very high value at the free edge and it falls approximately to zero at $y/b = 0.5$. As can be seen from Fig. 11.10, the perturbance owing to the free edge runs through a distance approximately equal to the laminate thickness. Thus, we may regard the interlaminar stresses as a *boundary layer phenomenon* restricted to the free edge and extending inward a distance equal to the laminate thickness. Pipes and Daniel (1971) confirmed these results experimentally. They used the Moiré technique to observe the surface displacements of a symmetric angle-ply laminate subjected to axial tension. Figure 11.11 shows that the agreement between experiment and theory is excellent.

An important aspect of this phenomenon of edge effects is that the laminate stacking sequence can influence the magnitude and nature of the interlaminar stresses [Pagano and Pipes, 1971; Pipes and Pagano, 1974; Pipes et al., 1973; Whitney, 1973; Oplinger et al., 1974]. It had been observed in some earlier work that identical angle-ply laminates stacked in two different sequences had different properties: the $[\pm 15°/\pm 45°]_s$ sequence had poor

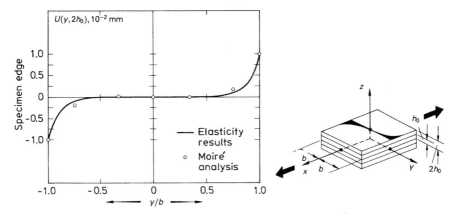

Fig. 11.11. Surface displacements of a symmetric angle-ply laminate subjected to axial tension. Experimental data points were determined by the Moiré technique. [From Pipes and Daniel (1971), used with permission.]

mechanical properties compared to the $[\pm 45°/\pm 15°]_s$ sequence. Pagano and Pipes [1971] showed that interlaminar normal stress σ_z changed from tension to compression as the ply sequence was inverted. A tensile interlaminar stress at the free edge would initiate delamination there, which would account for the observed difference in the mechanical properties. Whitney [1973] observed the same effect in carbon/epoxy composites in fatigue testing; namely, a specimen having a stacking sequence causing a tensile interlaminar stress at the free edge showed delaminations well before the fracture, while a specimen with a stacking sequence causing compressive interlaminar stress at the free edge showed little incidence of delaminations.

We can summarize the edge effects in laminated composites as follows:

1. The classical lamination theory of plates in plane stress is valid in the laminate interior, provided the laminate is sufficiently wide (i.e., $b/4h_0 \gg 2$).
2. Interlaminar stresses are confined to narrow regions of dimensions comparable to the laminate thickness and adjoining the free edges (i.e., $y = \pm b$).
3. The ply stacking sequence in a laminate affects the magnitude as well as the sign of the interlaminar stresses, which in turn affects the mechanical performance of the laminate. Specifically, a tensile interlaminar stress at the free edge is likely to cause delaminations.

References

R.M. Christensen (1979). *Mechanics of Composite Materials*, John Wiley & Sons, New York.

I.M. Daniel and O. Ishai (1994). *Engineering Mechanics of Composite Materials*, Oxford University Press, New York.

J.C. Halpin (1984). *Primer on Composite Materials*, second ed., Technornic, Lancaster, PA.

R.M. Jones (1975). *Mechanics of Composite Materials*, Scripta Book Co., Washington, DC.

A.E.H. Love (1952). *A Treatise on the Mathematical Theory of Elasticity*, fourth ed., Dover, New York.

R.L. McCullough (1971). *Concepts of Fiber-Resin Composites*, Marcel Dekker, New York, p. 16.

J.F. Nye (1969). *Physical Properties of Crystals*, Oxford University Press, London.

D.W. Oplinger, B.S. Parker, and F.P. Chiang (1974). *Exp. Mech.*, **14**, 747.

N.J. Pagano and R.B. Pipes (1971). *J. Composite Mater.*, **5**, 50.

R.B. Pipes and I.M. Daniel (1971). *J. Composite Mater.*, **5**, 255.

R.B. Pipes, B.E. Kaminski, and N.J. Pagano (1973). In *Analysis of the Test Methods for High Modulus Fibers and Composites*, **ASTM STP 521**, ASTM, Philadelphia, p. 218.

R.B. Pipes and N.J. Pagano (1970). *J. Composite Mater.*, **4**, 538.

R.B. Pipes and N.J. Pagano (1974). *J. Appl. Mech.*, **41**, 668.

S. Timoshenko and J.N. Goodier (1951). *Theory of Elasticity*, McGraw-Hill, New York.

S.W. Tsai and H.T. Hahn (1980). *Introduction to Composite Materials*, Technomic, Westport, CT.

J.M. Whitney (1973). In *Analysis of the Test Methods for High Modulus Fibers and Composites*, **ASTM SW 521**, ASTM, Philadelphia, p. 167.

Suggested Reading

L.R. Calcote (1969). *Analysis of Laminated Composite Structures*, Van Nostrand Reinhold, New York.

I.M. Daniel and O. Ishai (1994). *Engineering Mechanics of Composite Materials*, Oxford University Press, New York.

C.T. Herakovich (1998). *Mechanics of Fibrous Composites*, John Wiley & Sons, New York.

M.W. Hyer (1997). *Stress Analysis of Fiber-Reinforced Composite Materials*, McGraw-Hill, New York.

R.M. Jones (1999). *Mechanics of Composite Materials*, 2nd ed., Taylor & Francis, Philadelphia.

S.W. Tsai and H.T. Hahn (1980). *Introduction to Composite Materials*, Technomic, Westport, CT.

CHAPTER 12

Monotonic Strength and Fracture

In this chapter we describe the monotonic strength and fracture behavior of composites at ambient temperatures. The term *monotonic behavior* means behavior under an applied stress that increases in one direction, i.e., not a cyclic loading condition. We discuss the behavior of composites under fatigue or cyclic loading as well as under conditions of creep in Chapter 13.

12.1 Tensile Strength of Unidirectional Fiber Composites

In Chapter 10 we discussed the prediction of elastic and thermal properties when the component properties are known. A particularly simple but crude form of this is the rule-of-mixtures, which works reasonably well for predicting the longitudinal elastic constants. Unfortunately, the same cannot be said for the strength of a fiber composite. It is instructive to examine why the rule-of-mixtures approach does not work for strength properties. For a composite containing continuous fibers, unidirectionally aligned and loaded in the fiber direction (isostrain condition), we can write for the stress in the composite

$$\sigma_c = \sigma_f V_f + \sigma_m(1 - V_f) \tag{12.1}$$

where σ is the axial stress, V_f is the volume fraction, and the subscripts c, f, and m refer to composite, fiber, and matrix, respectively. The important question here is: What is the value of the matrix stress, σ_m? Ideally, it should be the in situ value of the matrix flow stress at a given strain. The main reason that the rule-of-mixtures does not always work for predicting the strength of a composite, while it works reasonably well for Young's modulus in the longitudinal direction, is that the elastic modulus is a relatively microstructure-insensitive property while strength is a highly microstructure-sensitive property. For example, the grain size of a polycrystalline material affects its strength but not its modulus. Thus, the response of a composite for elastic modulus is nothing but the volume-weighted average of the individual responses of the isolated components. Because the strength is an extremely structure-sensitive property, synergism can occur in regard to composite strength. Consider the various factors that may influence the composite strength properties. First, matrix or fiber structure may be altered during fabrication. Second, composite materials consist of two components whose thermomechanical properties are generally quite different and thus may have residual stresses and/or undergo structure alterations owing to the internal stresses. We discussed at length in Chapter 10 the effects of differential contraction during cooling from the fabrication temperature to ambient temperature, which leads to thermal stresses large enough to deform the matrix plastically and work-harden it (Chawla and Metzger, 1972; Chawla, 1973a, 1973b; Arsenault and Fisher, 1983; and Vogelsang et al., 1986).

Yet another source of microstructural modification of a component is a phase transformation induced by the fabrication process. In a metallic laminate composite made by roll-bonding aluminum and austenitic stainless steel (type 304), it was observed that the fabrication procedure work-hardened the steel and partially transformed the austenite to martensite (Chawla et al., 1986).

The matrix stress state may also be influenced by rheological interaction between the components during straining (Ebert and Gadd, 1965; Kelly and Lilholt, 1969). The plastic constraint on the matrix owing to the large differences in the Poisson's ratios of the matrix and the fiber, especially in the stage wherein the fiber deforms elastically while the matrix deforms plastically, can considerably alter the stress state in the composite. Thus, microstructural changes in one or both of the components, or rheological interaction between the components during straining can lead to the phenomenon of synergism in the strength properties. In view of this, the rule-of-mixtures should be regarded, in the best of circumstances, as an order of magnitude indicator of the strength of a composite. Nevertheless, it is instructive to consider this lower bound on the composite strength behavior. We ignore any negative deviations from the rule of mixtures caused by any fiber misalignment or the formation of a reaction product between the fiber and matrix. We also assume that the components do not interact during straining and that their properties in the composite state are the same as those in the isolated state.

We can show, following Kelly and Davies (1965), that a composite must have a certain minimum fiber (continuous) volume fraction V_{min} for the composite to show a real fiber reinforcement. Assuming that the fibers are all identical and uniform, that is, all fibers have the same ultimate tensile strength, we can calculate the composite ultimate strength σ_{cu} that will be ideally attained at a strain at which fibers fracture. Thus, we can write from Eq. (12.1)

$$\sigma_{cu} = \sigma_{fu} V_f + \sigma'_m (1 - V_f) \quad V_f > V_{min} \qquad (12.2)$$

where σ_{fu} is the fiber ultimate tensile stress in the composite, σ'_m is the matrix stress at the strain corresponding to the fiber ultimate tensile strength, and $V_f + V_m = 1$. At low fiber volume fractions, a work-hardened metallic

matrix can counterbalance the loss of load-carrying capacity as a result of
fiber fracture. At such low V_fs, the matrix controls the composite strength. If
all the fibers break at the same time, we must satisfy the following relationship in order to have real fiber strengthening:

$$\sigma_{cu} = \sigma_{fu} V_f + \sigma'_m (1 - V_f) \geq \sigma_{mu} (1 - V_f) \tag{12.3}$$

where σ_{mu} is the matrix ultimate tensile strength. The equality in this expression gives the minimum fiber volume fraction $V_{\min}$ that must be exceeded to
have real reinforcement, i.e., fiber strengthening. Thus,

$$V_{\min} = \frac{\sigma_{mu} - \sigma'_m}{\sigma_{fu} + \sigma_{mu} - \sigma'_m} \tag{12.4}$$

Note that the value of $V_{\min}$ increases with decreasing fiber strength.

In reality, we want the composite strength to be more than the matrix
ultimate strength in isolation. For this to be true, we can define a critical
fiber volume fraction V_{crit} that must be exceeded. Thus,

$$\sigma_{cu} = \sigma_{fu} V_f + \sigma'_m (1 - V_f) \geq \sigma_{mu} \tag{12.5}$$

The equality in Eq. (12.5)

$$V_{crit} = \frac{\sigma_{mu} - \sigma'_m}{\sigma_{fu} - \sigma'_m} \tag{12.6}$$

V_{crit} increases with increasing degree of matrix work-hardening $(\sigma_{mu} - \sigma'_m)$.
Figure 12.1 shows graphically the determination of $V_{\min}$ and V_{crit}. Note that
V_{crit} is always greater than $V_{\min}$.

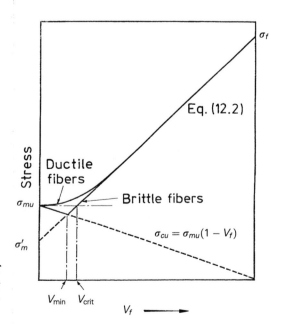

Fig. 12.1. Determination of
minimum and critical fiber
volume fractions for fiber reinforcement.

12.2   Compressive Strength of Unidirectional Fiber Composites

Fiber composites under compressive loading can be regarded, as a first approximation, as elastic columns under compression. Thus, the main failure modes in the failure of a composite are the ones that occur in the buckling of columns. Buckling occurs when a slender column under compression becomes unstable against lateral movement of the central portion. The critical stress corresponding to failure by buckling of a column is given by

$$\sigma_c = \frac{\pi^2 E}{16} \left(\frac{d}{l}\right)^2$$

where d is the diameter and l is the length of the column. It is easy to see that a high aspect ratio (l/d) will result in a low σ_c. Of course, in a fiber-reinforced composite we do not load a fiber directly. The matrix provides some stability in the lateral direction. Rosen (1965a) showed by means of photoelasticity that fiber composites fail by periodic buckling of the fibers, with the buckling wavelength being proportional to the fiber diameter. This is not surprising in view of the fact that in the analysis of a column on an elastic foundation, it is observed that the buckling wavelength depends on the column diameter. Figure 12.2 shows schematically the three situations: an unbuckled fiber composite, in-phase buckling, and out-of-phase buckling. The in-phase buckling of fibers involves shear deformation of the matrix. In such a case the composite strength in compression is proportional to the matrix shear modulus G_m; that is, $\sigma_c = G_m/V_m$, where V_m is the matrix volume fraction. For an isotropic matrix we have $G_m = E_m/2(1 + v_m)$, where E_m and v_m are the matrix Young's modulus and Poisson's ratio, respectively. Thus,

$$\sigma_c = \frac{E_m}{2(1 + v_m) V_m} \qquad (12.7)$$

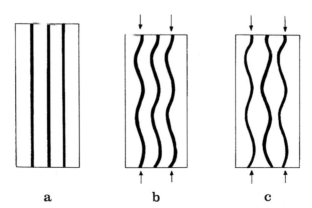

Fig. 12.2. a Unbuckled fiber composite, **b** inphase buckling of fibers, and **c** out-of-phase buckling of fibers.

Out-of-phase buckling of fibers involves transverse compressive and tensile strains. The compressive strength in such a case is proportional to the geometric mean of the fiber and matrix Young's moduli (Rosen, 1965a):

$$\sigma_c = 2V_f \left(\frac{V_f E_m E_f}{3V_m}\right)^{1/2} \tag{12.8}$$

where V and E denote the volume fraction and Young's modulus, respectively, and the subscripts f and m denote fiber and matrix, respectively.

From Eqs. (12.7) and (12.8), we can see that the two failure modes in compression have a different dependence on the moduli of the components, that is,

$$\sigma_c \propto (E_m E_f)^{1/2} \quad \text{out-of-phase}$$
$$\sigma_c \propto G_m \quad \text{in-phase} \tag{12.9}$$

Thus, if we were to put the same fiber in two different matrices (i.e., with different matrix moduli), we should be able to distinguish between these two compressive failure modes. Lager and June (1969) did just that with boron fibers in two different polymer matrices. An out-of-phase buckling mode predominated at low fiber volume fractions. At high fiber volume fractions, fibers exerted more influence on each other and a coupled or in-phase buckling mode prevailed. The approximate nature of Eqs. (12.7) and (12.8) is easy to see. Both imply that as $V_m \rightarrow 0$ (or $V_f \rightarrow 1$), $\sigma_c \rightarrow \infty$, that is, the fibers are infinitely strong. Of course, no fibers are infinitely strong. Fiber/matrix adhesion (Hancox, 1975) and matrix yielding (Piggott and Harris, 1980) also affect the compressive strength of fiber composites. In the case of a laminated composite, poor interlaminar bonding can result in easy buckling of fibers (Piggott, 1984). Heat and moisture as well as the ply stacking sequence can also affect the compressive strength. Budiansky and Fleck (1993) proposed a fiber microbuckling model which predicts a kinking stress as a function of the shear modulus of the composite, shear yield strain, and the strain hardening exponent. Gupta and coworkers (Gupta et al., 1994; Anand et al., 1994; Grape and Gupta, 1995a, 1995b) examined the behavior under uniaxial and biaxial compression of laminated carbon/carbon and carbon/polyimide composites. Carbon/carbon laminates under uniaxial showed the formation of a diagonal shear fault, which consisted of a mixture of fiber bundle kinks and interply delaminations. Carbon/polyimide composites under in-plane biaxial loading showed a new failure mechanism where the shear in the off-axis plies led to axial interply delaminations which were very similar to wing cracks observed in deformation of brittle materials under compression. Compressive strength of MMCs is generally higher than that of PMCs. Dève (1997) observed a compressive strength ≥ 4 GPa for aluminum alloy matrix containing high fiber volume fraction (55–65%) Nextel 610 alumina fibers. Fiber microbuckling was the observed failure mode. For a review of various models and experimental techniques related to compressive failure of fiber reinforced composites, see Schultheisz and Waas (1996) and Waas and Schultheisz (1996).

12.3   Fracture Modes in Composites

A great variety of deformation modes can lead to failure of the composite. The operative failure mode depends, among other things, on loading conditions and the microstructure of a particular composite system. By *microstructure*, we mean fiber diameter, fiber volume fraction, fiber distribution, and damage resulting from thermal stresses that may develop during fabrication and/or in service. In view of the fact that many factors can and do contribute to the fracture process in composites, it is not surprising that a multiplicity of failure modes is observed in a given composite system.

12.3.1   Single and Multiple Fracture

In general, the fiber and matrix will have different values of strain at fracture. When the component that has the smaller breaking strain fractures, for example, a brittle fiber or a brittle ceramic matrix, the load carried by this component is thrown onto the other one. If the component with a higher strain of fracture can bear this additional load, the composite will show multiple fracture of the brittle component. A manifestation of this phenomenon is fiber bridging of the ceramic matrix (see Chapter 7). Eventually, a particular transverse section of the composite becomes so weak that the composite is unable to carry the load any further and it fails.

Consider the case of a fiber reinforced composite in which the fiber fracture strain is less than that of the matrix, for example, a ceramic fiber in a metallic matrix. The composite will then show a single fracture (Hancox, 1975) when

$$\sigma_{fu} V_f > \sigma_{mu} V_m - \sigma'_m V_m \tag{12.10}$$

where σ'_m is the matrix stress corresponding to the fiber fracture strain and σ_{fu} and σ_{mu} are the ultimate tensile stresses of the fiber and matrix, respectively. Equation (12.10) states that when the fibers break, the matrix will not be able to support the additional load. This is commonly the case with composites containing a large quantity of brittle fibers in a ductile matrix. All the fibers break in more or less one plane and the composite also fails in that plane.

If we have a composite that satisfies the condition

$$\sigma_{fu} V_f < \sigma_{mu} V_m - \sigma'_m V_m \tag{12.11}$$

then the fibers will be broken into small segments until the matrix fracture strain is reached.

In the case where the fibers have a fracture strain greater than that of the matrix (e.g., a ceramic matrix reinforced with ductile fibers), we will have multiple fractures in the matrix. We can write the expression for this as (Hale and Kelly, 1972)

$$\sigma_{fu} V_f < \sigma_{mu} V_m + \sigma_f' V_f \qquad (12.12)$$

where σ_f' is the stress in the fiber at the matrix fracture strain.

12.3.2   Debonding, Fiber Pullout, and Delamination Fracture

Debonding, fiber pullout, and delamination fracture are some of the features that are commonly observed in fiber reinforced composites and that are not observed in monolithic materials. Consider the situation wherein a crack originates in the matrix and approaches the fiber/matrix interface. In a short fiber composite with a critical length l_c, fibers with extremities within a distance $l_c/2$ from the plane of the crack will debond and pull out of the matrix (see Fig. 10.13). These fibers will not break. In fact, the fraction of fibers pulling out will be l_c/l. Continuous fibers ($l > l_c$) invariably have flaws distributed along their length. Thus, some of them may fracture in the plane of the crack while others may fracture away from the crack plane. This is treated in some detail later in this section. The final fracture of the composite will generally involve some fiber pullout. Consider a model composite consisting of a fiber of length l embedded in a matrix; see Figure 12.3. If this fiber is pulled out, the adhesion between the matrix and the fiber will produce a shear stress τ parallel to the fiber surface. The shear force acting on the fiber as a result of this stress is given by $2\pi r_f \tau l$, where r_f is the fiber radius. Let τ_i be the maximum shear stress that the interface can support and let σ_{fu} be the fiber fracture stress in tension. The maximum force caused by this normal stress on the fiber is $\pi r_f^2 \sigma_{fu}$. For maximum fiber strengthening, we would like the fiber to break rather than get pulled out of the matrix. From a toughness point of view, however, fiber pullout may be more desirable. We can then write from the balance of forces the following condition for the

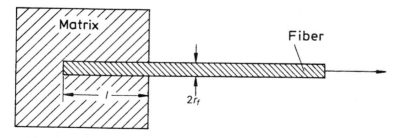

Fig. 12.3. A fiber of length l, embedded in a matrix, being pulled out.

fiber to be broken:

$$\pi r_f^2 \sigma_{fu} < 2\pi r_f \tau_i l$$

or

$$\frac{\sigma_{fu}}{4\tau_i} < \frac{l}{2r_f} = \frac{l}{d} \qquad (12.13)$$

where d is the fiber diameter and the ratio l/d is the aspect ratio of the fiber. On the other hand, for fiber pullout to occur, we can write

$$\frac{l}{d} \leq \frac{\sigma_{fu}}{4\tau_i} \qquad (12.14)$$

The equality in this expression gives us the critical fiber length l_c for a given fiber diameter. Thus,

$$\frac{l_c}{d} = \frac{\sigma_{fu}}{4\tau_i} \qquad (12.15)$$

This equation provides us with a means of obtaining the interface strength, namely, by embedding a single fiber in a matrix and measuring the load required to pull the fiber out. The load-displacement curve shows a peak corresponding to debonding, followed by an abrupt fall and wiggling about a constant stress level. Note that Eq. (12.15) gives l_c/d to be half that given by Eq. (10.65). This is because in the current case the fiber is being loaded from one end only.

A point that has not been mentioned explicitly so far is that real fibers do not have uniform properties but rather show a statistical distribution. Weak points are distributed along the fiber length. We will treat these statistical aspects of fiber strength in detail in Section 12.4. Suffice it to say that it is more than likely that a fiber would break away from the main fracture plane. Interfacial debonding occurs around the fiber breakpoint. The broken fiber parts are pulled out from their cylindrical holes in the matrix during further straining. Figure 12.4a shows schematically the fiber pullout in a continuous fiber composite, while a practical example of fiber pullout in a boron fiber/ aluminum matrix is shown in Figure 12.4b. Work is done in the debonding process as well as in fiber pullout against frictional resistance at the interface. Outwater and Murphy (1969) showed that the maximum energy required for debonding is given by

$$W_d = \left(\frac{\pi d^2}{24}\right)\left(\frac{\sigma_{fu}^2}{E_f}\right) x \qquad (12.16)$$

where x is the debond length.

Cottrell (1964) pointed out the importance of fiber pullout in regard to toughness. The length l should be large but close to l_c for maximizing the fiber pullout work and to prevent the composite from separating into two

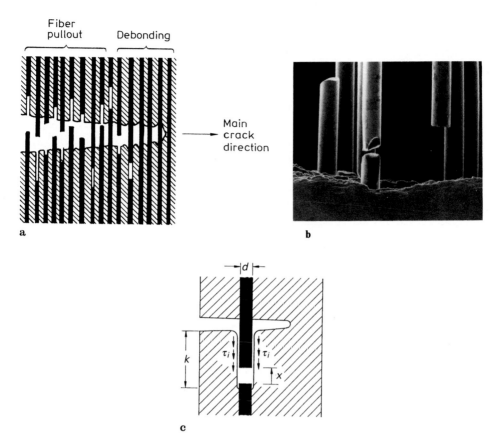

Fig. 12.4. a Schematic of fiber pullout in a continuous fiber composite. **b** Fiber pullout in a B/Al system. **c** Schematic of an isolated fiber pullout through a distance x against an interfacial shear stress, τ_i.

halves. It should be recognized at the same time that for $l < l_c$, the fiber will not get loaded to its maximum possible strength level and thus full fiber strengthening potential will not be realized.

We can estimate the work done in pulling out an isolated fiber in the following way (Fig. 12.4c). Let the fiber be broken a distance k below the principal crack plane, where $0 < k < l_c/2$. Now let the fiber be pulled out through a distance x against an interfacial frictional shear stress, τ_i. Then the total force at that instant on the debonded fiber surface, which is opposing the pullout, is $\tau_i \pi d(k - x)$. When the fiber is further pulled out a distance dx, the work done by this force is $\tau_i \pi d(k - x)\, dx$. We can obtain the total work done in pulling out the fiber over the distance k by integrating. Thus,

$$\text{Work of fiber pullout} = \int_0^k \tau_i \pi\, d(k - x)\, dx = \frac{\tau_i \pi\, dk^2}{2} \qquad (12.17a)$$

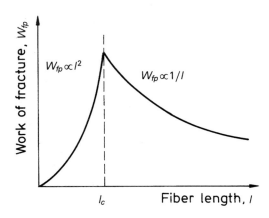

Fig. 12.5. Variation of fiber pull-out work with fiber length.

Now the pullout length of the fiber can vary between a minimum of 0 and a maximum of $l_c/2$. The average work of pullout per fiber is then

$$W_{fp} = \frac{1}{l_c/2} \int_0^{l_c/2} \frac{\tau_i \pi \, dk^2}{2} \, dk = \frac{\tau_i \pi \, dl_c^2}{24}$$

This analysis assumes that all fibers are pulled out. In a discontinuous fiber composite, it has been observed experimentally (Kelly, 1970) that fibers with ends within a distance $l_c/2$ of the main fracture plane and that cross this fracture plane suffer pullout. Thus, it is more likely that a fraction (l_c/l) of fibers will pull out. The average work done per fiber can be written

$$W_{fp} = \left(\frac{l_c}{l}\right) \frac{\pi \, d\tau_i l_c^2}{24} \tag{12.17b}$$

In general, fiber pullout provides a more significant contribution to composite fracture toughness than fiber/matrix debonding. The reader should appreciate the fact that debonding must precede pullout. Figure 12.5 shows schematically the variation of work of fracture with fiber length. For $l < l_c$, W_{fp} increases as l^2 (substitute l for k in Eq. 12.17a). Physically, this makes sense because as the fiber length increases, an increasing fiber length will be pulled out. For $l > l_c$, as pointed out earlier, some fibers will fracture in the plane of fracture of the composite and thus their contribution to W_{fp} will be nought. Those fibers whose ends are within a distance of $l_c/2$ from the fracture plane, they will undergo the process of fiber pullout. In this case, the average work of fracture is given by Eq. 12.17b, i.e., W_{fp} varies as $1/l$ because with increasing fiber length, fiber breaks intervene and the fiber pullout decreases. The work of fracture, W_{fp} peaks at $l = l_c$. The reader can easily show that for $l = l_c$, the average stress in the fiber will only be half that in an infinitely long fiber.

One of the attractive characteristics of composites is the possibility of obtaining an improved fracture toughness behavior together with high strength. Fracture toughness can loosely be defined as resistance to crack propagation. Consider a fibrous composite containing a crack transverse to the fibers. We can then increase the crack propagation resistance by one of the following means, each of which involves additional work:

1. Plastic deformation of the matrix (applicable in a metal matrix).
2. Fiber pullout.
3. Presence of weak interfaces, fiber/matrix separation, and deflection of the crack.

For a metal matrix composite the work of fracture is mostly the work done during plastic deformation of the matrix. The work of fracture is proportional to $d(V_m/V_f)^2$ (Cooper and Kelly, 1967), where d is the fiber diameter and V_m and V_f are the matrix and fiber volume fractions, respectively. This is understandable inasmuch as in the case of large-diameter fibers, for a given V_f, the advancing crack will have to pass through a greater plastic zone of the matrix and will result in a larger work of fracture.

Fiber pullout increases the work of fracture by causing a large deformation before fracture. In this case, the controlling parameter for work of fracture is the ratio d/τ_i, where d is the fiber diameter and τ_i is the interface shear strength. In the case of short fibers, the work of fracture resulting from fiber pullout also increases with the fiber length, reaching a maximum at l_c. In the case of continuous fibers, the work of fracture increases with an increase in spacing between the defects (Kelly, 1971; Cooper, 1970). It would thus appear that one can increase the work of fracture by increasing the fiber diameter. This was discussed in Section 6.5 in regard to the toughness of metal matrix composites.

Crack deflection along an interface frequently follows separation of the fiber/matrix interface. This provides us with a potent mechanism of increasing the crack propagation resistance in composites; we discussed this topic in Chapter 7. This improvement in fracture toughness owing to the presence of weak interfaces has been confirmed experimentally. This crack deflection mechanism can be a major source of toughness in ceramic matrix composites (Chawla, 1993). Yet another related failure mode in laminated composites is the delamination failure associated with the plies and the fiber/matrix interface. This fracture mode is of importance in structural applications involving long-term use, for example, under fatigue conditions and where environmental effects are important. Highly oriented fibers such as aramid can also contribute to the work of fracture. Figure 12.6 shows a delamination-type fracture in a Kevlar aramid/epoxy composite (Saghizadeh and Dharan, 1985) and the characteristic fibrillation of the Kevlar fiber, which stems from its structure as described in Chapter 2. Carbon/epoxy composites, when tested for delamination fracture, showed clean exposed fiber surfaces (Saghizadeh and Dharan, 1985). For details regarding this subject of delamination, the reader is referred to Johnson (1985) in Suggested Reading.

Fig. 12.6. A delamination-type fracture of Kevlar/epoxy composite. Note the characteristic fibrils. [From Saghizadeh and Dharan (1985), used with permission.]

12.4 Effect of Variability of Fiber Strength

Most high-performance fibers are brittle. Thus, their strength must be characterized by a statistical distribution function. Figure 12.7a shows the strength distribution for a material showing statistical variation in strength, while Figure 12.7b shows what is called a *Dirac delta distribution*, where the variability of strength is insignificant. We can safely put most metallic wires in the latter category. Most high-strength and high-stiffness fibers (aramid, polyethylene, B, C, SiC, Al_2O_3, etc.), however, follow some kind of statistical distribution of strength.

The Weibull statistical distribution function has been found to characterize the strength of brittle materials fairly well. For high-strength fibers also, the Weibull treatment of strength has been found to be quite adequate (Coleman, 1958). Here we follow the treatment, due to Rosen (1965a, 1965b, 1983), of this fiber-strength variability problem. We can express the dependence of fiber strength on its length in terms of the following distribution function:

$$f(\sigma) = L\alpha\beta\sigma^{\beta-1}\exp(-L\alpha\sigma^\beta) \tag{12.18}$$

where L is the fiber length, σ is the fiber strength, and α and β are statistical

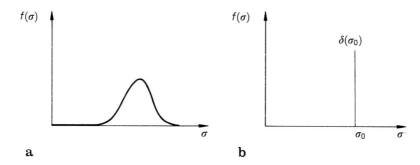

Fig. 12.7. a Strength distribution for a brittle material. **b** Strength distribution (Dirac delta) for material with insignificant variability of strength.

parameters. $f(\sigma)$, a probability density function, gives the probability that the fiber strength is between σ and $\sigma + d\sigma$.

We define the kth moment M_k of the statistical distribution function as

$$M_k = \int_0^\infty \sigma^k f(\sigma)\, d\sigma \qquad (12.19)$$

Knowing that the mean strength of the fiber is given by $\bar\sigma = \int_0^\infty \sigma f(\sigma)\, d\sigma$, we can write

$$\bar\sigma = M_1 \qquad (12.20)$$

and the standard deviation s can be expressed as

$$s = (M_2 - M_1^2)^{1/2} \qquad (12.21)$$

From the Weibull distribution [Eq. (12.18)] and Eqs. (12.20) and (12.21), we obtain

$$\bar\sigma = (\alpha L)^{-1/\beta} \Gamma\left(1 + \frac{1}{\beta}\right) \qquad (12.22)$$

and

$$s = (\alpha L)^{-1/\beta} \left[\Gamma\left(1 + \frac{2}{\beta}\right) - \Gamma^2\left(1 + \frac{1}{\beta}\right) \right]^{1/2} \qquad (12.23)$$

where $\Gamma(n)$ is the gamma function given by $\int_0^\infty \exp(-x) x^{n-1}\, dx$. The coefficient of variation μ for this distribution then follows from

$$\mu = \frac{s}{\bar\sigma} = \frac{[\Gamma(1 + 2/\beta) - \Gamma^2(1 + 1/\beta)]^{1/2}}{\Gamma(1 + 1/\beta)} \qquad (12.24)$$

We note that μ is a function only of the parameter β. Rosen has shown that for $0.05 \le \mu \le 0.5$, $\mu \simeq \beta^{-0.92}$ or $\mu \simeq 1/\beta$. In other words, β is an inverse measure of the coefficient of variation μ. For fibers characterized by the

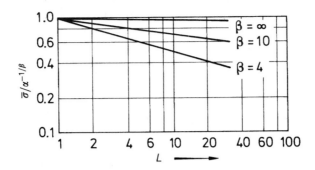

Fig. 12.8. Normalized mean fiber strength versus fiber length L.

Weibull distribution [Eq. (12.18)], $\beta > 1$. For glass fibers, μ can be about 0.1, which would correspond to $\beta = 11$. For boron and SiC fibers, μ can be between 0.2 and 0.4 and β will be between 2.7 and 5.8.

From Eq. (12.22), we can write for a unit length of fiber

$$\bar{\sigma}_1 = k\alpha^{-1/\beta} \tag{12.25}$$

where

$$k = \Gamma\left(1 + \frac{1}{\beta}\right) \tag{12.26}$$

For $\beta > 1$, we have $0.88 \le k \le 1.0$. Thus, we can regard the quantity $\alpha^{-1/\beta}$ as the reference strength level. We can plot Eq. (12.22) in the form of curves of $\bar{\sigma}/\alpha^{-1/\beta}$ (a normalized mean strength) against fiber length L for different β values. In Figure 12.8, $\beta = \infty$ corresponds to a spike distribution function, that is, the Dirac delta function. In such a case, all the fibers have identical strength and there is no fiber length dependence. For $\beta = 10$, which corresponds to a $\mu \simeq 12\%$, an order-of-magnitude increase in fiber length produces a 20% fall in average strength. For $\beta = 4$, the corresponding fall in strength is about 50%.

If we differentiate Eq. (12.18) and equate it to zero, we obtain the statistical mode σ^*, which is the most probable strength value. Thus,

$$\sigma^* \simeq \left(\frac{\beta - 1}{\beta}\right)^{1/\beta} (\alpha L)^{-1/\beta}$$

For large β,

$$\sigma^* \simeq (\alpha L)^{-1/\beta}$$

$(\alpha L)^{-1/\beta}$, as mentioned earlier, is a reference stress level. The values of α and β can be obtained from experimental $\bar{\sigma}$ and μ values.

There is yet another important statistical point with regard to this variability of fiber strength. This has to do with the fact that in a unidirectionally aligned fiber composite, the fibers act in a bundle in parallel. It turns out that the strength of a bundle of fibers whose elements do not possess a uniform strength is not the average strength of the fibers. Coleman (1958) has investigated this nontrivial problem.

In the simplest case, we assume that all fibers have the same cross section and the same stress-strain curve but with different strain-to-fracture values. If the strength distribution function is $f(\sigma)$, the probability that a fiber will break before a certain value σ is attained is given by the cumulative strength distribution function $F(\sigma)$. We can write

$$F(\sigma) = \int_0^\sigma f(\sigma) \, d\sigma \tag{12.27}$$

One makes a large number of measurements of strength of individual packets or bundles of fibers. Each bundle has the same large number of fibers of identical cross section and they are loaded from their extremities. From this we can find the mean fiber strength in the bundle. Daniels (1945) showed that, for a very large number of fibers in the bundle, the distribution of the mean fiber strength at bundle failure σ_B attains a normal distribution, with the expectation value being given by

$$\bar{\sigma}_B = \sigma_{fu}[1 - F(\sigma_{fu})] \tag{12.28}$$

The maximum fiber strength σ_{fu} corresponds to the condition where the bundle supports the maximum load. Thus, σ_{fu} is obtained from

$$\frac{d}{d\sigma}\{\sigma[1 - F(\sigma)]\}_{\sigma=\sigma_{fu}} = 0 \tag{12.29}$$

σ_B values are characterized by the following density function for normal distribution function:

$$\omega(\sigma_B) = \frac{1}{\Psi_B\sqrt{2\pi}} \exp\left[-\frac{1}{2}\left(\frac{\sigma_B - \bar{\sigma}_B}{\Psi_B}\right)^2\right] \tag{12.30}$$

where Ψ_B is the standard deviation given by

$$\Psi_B = \sigma_{fu}\{F(\sigma_{fu})[1 - F(\sigma_{fu})]\}^{1/2}N^{-1/2} \tag{12.31}$$

where N is the number of fibers in the bundle.

As N becomes very large, not unexpectedly, the standard variation Ψ_B becomes small. That is, the larger the number of fibers in the bundle, the more reproducible is the bundle strength. For bundles characterized by Eq. (12.30), we can define a cumulative distribution function $\Omega(\sigma_B)$. Thus,

$$\Omega(\sigma_B) = \int_0^{\sigma_B} \omega(\sigma_B) \, d\sigma_B \tag{12.32}$$

Considering the Weibull distribution [Eq. (12.18)], we have [from Eq.

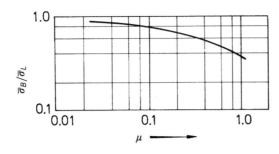

Fig. 12.9. Normalized fiber bundle strength versus variance μ of the fiber population.

(12.29)]

$$\sigma_{fu} = (L\alpha\beta)^{-1/\beta} \tag{12.33}$$

and from Eq. (12.28), it follows that

$$\bar{\sigma}_B = (L\alpha\beta e)^{-1/\beta} \tag{12.34}$$

Comparing this mean fiber bundle strength value [Eq. (12.34)] to the mean value of the fiber strength obtained from equal-length fibers tested individually [Eq. (12.22)], we note that when there is no dispersion in fiber strength, the mean bundle strength equals the mean fiber strength; see Fig. 12.9. As the coefficient of variation of fiber strength increases above zero, the mean bundle strength decreases and, in the limit of an infinite dispersion, tends to zero. For a 10% variance, the mean bundle strength is about 80% of the mean fiber strength, while for a variance of 25%, the bundle strength is about 60% of the mean fiber strength.

In view of the statistical distribution of fiber strength, it is natural to extend these ideas to composite strength. We present here the treatment introduced by Rosen and coworkers (1965b, 1983). On straining a fiber composite, fibers fracture at various points before a complete failure of the composite. There occurs an accumulation of fiber fractures with increasing load. At a certain point, one transverse section will be weakened as a result of the statistical accumulation of fiber fractures, hence the name *cumulative weakening model of failure*. Because the fibers have nonuniform strength, it is expected that some fibers will break at very low stress levels. Figure 12.10 shows the perturbation in stress state when a fiber fracture occurs. At the point of fiber fracture, the tensile stress in the fiber drops to zero. From this point the tensile stress in the fiber increases along the fiber length along the two fiber segments as per the load transfer mechanism by interfacial shear described in Chapter 10. These local stresses can be very large. As a result, either the matrix would yield in shear or an interfacial failure would occur. Additionally, this local drop in stress caused by a fiber break will throw the load onto adjoining fibers, causing stress concentrations there (see Fig. 12.11). Upon continued straining, progressive fiber fractures cause a cumu-

Fig. 12.10. Perturbation of stress state caused by a fiber break.

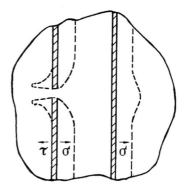

lative weakening and a redistribution of the load in the composite. After the first fiber fracture, the interfacial shear stresses may cause delamination between this broken fiber and the matrix, as shown in Figure 12.11a. When this happens, the broken and delaminated fiber becomes totally ineffective and the composite behaves as a bundle of fibers. The second alternative is that a crack, starting from the first fiber break, propagates through the other fibers in a direction normal to the fibers; see Fig.12.11b. Such a situation will occur only if the fiber and matrix are very strongly bonded and if the major component is very brittle. In the absence of these two modes, cumulative damage results. With increased loading, additional fiber fractures occur and a statistical distribution is obtained; see Fig. 12.11c.

Rosen (1983) considers in this model that the composite strength is controlled by a statistical accumulation of failures of individual volume elements that are separated by barriers to crack propagation; thus, these elements fail

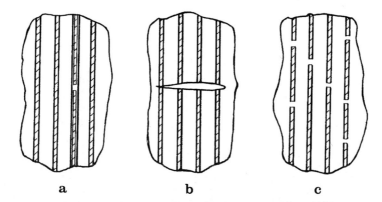

a   b   c

Fig. 12.11. Fracture models: **a** interface delamination where the composite acts as a fiber bundle, **b** first fiber fracture turns into a complete fracture, **c** cumulative damage—statistical.

independently. The load on the matrix is ignored. Increased loading leads to individual fiber breaks at loads less than the ultimate fracture load of the composite. An individual fiber break does not make the whole fiber length ineffective, it only reduces the capacity of the fiber being loaded in the vicinity of fiber failure. The stress distribution in the fiber is one of full load over its entire length less a length near the break over which the load is zero. This length is called the *ineffective length*. Thus, the composite is considered to be made up of a series of layers, each layer consisting of a packet of fiber elements embedded in the matrix. The length of the fiber element or the packet height is equal to the ineffective length.

As the load is increased, fiber breaks accumulate until at a critical load a packet of elements is unable to transmit the applied load and the composite fails. Thus, composite failure occurs because of this weakened section. A characteristic length δ corresponding to the packet height must be chosen. The term δ is defined as the length over which the stress attains a certain fraction ϕ of the unperturbed stress in the fiber. Rosen took ϕ equal to that length over which the stress increases elastically from zero (at a fiber end) to 90% of the unperturbed level. One then derives the stress distribution in the fiber elements and packets. The theory of the weakest link is then applied to obtain the composite strength. In the case of fibers characterized by the Weibull distribution, the stress distribution in the fiber elements is

$$w(\sigma) = \delta\alpha\beta\sigma^{\beta-1}\exp(-\delta\alpha\sigma^{\beta}) \tag{12.35}$$

where δ is the fiber element length (i.e., the ineffective length).

The composite is now a chain, the strength of whose elements is given by a normal distribution function $w(\sigma_B)$. The strength of a chain having m links of this population is characterized by a distribution function $\lambda(\sigma_c)$:

$$\lambda(\sigma_c) = mw(\sigma_c)[1 - \Omega(\sigma_c)]^{m-1} \tag{12.36}$$

where

$$\Omega(\sigma_c) = \int_0^{\sigma_c} w(\sigma_c)\, d\sigma \tag{12.37}$$

Suppose now that the number of elements N in a composite is so large that the standard deviation of the packet tends to zero. Then the statistical mode of composite strength is equal to $\bar{\sigma}_B$ [Eq. (12.28)].

In the case of a Weibull distribution, it follows from Eq. (12.34) that

$$\sigma_c^* = (\delta\alpha\beta e)^{-1/\beta} \tag{12.38}$$

and the statistical mode of the tensile strength of the composite is

$$\sigma^* = V_f(\delta\alpha\beta e)^{-1/\beta} \tag{12.39}$$

where V_f is the fiber volume fraction.

It should be pointed out that δ will be of the order of 10–100 fiber diameters. and thus much smaller than the gage length used for individual tests.

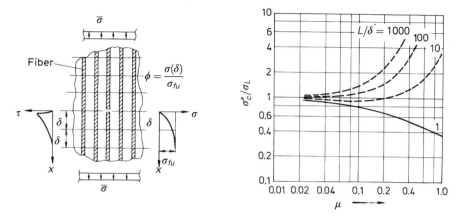

Fig. 12.12. Normalized composite strength $\sigma_C^*/\bar{\sigma}_L$ versus variance μ. [From Fiber Composite Materials, ASM, 1965, pp. 39, 38, used with permission.]

If we compare the (cumulative) average strength of a group of fibers of length L with the expected fiber strength from Eqs. (12.24) and (12.38), we get

$$\frac{\sigma_c^*}{\bar{\sigma}_L} = \left(\frac{L}{\delta\beta e}\right)^{1/\beta} \frac{1}{\Gamma(1+1/\beta)} \tag{12.40}$$

For $\beta = 5$ and $L/\delta = 100$, we have $\sigma_c^*/\bar{\sigma}_L = 1.62$; that is, the composite will be much stronger than what we expect from individual fiber tests. We can plot the composite strength σ_c^*, normalized with respect to the average strength $\bar{\sigma}_L$, of individual fibers of length L against μ, the variance of individual fiber strengths; see Fig. 12.12. The curves shown are for different values of the ratio L/δ. For $L/\delta = 1$, that is, the fiber length is equal to the ineffective length, the statistical mode of composite strength is less than the average fiber strength. This difference between the two increases with an increase in μ of the fibers. For a more realistic ratio, for example, $L/\delta > 10$, we note that the composite strength is higher than the average fiber strength.

A modification of this cumulative weakening model has been proposed by Zweben and Rosen (1970), which takes into account the redistribution of stress that results at each fiber break; that is, there is greater probability that fracture will occur in immediately adjacent fibers because of a stress magnification effect.

12.5 Strength of an Orthotropic Lamina

We saw in Chapters 10 and 11 that fiber reinforced composites are aniso-tropic in elastic properties. This results from the fact that the fibers are aligned in the matrix fibers. Additionally, the fibers are, generally, a lot stiffer and stronger than the matrix. Therefore, not unexpectedly, fiber reinforced composites also show anisotropy in strength properties. Quite frequently, the strength in the longitudinal direction is as much as an order of magnitude greater than that in the transverse direction. It is of great importance for design purposes to be able to predict the strength of a composite under the loading conditions prevailing in service. The use of a failure criterion gives us information about the strength under combined stresses. We assume, for simplicity, that the material is homogeneous; that is, its properties do not change from point to point. In other words, we treat a fiberreinforced lamina as a homogeneous, orthotropic material (Rowlands, 1985). We present a brief account of different failure criteria.

12.5.1 Maximum Stress Theory

Failure will occur when any one of the stress components is equal to or greater than its corresponding allowable or intrinsic strength. Thus, failure will occur if

$$\sigma_1 \geq X_1^T \quad \sigma_1 \leq -X_1^C$$
$$\sigma_2 \geq X_2^T \quad \sigma_2 \leq -X_2^C \tag{12.41}$$
$$\sigma_6 \geq S \quad\quad \sigma_6 \leq S$$

where X_1^T is the ultimate uniaxial tensile strength in the fiber direction, X_1^C is the ultimate uniaxial compressive strength in the fiber direction, X_2^T is the ultimate uniaxial tensile strength transverse to the fiber direction, X_2^C is the ultimate uniaxial compressive strength transverse to the fiber direction, and S is the ultimate planar shear strength. When any one of the inequalities indicated in Eq. (12.41) is attained, the material will fail by the failure mode related to that stress inequality. No interaction between different failure modes is permitted in this criterion. Consider an orthotropic lamina, that is, a unidirectional fiber reinforced prepreg subjected to a uniaxial tensile stress σ_x in a direction making an angle θ with the fiber direction. We then have, for the stress components in the 1-2 system,

$$\begin{bmatrix} \sigma_1 \\ \sigma_2 \\ \sigma_6 \end{bmatrix} = [T]_\sigma \begin{bmatrix} \sigma_x \\ \sigma_y \\ \sigma_s \end{bmatrix} \tag{12.42}$$

where

$$[T]_\sigma = \begin{bmatrix} m^2 & n^2 & 2mn \\ n^2 & m^2 & -2mn \\ -mn & mn & m^2 - n^2 \end{bmatrix} \quad m = \cos\theta, \ n = \sin\theta$$

Using the fact that only σ_x is nonzero, we obtain

$$\sigma_1 = \sigma_x m^2$$

$$\sigma_2 = \sigma_x n^2$$

$$\sigma_6 = \sigma_x mn$$

According to the maximum stress criterion, the three possible failure modes are:

$$\sigma_x = \frac{X_1^T}{m^2} \quad \text{(longitudinal tensile)}$$

$$\sigma_x = \frac{X_2^T}{n^2} \quad \text{(transverse tensile)} \qquad (12.43)$$

$$\sigma_x = \frac{S}{mn} \quad \text{(planar shear)}$$

With varying θ, the failure mode will change from longitudinal tension to planar shear to transverse tension, as shown by the dashed lines in Figure 12.13. Poor agreement with experiment, particularly around $\theta = \pi/4$, is clear from this figure. This indicates that at intermediate angles failure mode interactions do occur.

12.5.2 Maximum Strain Criterion

This criterion is analogous to the maximum stress criterion. Failure occurs when any one of the strain components is equal to or greater than its corre-

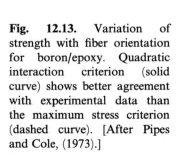

Fig. 12.13. Variation of strength with fiber orientation for boron/epoxy. Quadratic interaction criterion (solid curve) shows better agreement with experimental data than the maximum stress criterion (dashed curve). [After Pipes and Cole, (1973).]

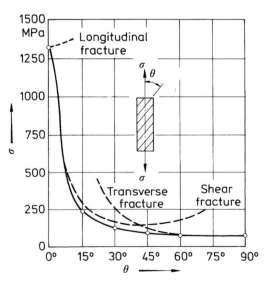

sponding allowable strain. Thus,

$$\varepsilon_1 \geq e_1^T \quad \varepsilon_1 \leq -e_1^C$$
$$\varepsilon_2 \geq e_2^T \quad \varepsilon_2 \leq -e_2^C \tag{12.44}$$
$$\varepsilon_6 \geq e_6 \quad \varepsilon_6 \leq e_6$$

where e_1^T is the ultimate tensile strain in the fiber direction, e_1^C is the ultimate compressive strain in the fiber direction, e_2^T is the ultimate tensile strain in the transverse direction, e_2^C is the ultimate compressive strain in the transverse direction, and e_6 is the ultimate planar shear strain. This criterion is also not very satisfactory.

12.5.3   Maximum Work (or the Tsai-Hill) Criterion

According to the Tsai-Hill criterion, failure of an orthotropic lamina will occur under a general stress state when

$$\frac{\sigma_1^2}{X_1^2} - \frac{\sigma_1\sigma_2}{X_1^2} + \frac{\sigma_2^2}{X_2^2} + \frac{\sigma_6^2}{S^2} \leq 1 \tag{12.45}$$

where X_1, X_2, and S are the longitudinal tensile failure strength, the transverse tensile failure strength, and the in-plane shear failure strength, respectively. If compressive stresses are involved, then the corresponding compressive failure strengths should be used.

Consider again a uniaxial stress σ_x applied to an orthotropic lamina. Then, following Eq. (12.42), we can write

$$\sigma_1 = \sigma_x m^2$$
$$\sigma_2 = \sigma_x n^2 \tag{12.46}$$
$$\sigma_6 = \sigma_x mn$$

where $m = \cos\theta$ and $n = \sin\theta$. Substituting these values in Eq. (12.45), we have

$$\frac{m^4}{X_1^2} + \frac{n^4}{X_2^2} + m^2 n^2 \left(\frac{1}{S^2} - \frac{1}{X_1^2}\right) < \frac{1}{\sigma_x^2} = \frac{1}{\sigma_\theta^2} \tag{12.47}$$

12.5.4   Quadratic Interaction Criterion

As the name indicates, this criterion takes into account the stress interactions. Tsai and Wu (1971) proposed this modification of the Hill theory for a lamina by adding some additional terms. Tsai and Hahn (1980) provide a good account of this criterion. According to this theory, the failure surface in stress space can be described by a function of the form

$$f(\sigma) = F_i\sigma_i + F_{ij}\sigma_i\sigma_j = 1 \quad i, j = 1, 2, 6 \tag{12.48}$$

where F_i and F_{ij} are the strength parameters. For the case of plane stress, i, $j = 1, 2, 6$ and we can expand Eq. (12.48) as follows:

$$F_1\sigma_1 + F_2\sigma_2 + F_6\sigma_6 + F_{11}\sigma_1^2 + F_{22}\sigma_2^2 + F_{66}\sigma_6^2 + 2F_{12}\sigma_1\sigma_2$$

$$+ 2F_{16}\sigma_1\sigma_6 + 2F_{26}\sigma_2\sigma_6 = 1 \tag{12.49}$$

For the orthotropic lamina, sign reversal for normal stresses, whether tensile or compressive, is important. The linear stress terms provide for this difference. For the shear stress component, the sign reversal should be immaterial. Thus, terms containing the first-degree shear stress must vanish. These terms are $F_{16}\sigma_1\sigma_6$, $F_{26}\sigma_2\sigma_6$, and $F_6\sigma_6$. The stress components in general are not zero. Therefore, for these three terms to vanish, we must have

$$F_{16} = F_{26} = F_6 = 0$$

Equation (12.49) is now simplified to

$$F_1\sigma_1 + F_2\sigma_2 + F_{11}\sigma_1^2 + F_{22}\sigma_2^2 + F_{66}\sigma_6^2 + 2F_{12}\sigma_1\sigma_2 = 1 \tag{12.50}$$

There are six strength parameters in Eq. (12.50). We can measure five of these by the following simple tests.

Longitudinal (Tensile and Compressive) Tests

If $\sigma_1 = X_1^T$, then $F_{11}(X_1^T)^2 + F_1 X_1^T = 1$.
If $\sigma_1 = -X_1^C$, then $F_{11}(X_1^C)^2 - F_1 X_1^C = 1$.
From these we get

$$F_{11} = \frac{1}{X_1^T X_1^C} \tag{12.51}$$

and

$$F_1 = \frac{1}{X_1^T} - \frac{1}{X_1^C} \tag{12.52}$$

Transverse (Tensile and Compressive) Tests

If X_2^T and X_2^C are the transverse tensile and compressive strengths, respectively, then proceeding as earlier, we get

$$F_{22} = \frac{1}{X_2^T X_2^C} \tag{12.53}$$

and

$$F_2 = \frac{1}{X_2^T} - \frac{1}{X_2^C} \tag{12.54}$$

Longitudinal Shear Test

If S is the shear strength, we have

$$F_{66} = \frac{1}{S^2} \qquad (12.55)$$

Thus, we can express all the failure constants except F_{12} in terms of the ultimate intrinsic strength properties. F_{12} is the only remaining parameter and it must be evaluated by means of a biaxial test, not a small inconvenience. Many workers (Hoffman, 1967; Cowin, 1979) have proposed variations of the Tsai-Wu criterion involving F_{12} explicitly in terms of uniaxial strengths. Tsai and Hahn (1980) suggest that, in the absence of other data, $F_{12} = -0.5\sqrt{F_{11}F_{22}}$. It turns out, however, that small changes in F_{12} can significantly affect the predicted strength (Tsai and Hahn, 1980). Figure 12.13 shows for the boron/epoxy system the variation of strength with orientation assuming $F_{12} = 0$ (Pipes and Cole, 1973). The intrinsic properties of this system are as follows:

$$X_1^T = 27.3\,\text{MPa} \quad X_2^T = 1.3\,\text{MPa} \quad S = 1.4\,\text{MPa}$$
$$X_1^C = 52.4\,\text{MPa} \quad X_2^C = 6.5\,\text{MPa}$$

Note the excellent agreement between the curve computed using the quadratic interaction criterion and the experiment. The agreement with the maximum stress criterion (dashed curve) is poor.

Comparison of Failure Theories

Important attributes of the four main failure theories, namely, maximum stress, maximum strain, maximum work, and quadratic interaction are compared in Table 12.1. The applicability of a given theory depends on material properties and the failure modes (Daniel and Ishai, 1994). As expected, the maximum stress and maximum strain criteria are generally valid with brittle materials. They do require three subcriteria but are conceptually quite simple and experimental determination of parameters is also quite simple and straightforward. The two interactive theories, maximum work and quadratic interaction, are more suitable for computational purposes. In particular, the quadratic interaction criterion is quite general and comprehensive. Both require more complicated experimental characterization. According to Daniel and Ishai (1994), when material behavior and failure modes are not known and when a conservative approach is required, all four criteria should be evaluated and use the most conservative envelope in each quadrant. Figure 12.14 shows the four criteria and the shaded part of the envelopes conforms to this conservative approach.

Table 12.1. Comparison of failure theories. [After Daniel and Ishai (1994).]

Criterion	Physical Basis	Computational Aspects	Experimental Characterization
Maximum stress	Tensile behavior of brittle material, no stress interaction	Inconvenient	Simple
Maximum strain	Tensile behavior of brittle material, no stress interaction	Inconvenient	Simple
Maximum work	Valid for ductile anisotropic materials, curve fitting for heterogenous brittle composites	Can be programmed, different functions required for tensile and compressive strengths	Biaxial testing needed
Quadratic interaction	Mathematically consistent, reliable curve fitting	Simple, comprehensive	Complicated, numerous parameters needed

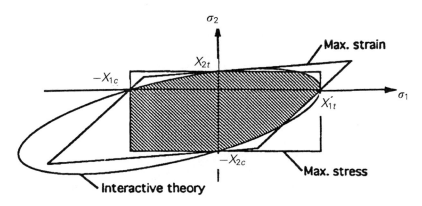

Fig. 12.14. Failure envelopes due to different failure criteria. The shaded past indicates the conservative failure region. [After Daniel and Ishai (1994).]

References

K. Anand, V. Gupta, and D. Dartford (1994). *Acta Metallurgica et Materialia,* **42,** 797.

R.J. Arsenault and R.M. Fisher (1983). *Scripta Met.,* **17,** 67.

B. Budiansky and N. Fleck (1993). *J. Mech. Phys. Solids,* **41,** 183.

K.K. Chawla (1973a). *Metallography,* **6,** 155.

K.K. Chawla (1973b). *Philos. Mag.*, **28**, 401.

K.K. Chawla (1993). *Ceramic Matrix Composites*, Chapman & Hall, London.

K.K. Chawla and M. Metzger (1972). *J. Mater. Sci.*, **7**, 34.

K.K. Chawla, J. Singh, and J.M. Rigsbee (1986). *Metallography*, **19**, 119.

B.D. Coleman (1958). *J. Mech. Phys. Solids*, **7**, 60.

J. Cook and J.E. Gordon (1964). *Proc. R. Soc. London*, **A228**, 508.

G.A. Cooper (1970). *J. Mater. Sci.*, **5**, 645.

G.A. Cooper and A. Kelly (1967). *J. Mech. Phys. Solids*, **15**, 279.

A.H. Cottrell (1964). *Proc. R. Soc.*, **282A**, 2.

S.C. Cowin (1979). *J. Appl. Mech.*, **46**, 832.

I.M. Daniel and O. Ishai (1994). *Engineering Mechanics of Composite Materials*, Oxford University Press, New York, p. 126.

H.E. Daniels (1945). *Proc. R. Soc.*, **A183**, 405.

H.E. Dève (1997). *Acta Mater.*, **45**, 5041.

L.J. Ebert and J.D. Gadd (1965). In *Fiber Composite Materials*, ASM, Metals Park, OH, p. 89.

V. Gupta, K. Anand, and M. Kryska (1994). *Acta Metallurgica et Materialia*, **42**, 781.

J. Grape and V. Gupta (1995a). *Acta Metallurgica et Materialia*, **43**, 2657.

J. Grape and V. Gupta (1995b). *J. Composite Mater.*, **29**, 1850.

D.K. Hale and A. Kelly (1972). *Ann. Rev. Mater. Sci.*, **2**, 405.

N.L. Hancox (1975). *J. Mater. Sci.*, **10**, 234.

O. Hoffman (1967). *J. Composite Mater.*, **1**, 200.

A. Kelly (1970). *Proc. R. Soc. London*, **A319**, 95.

A. Kelly (1971). In *The Properties of Fibre Composites*, IPS Science & Technology Press, Guildford, Surrey, U.K., p. 5.

A. Kelly and G.J. Davies (1965). *Metallurgical Rev.*, **10**, 1.

A. Kelly and H. Lilholt (1969). *Philos. Mag.*, **20**, 311.

J.R. Lager and R.R. June (1969). *J. Composite Mater.*, **3**, 48.

J.O. Outwater and M.C. Murphy (1969). In *Proceedings of the 24th SPI/RP Conference*, paper 11-6, Society of Plastics Industry, New York.

M.R. Piggott (1984). In *Developments in Reinforced Plastics—4*, Elsevier Applied Science Publishers, London, p. 131.

M.R. Piggott and B. Harris (1980). *J. Mater. Sci.*, **15**, 2523.

R.B. Pipes and B.W. Cole (1973). *J. Composite Mater.*, **7**, 246.

B.W. Rosen (1965a). In *Fiber Composite Materials*, American Society for Metals, Metals Park, OH, p. 58.

B.W. Rosen (1965b). In *Fiber Composite Materials*, American Society for Metals, Metals Park, OH, p. 37.

B.W. Rosen (1983). In *Mechanics of Composite Materials: Recent Advances*, Pergamon Press, Oxford, p. 105.

R.E. Rowlands (1985). In *Failure Mechanics of Composites*, Vol. 3 of the series Handbook of Composites, North-Holland, Amsterdam, p. 71.

H. Saghizadeh and C.K.H. Dharan (1985). American Society of Mechanical Engineering. Paper #85WA/Mats-15, presented at the Winter Annual Meeting, Miami Beach, FL.

C.R. Schultheisz and A.M. Waas (1996). *Prog. Aerospace Sci.*, **32**, 1.

S.W. Tsai and H.T. Hahn (1980). *Introduction to Composite Materials*, Technomic, Westport, CT.

S.W. Tsai and E.M. Wu (1971). *J. Composite Mater.*, **5**, 58.

M. Vogelsang, R.J. Arsenault, and R.M. Fisher (1986). *Met. Trans. A*, **17A**, 379.

A.M. Waas and C.R. Schultheisz (1996). *Prog. Aerospace Sci.*, **32**, 43.

C. Zweben and B.W. Rosen (1970). *J. Mech. Phys. Solids*, **18**, 189.

Suggested Reading

I.M. Daniel and O. Ishai (1994). *Engineering Mechanics of Composite Materials*, Oxford University Press, New York.

W.S. Johnson (Ed.) (1985). *Delamination and Debonding of Materials*, ASTM STP 876, American Society of Testing and Materials, Philadelphia.

M.N. Nahas, *J. Composites Tech. & Res.*, 8, 138.

CHAPTER 13

Fatigue and Creep

In Chapter 12 we described the monotonic behavior of a composite under ambient temperature conditions of loading. There are many applications of composites where cyclic fatigue and high-temperature creep conditions are very important. Accordingly, in this chapter we go further in complexity and describe the fatigue and creep behavior of composites. *Fatigue* is the phenomenon of mechanical property degradation leading to failure of a material or a component under cyclic loading. The operative word in this definition is *cyclic*. This definition thus excludes the so-called phenomenon of static fatigue, which is sometimes used to describe stress corrosion cracking in glasses and ceramics in the presence of moisture. Creep refers to time-dependent deformation in a material, which becomes important at relatively high temperatures ($T > 0.4\,T_m$, where T_m is the melting point in kelvin). We first describe fatigue and then creep of composites.

13.1 Fatigue

Degradation of mechanical properties of a material or a component under cyclic loading is called fatigue. Understanding the fatigue behavior of composites of all kinds is of vital importance, because without such an understanding it would be virtually impossible to gain acceptance of the design engineers. Many high volume applications of composite materials involve cyclic loading situations, e.g., automobile components. It would be a fair admission that this understanding of the fatigue behavior of composites has lagged that of other aspects such as the elastic stiffness or strength. The major difficulty in this regard is that the application of conventional approaches to fatigue of composites, for example, the stress vs. cycles (S-N) curves or the application of linear elastic fracture mechanics (LEFM), is not straightforward. The main reasons for this are the inherent heterogeneity and anisotropic nature of the composites. This results in damage mechanisms in composites being very different from those encountered in conventional,

Fig. 13.1. Damage zone in a conventional, homogeneous, monolithic material (isotropic).

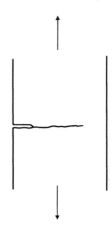

homogeneous, or monolithic materials. The fracture behavior of composites is characterized by a multiplicity of damage modes, such as matrix crazing (in a polymeric matrix), matrix cracking (in a brittle matrix), fiber fracture, delamination, debonding, void growth, and multidirectional cracking, and these modes appear rather early in the fatigue life of composites. Progressive loss of stiffness during fatigue of a composite is a very important characteristic, and it is very different from the fatigue behavior of monolithic materials. The different types of damage zones formed in an isotropic material (e.g., a metal, or ceramic, or polymer) and a fiber reinforced composite, which is an anisotropic material, are shown schematically in Figures 13.1 and 13.2, respectively. In the case of the isotropic material, a single crack propagates in a direction perpendicular to the cyclic loading axis (mode I loading). In the fiber reinforced composite, on the other hand, a variety of subcritical damage mechanisms lead to a highly diffuse damage zone. Despite these limitations, conventional approaches have been used and are therefore described here briefly before we describe some more innovative approaches to the problem of fatigue in composites.

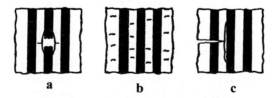

Fig. 13.2. Diffuse damage zone in a fiber reinforced composite (anisotropic): **a** fiber break and local debonding; **b** matrix cracking; **c** deflection of the principal crack along a weak fiber/matrix interface.

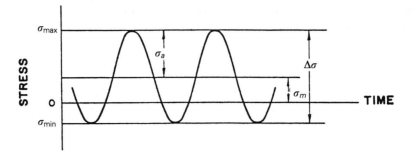

Fig. 13.3. Some useful parameters for fatigue.

Let us first define some useful parameters for our discussion of the fatigue phenomenon. (See Figure 13.3):

$$\text{cyclic stress range, } \Delta\sigma = \sigma_{max} - \sigma_{min}$$

$$\text{cyclic stress amplitude, } \sigma_a = (\sigma_{max} - \sigma_{min})/2$$

$$\text{mean stress, } \sigma_m = (\sigma_{max} + \sigma_{min})/2$$

$$\text{stress ratio, } R = \sigma_{min}/\sigma_{max}$$

13.1.1 S-N Curves

We saw in Chapter 10 that in a unidirectionally reinforced fiber composite, elastic modulus and strength improve in the direction of reinforcement. This also has its consequences in the fatigue behavior. S-N curves are commonly used with monolithic materials, especially metals and, to some extent, with polymers. It involves determination of the so-called S-N curves, where S is the stress amplitude and N is the number of cycles to failure. In general, for ferrous metals, one obtains a fatigue limit or endurance limit. For stress levels below this endurance limit, theoretically, the material can be cycled infinitely. In cases where such an endurance limit does not exist, one can arbitrarily define a certain number of cycles, say 10^6, as the cutoff value. Incorporation of fibers generally improves the fatigue resistance of any fiber-reinforced composite in the fiber direction. Not surprisingly, therefore, composites containing these fibers, aligned along the stress axis and in large volume fractions, do show high monotonic strength values that are translated into high fatigue strength values. Quite frequently, a rule-of-thumb approach in metal fatigue is to increase its monotonic strength, which concomitantly results in an increase in its cyclic strength. This rule-of-thumb assumes that the ratio fatigue strength/tensile strength is about constant. It should also be noted that the maximum efficiency in terms of stiffness and strength gains in fiber reinforced composites occurs when the fibers are continuous, uniaxially aligned and the properties are measured parallel to the fiber direction. As we go off-angle, the strength and stiffness drop sharply. Also, at off-angles, the role of the matrix becomes more important in the deformation and failure processes. One major drawback of this S-N approach to fatigue behavior of a material is that no distinction can be made between the crack initiation phase and the crack propagation phase.

The variety of operating mechanisms and the inadequacy of the S-N curve approach have been documented by many researchers. Owen et al. (1967, 1969), for example, studied the fatigue behavior of chopped strand mat glass/polyester composite and observed the following sequence of events in the fatigue failure process: (1) debonding, generally at fibers oriented transverse to the stress axis, (2) cracking in the matrix, and (3) final separation or fracture. Note that the debonding and cracking phenomena set in quite early in the fatigue life. Lavengood and Gulbransen (1969) investigated the importance of fiber aspect ratio and the role of matrix in the fatigue performance of composites. They studied the effect of cyclic loading on short boron fibers (50–55% volume fraction) in an epoxy matrix. They used a low frequency (3 Hz) to minimize the hysteretic heating effects and measured the number of cycles required to produce a 20% decrease in the composite elastic modulus, i.e., this was their arbitrary definition of fatigue life. The fatigue life increased with aspect ratio up to about 200, beyond which there was little effect. In all cases, the failure consisted of a combination of interfacial fracture and brittle failure of matrix at 45° to the fiber axis. Incorporation of fibers certainly improves the fatigue resistance of the fiber reinforced polymeric matrix composites in the fiber direction. Not surprisingly, therefore, composites containing these fibers, aligned along the stress axis and in large volume fractions, will show high monotonic strength values, which are translated into high fatigue strength values. An example of S-N curves of unreinforced polysulfone (PSF) and composites with a PSF matrix and different amount of short fibers (glass and carbon) is shown in Figure 13.4. These results were obtained from tests done at room temperature, a frequency between 5 and 20 Hz and at $R = 0.1$ (Mandell et al., 1983). Izuka et al. (1986) studied the fatigue behavior of two different types of carbon fibers (T800 and T300 carbon fiber) in an epoxy matrix. Both had 60% fiber volume fraction (V_f) but T800 carbon fiber has a maximum strength of 4.5 GPa and a Young's modulus of 230 GPa while T300 carbon fiber has 3.5 GPa and 210 GPa, respectively. As expected, the higher monotonic strength of T800 carbon fiber resulted in a superior S-N curve.

The polymeric matrices, however, show a viscoelastic behavior and are, generally, poor conductors of heat. Owing to the viscoelastic nature of the polymer matrix, there will be a phase lag between the stress and the strain, i.e., the strain lags the stress or vice versa, and that energy is stored in the material in the form of heat in each cycle. Because of the low thermal conductivity of the polymeric matrix, the heat generated is not dissipated quickly. This can cause a temperature difference between the interior and the surface of a PMC, and the fatigue behavior of PMCs becomes even more

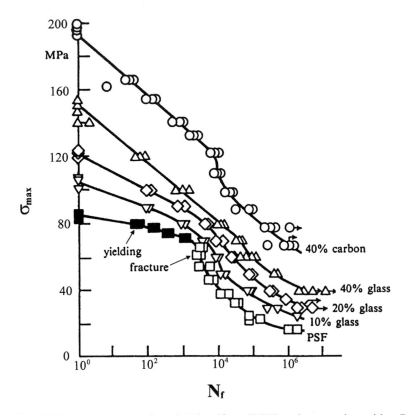

Fig. 13.4. S-N curves of unreinforced polysulfone (PSF) and composites with a PSF matrix and different amount of short fibers (glass and carbon). [After Mandell et al. (1983).]

complex due to such internal heating. The internal heating phenomenon, of course, depends on the cycling frequency. For frequencies less than 20 Hz, the internal heating effects are negligible. For a given stress level, this temperature difference increases with increasing frequency.

Typically, the carbon fibers are much more effective in improving the fatigue behavior of a given polymeric matrix than glass fibers. The reasons for this are the high stiffness and high thermal conductivity of carbon fibers vis à vis glass fibers. The higher thermal conductivity of carbon fibers will contribute to a lower hysteretic heating of the matrix at a given frequency.

Examples of S-N curves for MMCs are shown in Figure 13.5, which shows the S-N curves in tension-tension for unidirectionally reinforced boron (40% v/o)/Al6061, alumina (50% v/o)/Al, and alumina (50% v/o)/Mg composites (Champion et al., 1978). The cyclic stress is normalized with respect to the monotonic ultimate tensile stress. Note the rather flat S-N curves in all the cases and the fact that the unidirectional composites show

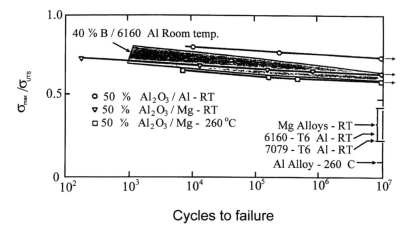

Fig. 13.5. S-N curves for MMCs in tension-tension for unidirectionally reinforced boron (40% v/o)/Al6061, alumina (50% v/o)/Al, and alumina (50% v/o)/Mg composites. [After Champion et al. (1978).]

better fatigue properties than the matrix when loaded parallel to the fibers. For example, at 10^7 cycles, the fatigue-to-tensile strength ratio of the composite is about 0.77, almost double that of the matrix.

Gouda et al. (1981) observed crack initiation early in the fatigue life at defects in boron fibers in unidirectionally reinforced B/Al composites. These cracks then grew along the fiber/matrix interface and accounted for a major portion of the fatigue life, as would be the case in a composite with a high fiber-to-matrix strength ratio. In composites with low fiber-to-matrix strength ratios, crack propagation may take up a major portion of fatigue life, but the crack would be expected to grow across the fibers and a poor fatigue resistance will result. This simply confirms the observation that in unidirectional composites the fatigue resistance will be maximum along the fiber direction and the greatest efficiency will be achieved if the fibers have uniform properties, as much as possible defect-free, and much stronger than the matrix. Similar results have been obtained by other researchers. For example, McGuire and Harris (1974) studied the fatigue behavior of tungsten fiber reinforced aluminum–4% copper alloy under tension-compression cycling ($R = \sigma_{min}/\sigma_{max} = -1$). They found that increasing the fiber volume fraction from 0 to 24% resulted in increased fatigue resistance. This was a direct result of increased monotonic strength of the composite as a function of the fiber volume fraction. The reader should note that due to the highly anisotropic nature of the fiber reinforced composites in general, the fatigue strength of any off-axis fibrous composite will be expected to decrease with increasing angle between the fiber axis and the stress axis. This has been confirmed by studies involving S-N behavior of alumina fiber reinforced magnesium composites (Hack et al., 1987; Page et al., 1987). It was found that the S-N behavior followed the tensile behavior. Increased fiber volume fractions resulted in enhanced fatigue life times in the axial direction but little or no improvement was observed in the off-axis directions. Fatigue crack initiation and propagation occurred primarily through the magnesium matrix. Thus, alloy additions to increase the strength of the matrix and fiber/matrix interface were tried. The alloy additions did improve the off-axis properties but decreased the axial properties. The reason for this was that while the alloy additions resulted in the matrix and interface strengthening, they decreased the fiber strength.

13.1.2   Fatigue Crack Propagation Tests

Fatigue crack propagation tests are generally conducted in an electro-hydraulic closed-loop testing machine on notched samples. The results are presented as log $[da/dN]$ (crack growth per cycle) vs. log ΔK (cyclic stress intensity factor). Crack growth rate, da/dN, is related to the cyclic stress intensity factor range, ΔK, according to the power law relationship formulated by Paris and Erdogan (1963):

$$da/dN = A(\Delta K)^m \tag{13.1}$$

where A and m depend on the material and test conditions. The applied cyclic stress intensity range is given by

$$\Delta K = Y\Delta\sigma\sqrt{\pi a}$$

where Y is a geometric factor, $\Delta\sigma$ is the cyclic stress range, and a is the crack length. The major problem in this kind of test is to make sure that there is one and only one dominant crack that is propagating. This is called the *self-similar* crack growth, i.e., the crack propagates in the same plane and direction as the initial crack. Fatigue crack propagation studies, under conditions of self-similar crack propagation, have been made on metallic sheet laminates (McCartney et al., 1967; Taylor and Ryder, 1976; Pfeiffer and Alic, 1978; Chawla and Liaw, 1979, Godefroid and Chawla, 1988) and unidirectionally aligned fiber reinforced MMCs (Saff et al., 1988). For the crack arrest geometry, if the interface is weak, then the crack on reaching the interface bifurcates and changes its direction and thus the failure of the composite is delayed. The improved fatigue crack propagation resistance in crack divider geometry has been attributed either to interfacial separation, which removes the triaxial state of stress, or to an interfacial holding back of crack in the faster crack-propagating component by the slower crack-propagating component. Generally, a relationship of the form of Eq. (13.1) describes the fatigue crack propagation behavior.

In general, the fibers provide a crack-impeding effect but the nature (morphology, rigidity, and fracture strain) of the fiber surface, the fiber/interface, and/or any reaction zone phases that might form at the interface can have great influence.

Fatigue crack propagation studies have also been done on aligned eutectic or in situ composites. Because many of these in situ composites are meant for high-temperature applications in turbines, their fatigue behavior has been studied at temperatures ranging from room temperature to 1100 °C. The general consensus is that the mechanical behavior of in situ composites, static and cyclic strengths, is superior to that of the conventional cast super-alloys (Stoloff, 1987).

It should be emphasized that only fatigue crack propagation rate data obtained under conditions of self-similar propagation can be used for comparative purposes. In a composite consisting of plies with different fiber orientation, in general, the self similar mode of crack propagation will not be obtained.

Fatigue of Composites under Compression

Fiber reinforced composites generally show lower fatigue resistance in compression than in tension. This may be due to the cooperative buckling of adjacent fibers and the accompanying matrix shear. In monotonic compression of unidirectionally reinforced fiber composites, a fiber kinking mechanism leads to failure. The failure in this case initiates at a weak spot, e.g., at a point where the fiber/matrix bonding is weak. This initial failure will, in turn, destabilize the neighboring fibers, causing more kink failure. Eventually, various kink failure sites can coalesce and lead to transverse tensile loading of the composite and longitudinal splitting. Pruitt and Suresh (1992) showed that in addition to kink band formation, in unidirectional carbon fiber/epoxy composites under cyclic compression, a single mode I crack can start and grow perpendicular to the fiber axis. Suresh finds this compression fatigue phenomenon macroscopically similar to that observed in metals, polymers, and ceramics. The origin of a mode I crack ahead of a stress concentration is the presence of residual tensile stresses as a result of a variety of permanent damage involving matrix, fiber, interface, etc.

Defects to simulate a delamination can be introduced in laminated composites by the following means:

1. Single circular inserts of different diameters located at different interfaces,
2. laminates containing a hole, and
3. laminates containing defects produced by controlled, low-velocity impacts.

Delamination growth under cyclic compression fatigue in a 38-ply T300 carbon/5208 epoxy composite laminate was studied by O'Brien (1984). The maximum strain decreased with cycling. The impacted laminates suffered the most severe degradation on compressive cycling while the laminates with a single implanted delamination suffered the least damage.

Fatigue Behavior of CMCs

We describe the fatigue behavior of CMCs under a separate subheading because of a special situation. Conventional wisdom had it that cyclic fatigue was unimportant so far as ceramics were concerned. However, in actuality the subject of cyclic fatigue in ceramics and ceramic matrix composites *is* an important one. Engineers and researchers began to appreciate the importance of cyclic fatigue in ceramics and ceramic matrix composites only in the 1970s. The fracture resistance of CMCs under cyclic conditions needs to be evaluated for design in a variety of potential structural applications. For example, it is not unusual to have a design requirement for a ceramic component in an automotive gas turbine to withstand more than 30,000 cycles of fatigue (low-cycle fatigue) (Helms and Haley, 1989). In the case of carbon fiber reinforced glass composites, no significant loss of strength was observed on cyclic loading (Phillips, 1983). However, the density and penetration of matrix cracks was more under cyclic loading than under static loading conditions. Also, under static loading this CMC showed higher work of fracture than under cyclic loading. Prewo et al. (1986) studied the tensile fatigue behavior of Nicalon-type silicon carbide fiber reinforced lithium aluminosilicate (LAS) glass-ceramic composite. They used two different types of LAS as the matrix material: one showed a linear tensile stress-strain curve to failure (LAS I) while the other showed a markedly nonlinear behavior due to extensive matrix cracking prior to ultimate failure (LAS II). It was observed that the level of tensile stress at which the inelastic behavior (proportional limit) of the composite began had an important bearing on the fatigue behavior of the CMC. The residual tensile strength and elastic modulus of the LAS I composite after fatigue was the same as that of as-fabricated composite. In the LAS II composite, cycling below the proportional limit produced the same result. However, on cycling to stress levels higher than the proportional limit, a second linear stress-strain region having a modulus less than the initial modulus was observed. Presumably, this change in behavior was due to matrix microcracking at stresses above the proportional limit.

As we said earlier, it was generally thought that the phenomenon of cyclic fatigue was unimportant in ceramics. Work by Suresh and coworkers (Suresh et al., 1988, 1991; Han and Suresh, 1989) on fatigue crack growth in a variety of brittle solids in compression, tension, and tension-compression fatigue shows that mechanical fatigue effects, i.e., due to cyclic loading, occur at room temperature in brittle solids as well. A variety of mechanisms such as microcracking, dislocation plasticity, stress- or strain-induced phase transformations, interfacial slip, and creep cavitation can promote an inelastic constitutive response in brittle solids of all kinds under compressive cycling. Particularly, in CMCs, the mechanisms of crack-tip deformation differ significantly under static and cyclic loading. They demonstrated that under pulsating compression, nucleation and growth of stable fatigue cracks occurred even at room temperature. Suresh et al. showed conclusively that cyclic compressive loading caused mode I fatigue crack growth in SiC whiskers/Si_3N_4 matrix composites. They also observed whisker pullout and breakage after fatigue cycling. Such behavior is generally not observed under monotonic loading. This mode I fatigue crack growth under far-field cyclic compression occurs because a residual zone of tensile stress is generated at the crack tip on unloading. Wang et al. (1991) investigated the behavior of a [0/90] carbon fiber reinforced silicon carbide composite under cyclic loading. They used tension-tension loading of smooth and notched samples and compression-compression loading. Damage in pulsating tension consisted of cumulative microcracking and spalling.

An important problem in high-temperature behavior of polycrystalline ceramics is the presence of intergranular glassy phases. Sintering and other processing aids can form glassy phases at the boundaries, which can result in rather conspicuous subcritical crack growth. Such subcritical crack growth can become very important in ceramic matrix composites because fibers such as silicon carbide can undergo oxidation. Han and Suresh (1989) examined the tensile cyclic fatigue crack growth in a silicon carbide whisker (33 vol. %)/alumina composite at 1400 °C. The composite showed subcritical fatigue crack growth at stress intensity values far below the fracture toughness. The fatigue behavior was characterized by the cyclic stress intensity factor, stress ratio, and frequency. They examined the crack tip region by optical and transmission electron microscopy and found that the nucleation and growth of flaws at the interface was the main damage mechanism. Diffuse microcracking in the wake of the crack and crack deflection/branching were observed. An increase in the test temperature (or the cyclic stress intensity or a reduction in the loading rate) can cause a rather significant increase in the size of the damage zone at the crack tip. Han and Suresh observed oxidation of silicon carbide whiskers to a silica-type glassy phase in the crack tip region at 1400 °C, in air. The alumina matrix can react with the main oxidation product, viz., SiO_2, to form aluminosilicates, SiC-rich or stoichiometric mullite, and the like. Viscous flow of glass can result in interfacial debonding, followed by the nucleation, growth, and coalescence of cavities. The important thing to note is that there is a difference in deformation and failure mechanisms under static and cyclic loadings, even in CMCs.

In a manner analogous to PMCs, there can be a hysteretic heating in CMCs under cyclic loading conditions due to interfacial friction (Holmes, 1991). Sørensen and Holmes (1995) observed that a lubricating layer may be beneficial in improving fatigue life of CFCMCs. A thicker coating, which would be expected to provide greater protection to the fiber against abrasion damage, resulted in less frictional heating because of less wear of the fibers during fatigue of a chemical vapor infiltrated (CVI) Nicalon/C/SiC composite (Chawla et al., 1997, 1998). The composite with a thinner coating exhibited much higher frictional heating. At higher frequencies, more heating was observed since the energy dissipated per unit time also increased. Substantial damage in terms of modulus was observed in fatigue of Nicalon/C/SiC, with most of the damage occurring during the first cycle. At a constant stress, the level of damage was not significantly dependent on frequency. At a given frequency, however, higher stresses induced more damage in both composites. A recovery in modulus of these woven composites was observed due to stretching and alignment of the plain-weave fabric during fatigue, creating a stiffer reinforcing architecture.

The laminate stacking sequence can affect the high-frequency fatigue behavior of CMCs. In SCS-6/Si_3N_4 composites, frictional heating in angle-ply laminates [±45] was substantially higher than that in cross-ply laminates [0/90] (Chawla, 1997). Because the angle ply had a lower stiffness, matrix microcracking in this composite was more predominant. Temperature rise in the specimens correlated very well with stiffness loss as a function of fatigue cycles in the composites (see Section 13.1.3).

Fatigue of Particle– and Whisker-Reinforced Composite

Ceramic particle reinforced metal matrix composites, such as silicon carbide or alumina particle reinforced aluminum alloy composites can have improved fatigue properties vis à vis unreinforced aluminum alloys, which can make these composites useful in applications where aluminum alloys would not be considered (Allison and Jones, 1993). Such systems have been studied by some researchers (Crowe and Hassen, 1982; Williams and Fine, 1985, 1987; Logsdon and Liaw, 1986; Shang et al., 1988; Davidson, 1989; Kumai et al., 1990; Bonnen et al., 1990; Christman and Suresh, 1988a, 1988b). In general, in terms of S-N curve behavior, the composite shows an improved fatigue behavior vis à vis the reinforced alloy. Such an improvement in stress-controlled cyclic loading or high cyclic fatigue is attributed to the higher stiffness of the composite. However, the fatigue behavior of the composite, evaluated in terms of strain amplitude versus cycles or low-cycle fatigue, was inferior to that of the unreinforced alloy (Bonnen et al., 1990). This was attributed to the generally lower ductility of the composite compared to unreinforced alloy.

Particle or short fibers can provide easy crack initiation sites. The detailed behavior can vary depending on the volume fraction of the reinforcement, shape, size, and most importantly on the reinforcement/matrix bond strength (Williams and Fine, 1987). For example, they observed fatigue crack initiation at the poles of SiC whiskers in 2124 aluminum. They also observed arrest of short cracks at the whisker/Al interfaces. Frequently in aluminum matrix composites, especially those made by casting, there are particles other than SiC, such as $CuAl_2$, $(Fe,Mn)_3 SiAl_{12}$, and $Cu_2Mg_5Si_6Al_5$ (Kumai and Knott, 1991). The phenomenon of particle pushing ahead of the solidification front results in SiC particles and the so-called constituent particles decorating the cell boundaries in the aluminum alloy matrix. Some possible reinforcement and crack tip interactions are shown in Figure 13.6, while Figure 13.7 shows schematic representations of crack growth versus cyclic intensity factor for a monolithic alloy and a particulate composite. Levin et al. (1989) observed superior resistance to fatigue crack growth in 15 volume % SiC/Al 6061 composite vis à vis Al 6061 alloy, which was attributed to a

Fig. 13.6. Some possible reinforcement and crack tip interactions.

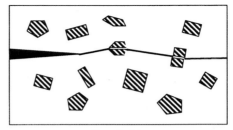

Coarse Particles
Strong Particle/Matrix Interface

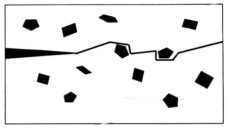

Fine Particles
Weak Particle/Matrix Interface

slower crack growth rate in the composite due to crack deflection caused by the SiC particles.

It would appear that choosing the optimum particle size and volume fractions, together with a clean matrix alloy, will result in a composite with improved fatigue characteristics. Shang et al. (1988) examined the effect of particle size on fatigue crack propagation as a function of cyclic stress intensity in a silicon carbide particle reinforced aluminum. They observed that for fine particle size, the threshold stress intensity, ΔK_{th}, for the composite was less than that for the unreinforced alloy, i.e., initial fatigue crack growth resistance of the composite was less than that of the unreinforced alloy. For coarse particles, the threshold intensity values were about the same for the two, while at very high values of the cyclic stress intensity, the fatigue crack growth of the composite was less than that of the unreinforced alloy.

13.1.3 Damage Mechanics of Fatigue

It was mentioned earlier that the complexities in composites lead to the presence of many modes of damage, such as matrix cracking, fiber fracture, delamination, debonding, void growth, and multidirectional cracking. These modes appear rather early in the fatigue life of composites, and these subcritical damage accumulation mechanisms come into play rather early in the

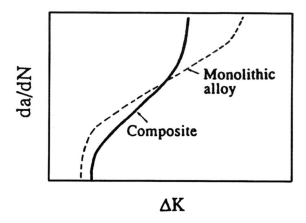

Fig. 13.7. Schematic representation of crack growth versus cyclic intensity factor for a monolithic alloy and a particulate MMC. [After Kumai et al. (1990).]

fatigue life—well before the fatigue limit, as determined in an S-N test, and a highly diffuse damage zone is formed. One manifestation of such damage is the stiffness loss as a function of cycling. In general, one would expect the scatter in fatigue data of composites to be much greater than that in fatigue of monolithic, homogeneous materials. This is because of the existence of a variety of damage mechanisms in composites, to wit, random distribution of matrix microcracks, fiber/matrix interface debonding, and fiber breaks. With continued cycling, an accumulation of damage occurs. This accumulated damage results in a reduction of the overall stiffness of the composite laminate. Measurement of stiffness loss as a function of cycling has been shown to be quite a useful technique for assessing the fatigue damage in composites. Information useful to designers can be obtained from such curves. In MMCs, the fatigue behavior of boron fiber and silicon carbide fiber reinforced aluminum and titanium alloy matrix composite laminates with different stacking sequences has been examined using the stiffness loss measurement technique (Dvorak and Johnson, 1980; Johnson, 1988; Johnson and Wallis, 1986). It was observed that on cycling below the fatigue limit but above a distinct stress range, ΔS_{SD}, the plastic deformation and cracking (internal damage) in the matrix led to a reduced modulus. Figure 13.8 shows the response of a boron fiber/aluminum matrix composite subjected to a constant cyclic stress range (225 MPa) with varying values of S_{max}, the maximum stress. The modulus drop occurred only when S_{max} was shifted upward. Johnson (1988) proposed a model that envisioned that the specimen reached a saturation damage state (SDS) during constant-amplitude fatigue testing. Gomez and Wawner (1988) also observed stiffness loss on subjecting silicon carbide/aluminum composites to tension-tension fatigue ($R = 0.1$) at 10 Hz. Periodically, the cycling was stopped and the elastic modulus was measured. Figure 13.9 shows a typical modulus loss curve for unidirectional

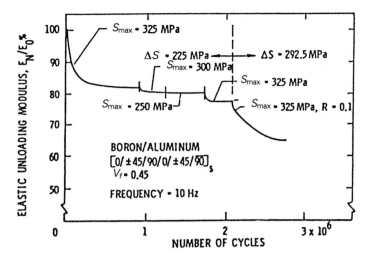

Fig. 13.8. Response of a boron fiber/aluminum matrix composite subjected to a constant cyclic stress range (225 MPa) with varying values of S_{max}, the maximum stress. Diffuse damage zone in a fiber reinforced composite (anisotropic): **a** fiber break and local debonding, **b** matrix cracking, deflection of the principal crack along a weak fiber/matrix interface. [After Johnson (1988).]

silicon carbide/aluminum composites. Modulus at N cycles, E_N, normalized with respect to the original modulus, E_0, is plotted against the log (number of cycles, N). These authors used a special type of silicon carbide fiber, called the SCS-8 silicon carbide fiber, which is a silicon carbide fiber with a modified surface to give a strong bond between the fiber and the aluminum matrix. The SCS coating broke off at high cycles and the fracture surface showed the coating clinging to the matrix. Karandikar and Chou (1992) used the approach of stiffness loss as a function of stress cycles with unidirectionally reinforced Nicalon fiber/calcium aluminosilicate (CAS) composites and obtained correlations between crack density and stiffness reduction.

Most of the so-called advanced composites, called *laminated composites*, involve the use of prepregs. Figure 13.10a shows schematically a comparison of damage accumulation as a function of fatigue cycles in a laminated composite made by appropriately stacking differently oriented plies and a monolithic, homogeneous material under constant stress amplitude fatigue (Hahn and Lorenzo, 1984). We plot damage ratio against cycle ratio. The damage ratio is the current damage normalized with respect to the damage at final failure. The cycle ratio, similarly, is the number of cycles at a given instant divided by the number of cycles to failure. In a homogeneous material, the term *damage* simply represents the crack length, and not surprisingly it increases monotonically with cycling. In the case of a laminate, we do not have a simple and unambiguous manifestation of damage, such as a crack length. Instead, damage means the crack density. Note that, unlike in

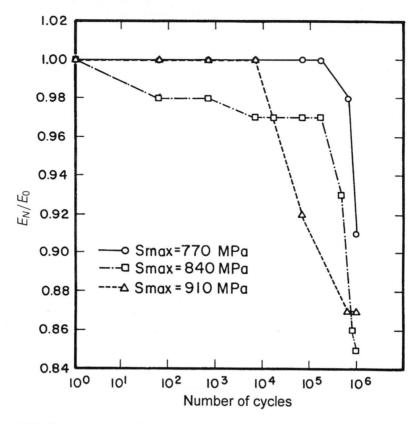

Fig. 13.9. Typical modulus loss curve for unidirectional silicon carbide/aluminum composites. E_N is the modulus after N cycles and E_0 is the modulus in the uncycled state. [After Gomez and Wawner (1988).]

homogeneous materials, the damage in laminates accelerates at first and then decelerates with cycling. This distinctive behavior is very important. Such a multiplicity of fracture modes is common to all composites. Figure 13.10b shows the fracture surface of B(W)/Al 6061 composite made by diffusion bonding. Note the ductile fracture in aluminum, brittle fracture in boron, fiber pullout (see the missing fiber in top-left-hand corner), and sheet delamination.

As was pointed out earlier, the fiber reinforced laminates can sustain a variety of subcritical damage (crazing and cracking of matrix, fiber/matrix decohesion, fiber fracture, ply cracking, delamination, and so on). For example, the cracking of a ply will result in a relaxation of stress in that ply, and with continued cycling no further cracking occurs in that ply. Ply cracking generally involves cracking in the matrix and along the fiber/matrix interface but rarely any fiber fracture. Other damage-accumulating mechanisms include the growth of existing cracks into interfaces leading to ply

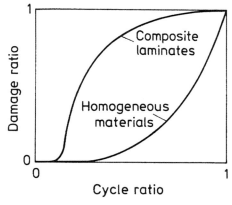

a

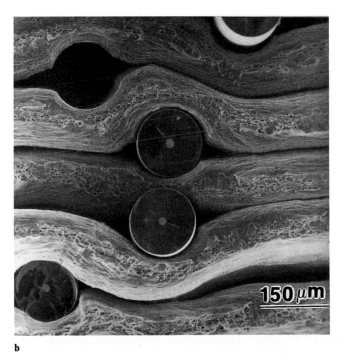

b

Fig 13.10. a Comparison of damage accumulation as a function of fatigue cycles in a laminated composite made by appropriately stacking differently oriented plies and a monolithic, homogeneous material under constant stress amplitude fatigue. [After Hahn and Lorenzo (1984).] **b** Fracture surface of B(W)/Al 6061 composite made by diffusion bonding. Note the ductile fracture in aluminum, brittle fracture in boron, fiber pullout (see the missing fiber in top-left-hand corner), and sheet delamination.

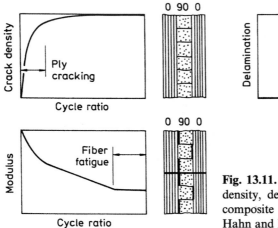

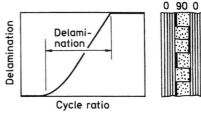

Fig. 13.11. Schematic of changes in crack density, delamination, and modulus in a composite laminate under fatigue. [After Hahn and Lorenzo (1984).]

delamination. The delamination of a ply results in a reduction of stress concentration on the neighboring plies. The subcritical damage can accumulate rather rapidly on cycling.

The various types of subcritical damage mentioned earlier result in a reduction of the load-carrying capacity of the laminate composite, which in turn manifests itself as a reduction of laminate stiffness and strength (Hahn and Kim, 1976; Highsmith and Reifsnider, 1982; Talreja, 1985; O'Brien and Reifsnider, 1981; Ogin et al., 1985). Figure 13.11 depicts schematically the changes in crack density, delamination, and modulus in a composite laminate under fatigue (Hahn and Lorenzo, 1984). Reifsnider et al. (1981) have modeled the fatigue development in laminate composites as occurring in two stages. In the first stage, homogeneous, noninteractive cracks appear in individual plies. In the second stage, the damage gets localized in zones of increasing crack interaction. The transition from stage one to stage two occurs at what has been called the *characteristic damage state* (CDS), which consists of a well-defined crack pattern characterizing saturation of the noninteractive cracking. Talreja (1985) used this model to determine the probability distribution of the number of cycles required to attain the CDS. Many researchers have experimentally related the stiffness changes in the laminated composites to the accumulated damage under fatigue (Hahn and Kim, 1976; Highsmith and Reifsnider, 1982; Talreja, 1985; Reifsnider, 1981). It can safely be said that the change in stiffness values is a good indicator of the damage in composites. An actual stiffness reduction curve for a $[0°/90°]s$ glass fiber reinforced laminated composite is shown in Figure 13.12 (Ogin et al., 1985). Under cyclic loading of a laminated composite, a variety of damage accumulation mechanisms can start at stress levels below those needed under monotonic conditions. Because of the availability of this

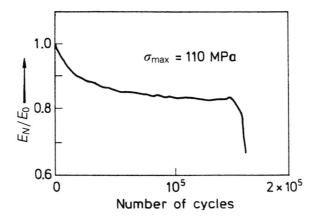

Fig. 13.12. Stiffness reduction curve for a $[0°/90°]_s$ glass fiber–reinforced laminated composite. [After Ogin et al. (1985).]

multiplicity of failure modes in a fibrous composite, it rarely fails in a simple manner as does a monolithic material. Following Beaumont (1989), we define the failure of the composite when a critical level of damage is reached or exceeded. Let D be the damage parameter that increases as a function of number of cycles, N. Then, dD/dN will be the damage growth rate. We can write for the damage growth rate

$$dD/dN = f(\Delta\sigma, R, D) \tag{13.2}$$

where $\Delta\sigma$ is the cyclic stress range, R is the stress ratio, and D is the current value of damage. Let N_f be the number of cycles to failure, i.e., the fatigue life corresponding to a critical level of damage. Then integrating Eq. (13.2) between limits of initial damage, D_i and final damage, D_f, we can get the number of cycles to failure

$$N_f = \int_{Di}^{Df} dD/f(\Delta\sigma, R, D) \tag{13.3}$$

The main problem is that the function f is not known. One measure of damage is the instantaneous load-bearing capacity or the stiffness, E. We can write

$$E = E_0 g(D) \tag{13.4}$$

where E_0 is the stiffness of the undamaged material and $g(D)$ is a function of damage, D.

Rewriting Eq. (13.4) as $E/E_0 = g(D)$ and differentiating, we get

$$(1/E_0)dE/dN = dg(D)/dN \tag{13.5}$$

Also,

$$D = g^{-1}(E/E_0)$$

where g^{-1} is the inverse of g. This allows us to rewrite Eq. (13.5) as

$$(1/E_0)\, dE/dN = g'[g^{-1}(E/E_0)]f(\Delta\sigma, R, D)$$

The function $g(D)$ can be obtained experimentally by obtaining data in terms of E/E_0 vs. N. We can then evaluate the function f as:

$$f(\Delta\sigma, R, D) = (1/g'[g^{-1}(E/E_0)])(1/E_0)\, dE/dN$$

We can evaluate the right-hand side of this expression for a range of $\Delta\sigma$, maintaining constant E/E_0, R, etc. If $2S$ is the average crack spacing, then crack density or damage $= 1/2S$. The following relationship was found experimentally for crack growth as a function of cycles, da/dN (Ogin, 1985):

$$da/dN \propto (\sigma_{max}^2 2S)^n$$

Total crack length a is proportional to damage or crack density, D, i.e.,

$$dD/dN \propto (\sigma_{max}^2/D)^n$$

Modulus of the damaged material E is given by

$$E = E_0(1 - cD) \tag{13.6}$$

where c is a material constant.

For a given value of E/E_0, we can write the modulus or stiffness reduction rate as

$$-1/E_0\, dE/dN = A[\sigma_{max}^2/E_0^2(1 - E/E_0)]^n \tag{13.7}$$

where A and n are cosntants. The left-hand side of this expression can be determined experimentally. Note that both sides are dimensionless quantities. The modulus reduction rate $(-1/E_0)\, dE/dN$, at a given value of E/E_0, is the tangent to the curve shown in Figure 13.12. This rate of Young's modulus reduction was well described over the range of peak fatigue stress between 110 and 225 MPa for the glass fiber/epoxy composite by Eq. (13.7); see Figure 13.13. We can integrate Eq. (13.7) to obtain a diagram relating stiffness reduction to number of cycles for different stress levels, as shown in Figure 13.14. Such a diagram gives us the number of cycles it will take, when the composite is cycled at a certain fraction of the monotonic ultimate strength, to attain a specific amount of stiffness reduction.

13.1.4 Hybrid Composites

Composites containing more than one type of fiber are called *hybrid composites*. Such composites, by using two or more type of fibers, extend the idea

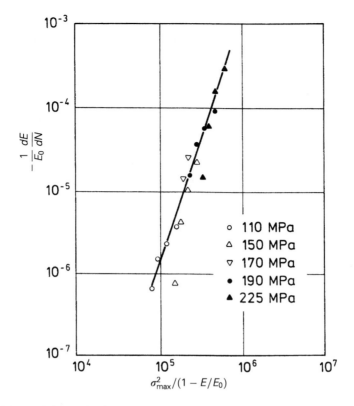

Fig. 13.13. Modulus reduction rate versus a parameter involving the peak stress. [After Ogin et al. (1985).]

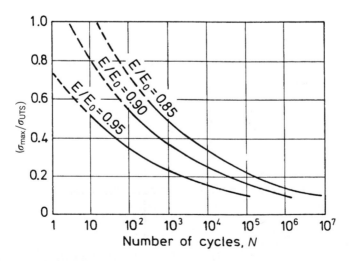

Fig. 13.14. Number of cycles required to attain a given stiffness reduction after cycling at different fractions of the ultimate tensile strength σ_{UTS}. [After Ogin et al. (1985).]

of tailor-making a composite material to meet specific property requirements. Partial replacement of expensive fibers by cheaper but adequate fiber types is another attractive feature of hybrid composites. Additionally, there is the possibility of obtaining a synergistic effect in the fatigue behavior of hybrid composites. Phillips (1976) observed enhanced fatigue strength in a carbon-glass hybrid system. Such synergistic results are by no means universal. For example, in the case of the flexural fatigue strength versus the number of cycles for 100% unidirectional carbon fiber/polyester, 100% chopped glass fiber/polyester, and unidirectional carbon fiber faces over chopped glass core, the hybrid composite curve was intermediate between the 100% carbon and 100% glass curves (Riggs, 1985).

An interesting type of hybrid composite is made of alternating layers of high-strength aluminum alloy sheets and unidirectional aramid fibers in an epoxy matrix. It is called *aramid aluminum laminates* (ARALL). Improved fatigue resistance of ARALL over that of monolithic aluminum structures is the main attractive feature. Cracks can grow only a short distance before being blocked by the araimd fibers spanning the crack tip. Figure 13.15 shows the slow fatigue crack growth characteristics of ARALL compared to two monolithic aluminum alloys (Mueller and Gregory, 1988). Applications of ARALL are envisaged in tension-dominated fatigue structures such as aircraft fuselage, lower wing, and tail skins. It should be mentioned that the use of ARALL will result in 15–30% weight savings over conventional construction. If a glass fiber reinforced epoxy sheet is used instead of an aramid/epoxy sheet, then the acronym GLARE is used.

Another version of hybrid laminated MMCs is a family of layered composites consisting of metallic outer skins with a viscoelastic core material (for

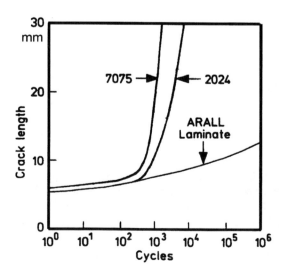

Fig. 13.15. Fatigue crack growth of ARALL compared to two monolithic aluminum alloys. [After Mueller and Gregory (1988).]

example, polyethylene, nylon, polypropylene, paper, or cork). Such composites will be useful where sound and vibration damping are required. The viscoelastic layer provides a high loss factor, i.e., a high capacity to convert vibrational energy to heat.

13.1.5 Thermal Fatigue

There exists a very fundamental physical incompatibility between the reinforcement (be that fiber, whisker, or particle) and the matrix, to wit, the difference in their thermal expansion (or contraction) coefficients. This problem of thermal expansion mismatch between the components of a composite is a very important one. Thermal stresses arise in composite materials due to the generally large differences in the thermal expansion coefficients (α) of the fiber and the matrix. It should be emphasized that thermal stresses in composites will arise even if the temperature change is uniform throughout the volume of the composite. Such thermal stresses can be introduced in composites during cooling from high fabrication, annealing, or curing temperatures or during any temperature excursions (inadvertent or by design) during service. Turbine blades, for example, are very much susceptible to thermal fatigue. The magnitude of thermal stresses in composites is proportional to $\Delta\alpha\,\Delta T$, where $\Delta\alpha$ is the difference in the expansion coefficients of the two components and ΔT is the amplitude of the thermal cycle (see Chapter 10).

In PMCs and MMCs, the matrix generally has a much higher coefficient of thermal expansion than the fiber. In CMCs, the thermal coefficients of the components are not that much different but the ceramic materials have very low strain-to-failure values, i.e., very low ductility. In general, the matrix has a much higher coefficient of thermal expansion than the fiber. Rather large internal stresses can result when fiber-reinforced composites are heated or cooled through a temperature range. When this happens in a repeated manner, we have what is called the phenomenon of *thermal fatigue*, because the cyclic stress is thermal in origin. Thermal fatigue can cause cracking in the brittle polymeric matrix or plastic deformation in a ductile metallic matrix (Chawla, 1973a, 1973b, 1975a, 1975b). Cavitation in the matrix and fiber/matrix debonding are the other forms of damage observed due to thermal fatigue in composites (Chawla, 1973a, 1973b, 1975b; Lee and Chawla, 1987; Lee et al., 1988). Xu et al. (1995) studied the damage evolution as a function of thermal cycles in terms of three metal matrix systems: Al_2O_3(fiber)/Mg alloy, B_4C_p(particle)/6061Al alloy, and SiC_p(particle)/8090Al alloy. The samples were thermally cycled between room temperature (22 °C) and 300 °C. The incidence of void formation at the fiber/matrix interface increased with the number of cycles. They observed that loss in stiffness and density could be used as damage parameters. The damage in density and elastic modulus caused by thermal cycling was more severe in the fiber reinforced composite than in the particle reinforced composites.

In view of the fact that CMCs are likely to find major applications at high temperatures, it is of interest to study their behavior under conditions of isothermal exposure as well as under conditions of thermal cycling. Wetherhold and Zawada (1991) studied the behavior of ceramic-grade Nicalon fiber in an aluminosilicate glass matrix under isothermal and thermal cycling. At 650 to 700 °C, isothermally exposed and thermally cycled samples showed rapid oxidation and loss in strength. Oxidation behavior overshadowed any thermal cycling effect for these test conditions. The embrittlement was attributed to oxygen infiltration from the surface, which destroyed the weak carbon-rich interface in this composite. At 800 °C, however, less embrittlement was observed and the fiber toughening effect remained. This decreased embrittlement at higher temperatures was attributed to smoothening of the sample surface by glass flow and slow oxygen infiltration. Boccaccini et al. (1997a, 1997b) studied the cyclic thermal shock behavior of Nicalon fiber reinforced glass matrix composites. The thermal mismatch between the fiber and the matrix in this system was almost nil. A decrease in Young's modulus and a simultaneous increase in internal friction as a function of thermal cycles were observed. The magnitude of internal friction was more sensitive to microstructural damage than Young's modulus. An interesting finding of theirs involved the phenomenon of crack healing when the glass matrix composite was cycled to a temperature above the glass transition temperature of the matrix where the glass flows.

It is possible to obtain a measure of the internal stresses generated on subjecting a composite to thermal cycling. Kwei and Chawla (1992) used a computer-controlled servohydraulic thermal fatigue system to perform tests on an alumina fiber/Al-Li alloy composite. Thermal fatigue testing in this case involved cycling the temperature of the sample while its gage length was kept constant. This constraint resulted in a stress on the sample, which was measured. Such a test provides the stress required to keep the specimen gage length constant as a function of thermal cycles i.e., a measure of internal stresses generated.

In general, one can reduce the damage in the matrix by choosing a matrix material that has a high yield strength and a large strain to failure (i.e., ductility). The eventual fiber/matrix debonding can only be avoided by choosing the components such that the difference in the thermal expansion characteristics of the fiber and the matrix is low.

13.2 Creep

Creep is defined as the time-dependent deformation in a material. It becomes important at relatively high temperatures, especially at temperatures greater than 0.4–0.5 T_H, where T_H is the homologous temperature equal to T/T_m, T is the temperature of interest in kelvin, and T_m is the melting point of the material in kelvin. The phenomenon of creep can cause small deformations under a sustained load over a long period of time. In a variety of situations or equipments (e.g., a pressure vessel or a rotating component), such slow deformations can lead to dimensional problems or even failure. Creep sets a limit on the maximum application temperature. In general, this limit increases with the melting point of a material. Without going into the theoretical and modeling details, suffice it to say that the basic governing equation of creep can be written in the following form:

$$\dot{\varepsilon} = A(\sigma/G)^n \exp(-\Delta Q/kT)$$

where $\dot{\varepsilon}$ is the creep strain rate, σ is the applied stress, n is an exponent, G is the shear modulus, ΔQ is the activation energy for creep, k is the Boltzmann's constant, and T is the temperature in kelvin. The stress exponent, n, typically varies between 3 and 7 in the dislocation climb regime and between 1 and 2 when diffusional mechanisms are operating. Pure dislocation creep, grain boundary sliding, vacancy motion in grains and in the grain boundaries, and dislocation can cause creep. The applied stress, grain size, porosity, and impurity content are important variables.

In polymers and PMCs such as Kevlar 49 aramid fiber/epoxy, one can observe creep even at room temperature (Ericksen, 1976). At a given temperature, cross-linked thermosets show less creep than thermoplastics. Creep in polymers is the same as defined above, viz., we apply a constant stress and observe the strain as a function of time. There is a related phenomenon called *stress relaxation*, which is also important in polymers. In stress relaxation, we impose a constant strain on the specimen and observe the drop in stress as a function of time. If we substitute a polymeric fiber such as aramid with a more creep-resistant fiber—say alumina, SiC, or even glass—we can make the composite more creep-resistant.

In very simple terms, creep in a PMC or MMC is likely to be dominated by the creep behavior of the matrix. As the matrix deforms in creep, the applied load is transferred to the load-bearing component, viz., fiber. Eventually, the fibers will carry all the load. McLean (1983, 1985) showed that for a composite containing a matrix that follows a power-law creep ($\dot{\varepsilon} = A\sigma^n$), the creep rate in the composite is given by:

$$\dot{\varepsilon}_c = \frac{A\sigma^n \left[1 - \dfrac{\dot{\varepsilon}}{\dot{\varepsilon}_\infty}\right]^n}{\left[1 + \dfrac{V_f E_f}{V_m E_m}\right] V_m^n}$$

where $\dot{\varepsilon}_\infty = (\sigma_c/V_f E_f)$ is the asymptotic creep strain.

A creep curve, strain versus time, for a 25% V_f silicon carbide whisker/ 2124 aluminum alloy matrix is shown in Figure 13.16 (Lilholt and Taya, 1987). The primary, secondary or steady-state, and tertiary stages are indi-

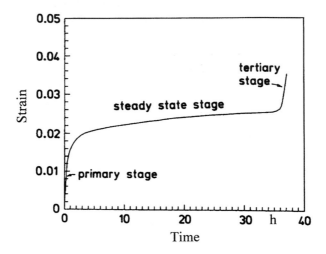

Fig. 13.16. A creep curve, strain versus time, for a 25% V_f silicon carbide whisker/ 2124 aluminum alloy matrix. [After Lilholt and Taya (1987).]

cated. The steady-state creep rate as a function of applied stress for the unreinforced silver matrix and tungsten fiber/silver matrix composite at 600 °C are shown in Figure 13.17 (Kelly and Tyson, 1966). The minimum creep rate as a function of the applied stress for Saffil short fiber–reinforced aluminum alloys and unreinforced aluminum alloys is shown in Figure 13.18 (Dlouhy et al., 1993). Note that the creep rate for the composite is lower than that of the control alloy, but the stress exponent or slope of the composite is much higher than that of the unreinforced alloy.

In some cases, creep behavior of MMCs shows a close correlation between the properties of the matrix and those of the composite (Kelly and Tyson, 1966; Kelly and Street, 1972a, 1972b; Dragone et al., 1991). In other cases, this is not observed. Creep experiments on an aluminum matrix containing ceramic particles or short fibers show a very high value of stress exponent, $n \sim 20$ and an activation energy for creep, $Q \sim 225$–400 kJ/mol. This is in contrast to the activation energy for self-diffusion in aluminum matrix, $Q \sim 150$ kJ/mol (Nieh, 1984; Nardone and Strife, 1987; Morimoto et al., 1988; Pandey et al., 1992; Dragone and Nix, 1992; Dlouhy et al., 1993, Eggler and Dlouhy, 1994). Various models have been proposed to rationalize these discrepancies. For a review of these models, the reader is referred to a review article by Dunand and Derby (1993). In CMCs too, the incorporation of fibers or whiskers can result in improved creep resistance. The creep rate, in four-point bending, of silicon carbide whisker (20 v/o) reinforced alumina was significantly reduced compared to that of the unreinforced alumina (Lin and Becher, 1990). For the creep tests done at 1200 and 1300 °C, the stress exponent, n in the relationship $\dot{\varepsilon} = A(\sigma)^n$, was 2 for the composite, not much different from the value of 2.3 for the unreinforced

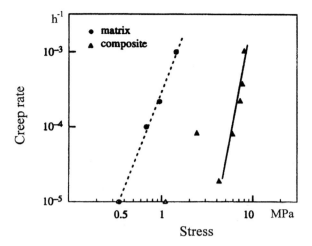

Fig. 13.17. The steady-state creep rate as a function of applied stress for the unreinforced silver matrix and tungsten fiber/silver matrix composite at 600 °C. [After Kelly and Tyson (1966).]

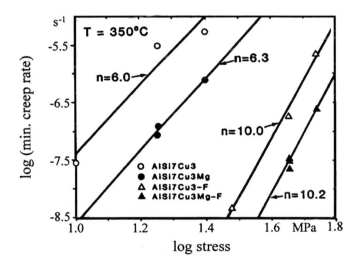

Fig. 13.18. The minimum creep rate as a function of the applied stress for Saffil short fiber–reinforced aluminum alloys and unreinforced aluminum alloys. Note that the creep rate for the composite is lower than that of the control alloy, but the stress exponent or slope of the composite is much higher than that of the unreinforced alloy.

alumina, indicating that the creep rate controlling process was similar in these two materials at these two temperatures. This improvement was attributed to a retardation of grain boundary sliding by SiC whiskers present at the grain boundaries. The creep curve at 1400 °C for the composites showed a marked change in the n value at a stress level of about 125 MPa indicating a change in the rate controlling process. A stress exponent value of about 2 is generally thought to be due to grain boundary sliding. The higher stress exponent and the higher creep rate at 1400 °C was attributed by the authors to extensive cavitation. There are, however, some differences in the creep behavior of CMCs vis à vis MMCs. Wiederhorn and Hockey (1991) have analyzed the creep behavior of CMCs, both particle and whisker reinforced. In two-phase ceramics, creep rate in tension was faster than in compression for identical stress and temperature conditions. At first sight, this might appear to be due to the ease of cavitation and microcracking during tension rather than in compression because tensile stresses assist cavitation while compressive stresses tend to close the cavities and micro-cracks. This is not so because Wiederhorn et al. (1988) observed the asymmetry in creep behavior in CMCs even under conditions where cavitation was absent.

Although continuous ceramic fibers can lead to substantial toughening of ceramics at room temperature, most of these fibers are not sufficiently creep-resistant. In fact, creep rates of many fibers are much higher than those of the corresponding monolithic ceramics (Lin and Becher, 1990; Routbort et al., 1990; Bender et al., 1991; Pysher et al., 1989). In the case of creep of ceramic matrix composites, one needs to consider the intrinsic creep resistance of the fiber, matrix, and interface region. Oxide fibers are fine-grained and generally contain some glassy phase. Nonoxide fibers are also fine-grained, multiphasic (with some glassy phase) and susceptible to oxidation (Bender et al., 1991). Nonoxide fiber/nonoxide matrix composites, such as SiC/SiC and SiC/Si$_3$N$_4$, generally show good low-temperature strength, but their poor oxidation resistance is a major limitation. Mah et al. (1984) observed that the strength of Nicalon-type SiC fiber was very sensitive to temperature above 1200 °C and its environment. Nonoxide fiber/oxide matrix composites or oxide fiber/nonoxide matrix composites, such as carbon/glass, SiC/glass, SiC/alumina, and Al$_2$O$_3$/SiC, generally do not possess high oxidation resistance because the permeability constant for the diffusion of oxygen is high, resulting in rapid oxygen permeation through the oxide matrix. Prewo et al. (1986) found that the glass matrix did not prevent the degradation of carbon fiber caused by oxidation. Holmes (1991) observed the formation of a complex glass layer (SiO$_2$-Y$_2$O$_3$-MgO) on the surface of hot-pressed SCS-6/Si$_3$N$_4$ subjected to creep. The glass layer forms by oxidation of silicon nitride and sintering aids. He also observed that the extent of fiber pullout decreased as the applied stress increased in a creep test. Rather pronounced separation along the fiber/matrix interface was observed after low stress (70 MPa) creep.

Creep behavior of laminated ceramic composites has also been studied.

The steady-state creep rate of monolithic silicon nitride matrix, [0] and [0/90] cross-plied SCS-6/Si$_3$N$_4$ composites as a function of applied stress at 1200 °C showed that the creep resistance of the composite was superior to that of the monolithic silicon nitride, while the creep resistance of the unidirectional composite was superior to that of the cross-plied composite because the fibers in the 90° direction contribute less to creep resistance than the fibers in the 0° direction (Yang and Chen, 1992).

From the preceding discussion of high-temperature behavior of nonoxide composites (even when one component is a nonoxide), it would appear that in situations where stability in air at high temperatures is a prime objective, oxide fiber/oxide matrix composites should be most promising because of their inherent stability in air. Some such systems have been investigated (Chawla et al., 1996). Among the oxide fibers, alumina-based and mullite fibers are the most widely used, while glass, glass-ceramics, alumina, and mullite are the most widely used oxide matrices. One can have two categories of oxide/oxide composites: oxide matrix reinforced with uncoated oxide fibers and oxide matrix reinforced with coated oxide fibers. In the first category of oxide/oxide composites, strength and modulus of the composite are generally better than the unreinforced oxide. The toughness characteristics of these composites are not substantially changed because of the strong chemical bonding at the fiber/matrix interface. Interface tailoring via fiber coating (the second category) is employed extensively in order to achieve the desired properties of the composites.

13.3 Closure

Let us summarize the important points of this chapter. In general, the fatigue resistance of a given material can be enhanced by reinforcing it with continuous fibers or by bonding two different metals judiciously selected to give the desired characteristics. Not unexpectedly, the improvement is greatest when the fibers are aligned parallel to the stress direction. While conventional approaches such as stress versus cycles (S-N) curves or fatigue crack propagation tests under conditions of self-similar crack propagation can be useful for comparative purposes and for obtaining information on the operative failure mechanisms, they do not provide information useful to designers. Fatigue crack propagation under mixed-mode cracking conditions should be analyzed analytically and experimentally. Novel approaches such as that epitomized by the measurement of stiffness reduction of the composite as a function of cycles seem to be quite promising. Because many applications of composites do involve temperature changes, it is important that thermal fatigue characteristics of these composites be evaluated in addition to their mechanical fatigue characteristics.

In regard to creep behavior of composites, introduction of creep-resistant reinforcement, especially fibers in a matrix that undergoes substantial creep, can result in a composite that is more creep-resistant than the matrix.

Experimental observations show that the stress exponent for the composite in the creep rate versus stress curve is frequently much higher than that for the unreinforced alloy.

References

J.E. Allison and J.W. Jones (1993). In *Fundamentals of Metal Matrix Composites*, S. Suresh, A. Mortensen, and A Needleman (Eds.), Butterworth-Heinemann, Boston, p. 269.

P.W.R. Beaumont (1989). In *Design with Advanced Composite Materials*, L.N. Phillips (Ed.), Springer-Verlag, Berlin, p. 303.

B.A. Bender, J.S. Wallace, and D.J. Schrodt (1991). *J. Mater. Sci.*, **12**, 970.

A.R. Boccaccini, D.H. Pearce, J. Janczak, W. Beier, and C.B. Ponton (1997a). *Materials Science and Technology*, **13**, 852.

A.R. Boccaccini, C.B. Ponton, and K.K. Chawla (1998). *Mat. Sci. Eng.* **A241**, 142.

J.J. Bonnen, C.P. You, J.E. Allison, and J.W. Jones (1990). In *Proc. Int. Conf. on Fatigue*, p. 887.

A.R. Champion, W.H. Krueger, H.S. Hartman, and A.K. Dhingra (1978). *Proc.: 1978 Intl. Conf. Composite Materials (ICCM/2)*, p. 883, TMS-AIME, New York.

K.K. Chawla (1973a). *Metallography*, **6**, 155.

K.K. Chawla (1973b). *Philos. Mag.*, **28**, 401.

K.K. Chawla (1975a). *Fibre Sci. & Tech.*, **8**, 49.

K.K. Chawla (1975b). *Grain Boundaries in Eng. Materials*, Proc. 4th Bolton Landing Conf., Claitor's Pub. Div., Baton Rouge, LA, p. 435.

K.K. Chawla and P.K. Liaw (1979). *J. Materials Sci.*, **14**, 2143.

K.K. Chawla, H. Schneider, Z.R. Xu, and M. Schmücker (1996). In *High Temperature Materials: Design & Processing Considerations*, Engineering Foundation Conf., Davos, Switzerland, TMS, Warrendale, PA, May 19–24, p. 235.

N. Chawla (1997). *Met. & Mater. Trans.*, **28A**, 2423.

N. Chawla, J.W. Holmes, and R.A. Lowden (1996). *Scripta Mater.*, **35**, 1411.

N. Chawla, Y.K. Tur, J.W. Holmes, J.R. Barber, and A. Szweda (1998). *J. Am. Ceram. Soc.*, **81**, 1221.

T. Christman and S. Suresh (1988a). *Acta Metall.*, **36**, 1691.

T. Christman and S. Suresh (1988b). *Mater. Sci. Eng.*, **102A**, 211.

C.R. Crowe and D.F. Hasson (1982). In *Proc. 6th Int. Conf. on the Strength of Metals and Alloys*, Pergamon, Oxford, p. 859.

D.L. Davidson (1989). *Eng. Fract. Mech.*, **33**, 965.

A. Dlouhy, N. Merk, and G. Eggeler (1993). *Acta. Metall. Mater.*, **41**, 3245.

T.L. Dragone and W.D. Nix (1992). *Acta. Metall. Mater.*, **40**, 2781.

T.L. Dragone, J.J. Schlautmann, and W.D. Nix (1991). *Metall. Trans*, **22A**, 1029.

D.C. Dunand and B. Derby (1993). In *Fundamentals of Metal Matrix Composites*, S. Suresh, A. Mortensen, and A Needleman (Eds.), Butterworth-Heinemann, Boston, p. 191.

G.J. Dvorak and W.S. Johnson (1980). *Intl. J. Fracture*, **16**, 585.

G. Eggeler and A. Dlouhy (1994). In *High Performance Composites: Commonalty of Phenomena*, K.K. Chawla, P.K. Liaw, and S.G. Fishman (Eds.), TMS, Warrendale, PA, p. 477.

R.H. Eriksen (1976). *Composites*, **7**, 189.

L.B. Godefroid and K.K. Chawla (1988). *3rd Latin American Colloquium on Fatigue and Fracture of Materials*, Rio de Janeiro, Brazil.

J.P. Gomez and F.W. Wawner (1988). Personal communication.

M. Gouda, K.M. Prewo, and A.J. McEvily (1981). *Fatigue of Fibrous Composite Materials*, ASTM STP 723, American Society of Testing and Materials, Philadelphia, p. 101.

J.E. Hack, R.A. Page, and G.R. Leverant (1987). *Met. Trans. A*, **15A**, 1389.

H.T. Hahn and R.Y. Kim (1976). *J. Composite Materials*, **10**, 156.

H.T. Hahn and L. Lorenzo (1984). In *Advances in Fracture Research, ICF6*, Pergamon Press, Oxford, Vol. 1, p. 549.

L.X. Han and S. Suresh (1989). *J. Amer. Ceram. Soc.*, **72**, 1233.

H.E. Helms and P.J. Haley (1989). In *Ceramic Materials and Components for Engines*, V.J. Tennery (Ed.), Amer. Ceram. Soc., Westerville, OH, p. 1347.

A.L. Highsmith and K.L. Reifsnider (1982). In *Damage in Composite Materials*, ASTM STP 775, Amer. Soc. of Testing and Mater., Philadelphia, p. 103.

J.W. Holmes (1991). *J. Mater. Sci.*, **26**, 1808.

Y. Izuka, T. Norita, T. Nishimura, and K. Fujisawa (1986). In *Carbon Fibers*, Noyes Pub., Park Ridge, NJ, p. 14.

W.S. Johnson (1988). *Mechanical and Physical Behavior of Metallic and Ceramic Composites*, 9th Risø Intl. Symp. on Metallurgy and Materials Science, Riso Nat. Lab., Roskilde, Denmark, p. 403.

W.S. Johnson and R.R. Wallis (1986). *Composite Materials: Fatigue and Fracture*, ASTM STP 907, ASTM, Philadelphia, p. 161.

P.G. Karandikar and T.-W. Chou (1992). *Ceram. Eng. Sci. Proc.*, **13**, 882.

A. Kelly and K.N. Street (1972a). *Proc. R. Soc. Lond. A*, **328**, 267.

A. Kelly and K.N. Street (1972b). *Proc. R. Soc. Lond. A*, **328**, 283.

A. Kelly and W.R. Tyson (1966). *J. Mech. Phys. Solids*, **14**, 177.

S. Kumai and J.F. Knott (1991). *Mater. Sci and Eng.*, A146, 317

S. Kumai, J.E. King, and J.F. Knott (1990). *Fatigue Fract. Eng. Mater. Struct.*, **13**, 511.

L.K. Kwei and K.K. Chawla (1992). *J. Materials Science*, **27**, 1101.

R.E. Lavengood and L.E. Gulbransen (1969). *Polymer Eng. Sci.*, **9**, 365.

C.S. Lee and K.K. Chawla (1987). In *Proc.: Industry-University Adv. Mater. Conf.*, TMS-AIME, Warrendale, PA, p. 289.

C.S. Lee, K.K. Chawla, J.M. Rigsbee, and M. Pfeifer (1988). *Cast Reinforced Metal Composites*, ASM Intl., Metals Park, OH, p. 301.

M. Levin, B. Karlsson, and J. Wasén (1989). In *Fundamental Relationships between Microstructures and Mechanical Properties of Metal Matrix Composites*, TMS, Warrendale, PA, p. 421.

H. Lilholt and M. Taya (1987). In *Proc.: ICCM/6*, Elsevier, p. 2.234-2.244.

H.-T. Lin and P.F. Becher (1990). *J. Amer. Ceram. Soc.*, **73**, 1378.

W.A. Logsdon and P.K. Liaw (1986). *Eng. Fract. Mech.*, **24**, 737.

T. Mah, N.L. Hecht, D.E. McCullum, J.R. Hoenigman, H.M. Kim, A.P. Katz, and H.A. Lipsitt (1984). *J. Mater. Sci.*, **19**, 1191.

J.F. Mandell, F.J. Mcgarry, D.D. Huang, and C.G. Li (1983). *Polymer Composites*, **4**, 32.

R.F. McCartney, R.C. Richard, and P.S. Trozzo (1967). *Trans. ASM*, **60**, 384.

M.A. McGuire and B. Harris (1974). *J. Phys. D:Appl. Phys.*, **7**, 1788.

M. McLean (1983). *Directionally Solidified Materials for High Temperature Service*, The Metals Soc., London.

M. McLean (1985). In *Proc.: 5th Intl. Conf. on Composite Materials (ICCM/V)*, TMS-AIME, Warrendale, PA, p. 639.

T. Morimoto, T. Yamaoka, H. Lilholt, and M. Taya (1988). *J. Eng. Mater. Tech. Trans. ASME*, **110**, 70.

L.R. Mueller and M. Gregory (1988). Paper presented at I Annual Metals and Metals Processing Conf. of SAMPE, Cherry Hill, NJ.

V.C. Nardone and J.R. Strife (1987). *Metall. Trans.*, **18A**, 109.

T.G. Nieh (1984). *Metall. Trans.*, **15A**, 139.

T.K. O'Brien (1984). *Interlaminar Fracture of Composites*, NASA TM-85768.

T.K. O'Brien and K.L. Reifsnider (1981). *J. Composite Materials*, **15**, 55.

S.L. Ogin, P.A. Smith, and P.W.R. Beaumont (1985). *Composites Sci. Tech.*, **22**, 23.

M.J. Owens and R. Dukes (1967). *J. Strain Analysis*, **2**, 272.

M.J. Owens, T.R. Smith, and R. Dukes (1969). *Plast. Polymers*, **37**, 227.

R.A. Page, J.E. Hack, R. Sherman, and G.R. Leverant (1987). *Met. Trans. A*, **15A**, 1397.

A.B. Pandey, R.S. Mishra, and Y.R. Mahajan (1992). *Acta. Metall. Mater.*, **40**, 2045.

P.C. Paris and F. Erdogan (1963). *J. Basic Eng. Trans. ASME*, **85**, 528.

N.J. Pfeiffer and J.A. Alic (1978). *J. Eng. Mater. Tech.*, **100**, 32.

D.C. Phillips (1983). In *Handbook of Composites*, Vol. 4, North-Holland, Amsterdam, p. 472.

L.N. Phillips (1976). *Composites*, **7**, 7.

K.M. Prewo (1987). *J. Materials Sci.*, **22**, 2695.

K.M. Prewo, J.J. Brennan, and G.K. Layden (1986). *Am. Ceram. Soc. Bull.*, **65**, 305.

L. Pruitt and S. Suresh (1992). *J. Mater. Sci. Lett.*, 1356.

D.J. Pysher, K.C. Goretta, R.S. Hodder, Jr., and R.E. Tressler (1989). *J. Amer. Ceram. Soc.*, **72**, 284.

J.P. Riggs (1985). In *Encyclopedia of Polymer Science Engineering*, 2nd ed., Vol. 2, John Wiley and Sons, New York, p. 640.

K.L. Reifsnider, E.G. Henneke, W.W. Stinchcomb, and J.C. Duke (1981). In *Mechanics of Composite Materials*, Pergamon Press, New York, p. 399.

J.L. Routbort, K.C. Goretta, A. Dominguez-Rodriguez, and A.R. de Arrellano-Lopez (1990). *J. Hard Materials*, **1**, 221.

C.R. Saff, D.M. Harmon, and W.S. Johnson (1988). *J. of Metals*, **40**, 58.

J.K. Shang, W. Yu, and R.O. Ritchie (1988). *Mater. Sci. Eng.*, **A102**, 181.

B.F. Sørensen and J.W. Holmes (1995). *Scripta Met. et Mater.*, **32**, 1393.

N.S. Stoloff (1987). In *Advances in Composite Materials*, Applied Sci. Pub., London, p. 247.

S. Suresh (1991). *J. Hard Materials*, **2**, 29.

S. Suresh, L.X. Han, and J.J. Petrovic (1988). *J. Am. Ceram. Soc.*, **71**, c158–c161.

R. Talreja (1985). *Fatigue of Composite Materials*, Technical University of Denmark, Lyngby, Denmark.

L.G. Taylor and D.A. Ryder (1976). *Composites*, **1**, 27.

Z. Wang, C. Laird, Z. Hashin, B.W. Rosen, and C.F. Yen (1991). *J. Mater. Sci.*, **26**, 5335.

R.C. Wetherhold and L.P. Zawada (1991). In *Fractography of Glasses and Ceramics*, V.D. Frechete and J.R. Varner (Eds.), *Ceramic Transactions*, Vol. 17, Amer. Ceram. Soc., Westerville, OH, p. 391.

S.M. Wiederhorn and B.J. Hockey (1991). *Ceramics Intl.*, **17**, 243.

S.M. Wiederhorn, W. Liu, D.F. Carroll, and T.-J. Chuang (1988). *J. Amer. Ceram. Soc.*, **12**, 602.

D.R. Williams and M.E. Fine (1985). In *Proc.: Fifth Intl Conf. Composite Materials (ICCM/V)*, TMS-AIME, Warrendale, PA, p. 639.

D.R. Williams and M.E. Fine (1987). In *Proc.: 6th Intl. Conf. on Composite Materials (ICCM/VI)*, Vol. 2, Elsevier Applied Science, London, p. 113.

Z.R. Xu, K.K. Chawla, A. Wolfenden, A. Neuman, G.M. Liggett, and N. Chawla (1995). *Mater. Sci. & Eng. A*, **A203**, 75.

J.-M. Yang and S.T. Chen (1992). *Adv. Composites Lett.*, **1**, 27.

Suggested Reading

R.W. Hertzberg and J.A. Manson (1980). *Fatigue of Engineering Plastics*, Academic Press, New York.

R. Talreja (1985). *Fatigue of Composite Materials*, Technical University of Denmark, Lyngby, Denmark.

R. Talreja (Ed.) (1994). *Damage Mechanics of Composite Materials*, Elsevier, Amsterdam.

CHAPTER 14

Designing with Composites

Understanding how to design with composites, especially fiber reinforced composites, is very important because composite materials do not represent just another new class of materials. Although there have been, over the years, ongoing efforts by researchers to improve the properties of different materials such as new alloys, composite materials represent a rather radical departure. Schier and Juergens (1983) analyzed the design impact of composites on fighter aircraft. The authors echoed the sentiments of many researchers and engineers in making the following statement: "Composites have introduced an extraordinary fluidity to design engineering, in effect forcing the designer-analyst to create a different material for each application...." A single component made of a laminated composite can have areas of distinctively different mechanical properties. For example, the wing-skin of an F/A-18 airplane is made up of 134 plies. Each ply has a specific fiber orientation and geometric shape. Computer graphics allow us to define each ply "in place" as well as its relationship to other plies. The reader can easily appreciate that storage and transmission of such engineering data via computer makes for easy communication between design engineers and manufacturing engineers. In this chapter, we discuss some of the salient points in regard to this important subject of designing with composites.

14.1 General Philosophy

We will make a few general philosophical points about designing with composites before going into the specifics. First of all, the whole idea behind the composite materials is that, in principle, one can make a material for any desired function. Thus, one should start with a function, not a material. A suitable composite material, say to meet some mechanical and/or thermal loading conditions, can then be made. An important corollary that follows is that one must exploit the anisotropy that is invariably present in fiber reinforced composites. Before discussing some specific design procedures, we review some of the advantages and fundamental characteristics of fiber-reinforced composites that make them so different from conventional monolithic materials.

14.2 Advantages of Composites in Structural Design

The main advantages of using composites in structural design are as follows.

Flexibility

- Ply lay-up allows for variations in the local detail design.
- Ply orientation can be varied to carry combinations of axial and shear loads.

Simplicity

- Large one-piece structures can be made with attendant reductions in the number of components.
- Selective reinforcement can be used.

Efficiency

- High specific properties, i.e., properties on a per-unit-weight basis.
- Savings in materials and energy.

Longevity

- Generally, properly designed composites show better fatigue and creep behavior than their monolithic counterparts.

14.3 Some Fundamental Characteristics of Composites

Composite materials come with some fundamental characteristics that are quite different from conventional materials. This is especially true of fiber reinforced composites. Among these important characteristics are the following:

- *Heterogeneity*: Composite materials, by definition, are heterogenous. There is large area of interface and the in situ properties of the components are different from those determined in isolation.
- *Anisotropy*: Composites in general, and fiber reinforced composites in particular, are anisotropic. For example, as we saw in Chapter 10, the modulus and strength are very sensitive functions of fiber orientation.
- *Coupling Phenomena*: Coupling between different loading modes, such as tension-shear, is not observed in conventional isotropic materials. We saw in Chapter 11 that in fiber reinforced composites such coupling phenom-

ena can be very important. These coupling phenomena make designing with composites more complex, as we shall see in Section 14.4.

- *Fracture Behavior*: Monolithic, conventional isotropic materials show what is called a *self similar crack propagation*. This means that the damage mode involves the propagation of a single dominant crack; one can then measure the damage in terms of the crack length. In composites, one has a multiplicity of fracture modes. A fiber reinforced composite, especially in the laminated form, can sustain a variety of subcritical damage (cracking of matrix, fiber/matrix decohesion, fiber fracture, ply cracking, delamination). For example, in a [0/90] laminate, the 90° ply will crack first. Such cracking of a ply, will result in a relaxation of stress in that ply, and with continued loading no further cracking occurs in that ply. Ply cracking could involve cracking in the matrix and maybe along the fiber/matrix interface. Other damage accumulating mechanisms include the growth of existing cracks into interfaces leading to ply delamination.

14.4   Design Procedures with Composites

As we said earlier, composite materials, particularly fiber reinforced composite materials, are not just another kind of new material. When designing with composites, one must take into account their special characteristics delineated in the preceding section. First, composite materials are inherently heterogeneous at a microstructural level, consisting as they do of two components that have different elastic moduli, different strengths, different thermal expansion coefficients, and so on. We saw in the micromechanical analysis (Chapter 10) that the structural and physical properties of composites are functions of (a) component characteristics, (b) geometric arrangement of one component in the other, and (c) interface characteristics. Even after selecting the two basic components, one can obtain a range of properties by manipulating the items (b) and (c). Second, the conventional monolithic materials are generally quite isotropic; that is, their properties do not show any marked preference for any particular direction, whereas fiber reinforced composites are highly anisotropic because of their very nature. The analysis and design of composites should take into account this strong directionality of properties, or, rather, this anisotropy of fiber composites must be exploited to the fullest advantage by the designer. The reader is referred to some figures in previous chapters: Figure 11.4 shows the marked influence of fiber orientation on the different elastic moduli of a composite. In a similar manner, Figure 12.14 shows the acute dependence of the composite strength on fiber orientation. Figure 10.7 shows the marked influence of fiber orientation on the coefficient of thermal expansion of a fiber reinforced composite.

In laminated composites, the ply stacking sequence can affect the properties of a composite. Recall Figure 8.5, which showed tensile creep strain at ambient temperature as a function of time for two different stacking sequences (Sturgeon, 1978). The laminate with carbon fibers at $\pm 45°$ showed more creep strain than one containing plies at $0°/90°/\pm 45°$. The reason for this was that in the $\pm 45°$ sequence, the epoxy matrix had creep strain contributions from (a) tension in the loading direction, (b) shear in the $\pm 45°$ directions, and (c) rotation of the plies in a scissor-like action. As we saw in Chapter 11, the $0°$ and $90°$ plies do not contribute to the scissor-like rotation due to the absence of tension-shear coupling in these specially orthotropic laminae. Thus the addition of $0°$ and $90°$ plies reduced the matrix shear deformation. Thus, for creep resistance, the $0°/90°/\pm 45°$ sequence is to be preferred over the $\pm 45°$ sequence.

For conventional materials, the designer needs only to consult a handbook or manual to obtain one unambiguous value of, say, modulus or any other property. For composite materials, however, the designer has to consult what are called *performance charts* or *carpet plots* representing a particular property of a given composite system. Figure 14.1 shows such plots for Young's modulus and the tensile strength, both in the longitudinal direction, for a 65% V_f carbon fiber/epoxy composite having a $0°/\pm 45°/90°$ ply sequence (Kliger, 1979). A conventional material, say aluminum, would be represented by just one point on such graphs. The important and distinctive point is that, depending on the components, their relative amounts, and the ply stacking sequence, we can obtain a range of properties in fiber composites. In other words, one can tailor-make a composite material as per the final objective. Figure 14.1 provides a good example of the versatility and flexibility of composites. Say our material specifications require, for an

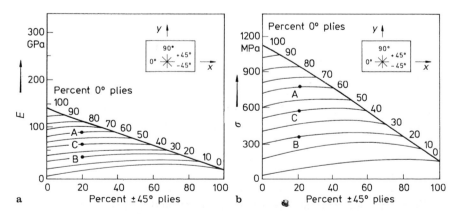

Fig. 14.1. Carpet plots for 65% V_f carbon/epoxy with a $0°/\pm 45°/90°$ stacking sequence: **a** Young's modulus, **b** tensile strength. [From Kliger (1979), used with permission.]

application, a material that is stronger than steel in the x direction ($\sigma_x = 500$ MPa) and a stiffness in the y direction equal to that of aluminum ($E = 70$ GPa). We can then pick a material combination that gives us a composite with these characteristics. Using Figure 14.1a and b, we can choose the following ply distribution:

60% at 0°
20% at 90°
20% at ±45°

Point A has a $\sigma_x = 725$ MPa and an $E_x = 90$ GPa. If we interchange the x and y coordinates and the amounts of the 0° and 90° plies, we get point B, which corresponds to a $\sigma_y = 340$ MPa and an $E_y = 40$ GPa. Thus, σ_x is now higher than required and E_y is too low. We now take some material from 0° and move it to 90°. Consider the following ply distribution:

40% at 0°
40% at 90°
20% at ±45°

This is represented by point C, having $\sigma_x = 580$ MPa and $E_y = 70$ GPa. Reversing the x and y coordinates again gives point C with $\sigma_x = 580$ MPa and $E_y = 70$ GPa. Point C then meets our initial specifications.

Stacking sequences other than $[0°/\pm 45°/90°]$ can also satisfy one's requirements. A $[0°/\pm 30°/\pm 60°/90°]$ arrangement gives quasi-isotropic properties in the plane. Carpet plots, however, work for three-ply combinations and the sequence $[0°/\pm 45°/90°]$ has some other advantages as described later. It should be pointed out that carpet plots for strengths in compression and shear, shear modulus, and thermal expansion coefficients can also be made. It is instructive to compare the expansion behavior of monolithic and fiber reinforced composites. Figure 14.2 compares the thermal expansion coefficients of some metals and composites as a function of fiber orientation (Fujimoto and Noton, 1973). It should be recalled that cubic materials are isotropic in thermal properties, even in single-crystal form. Noncubic monolithic materials in a polycrystalline form, with randomly distributed grains, will also be practically isotropic in thermal properties. Thus, titanium, steel, and aluminum are each represented by a single point on this figure. Not unexpectedly, fiber reinforced composites such as boron/epoxy and carbon/epoxy show highly anisotropic thermal expansion behavior. Both aramid and carbon fibers have negative expansion coefficients parallel to the fiber axis. One can exploit such anisotropy quite usefully to make components that show well-controlled expansion characteristics. Although aerospace applications of composite materials have been very much in the forefront, one should appreciate the fact that composites, in one form or another, have been in use in the electrical and electronic industry for quite a long time. These applications range from cables to printed circuit boards. There are many applications of composites in the

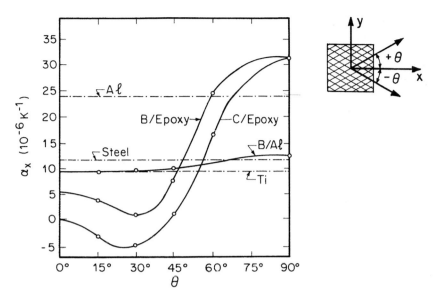

Fig. 14.2. Variation of thermal expansion coefficients with orientation θ. [From Fujimoto and Noton (1973), used with permission.]

electronic packaging industry (Seraphim et al., 1986). An interesting example is that of the leadless chip carrier (LCC) that allows electronic designers to pack more electrical connections into less space than is possible with conventional flat packs or dual in-line packages, which involve E-glass/epoxy resin laminates. Glass/epoxy laminate, however, is not very suitable for LCC as a substrate because it expands and contracts much more than the alumina chip carrier (the coefficient of expansion, α, of E-glass/epoxy can vary in the range of 14 to $17 \times 10^{-6}\,K^{-1}$, while that of alumina is $\sim 8 \times 10^{-6}\,K^{-1}$). Aramid fiber (40–50 v/o)/epoxy composite provides the desired matching expansion characteristics with those of the alumina chip carrier. Thus, the thermal fatigue problem associated with the E-glass/epoxy composites and the ceramic chip carrier is avoided. It should be pointed out that, in general, thermal expansion of particulate composites, assuming an isotropic distribution of particles in the matrix, will be more or less isotropic.

We saw in Chapter 11 that laminated composites show coupling phenomena: tension-shear, tension-bending, and bending-twisting. We also saw in Chapter 10 that certain special ply sequences can simplify the analyses. Thus, using a balanced arrangement (for every $+\theta$ ply there is a $-\theta$ ply in the laminate), we can eliminate shear coupling. In such an arrangement, the shear distortion of one layer is compensated by an equal and opposite shear distortion of the other layer. Yet another simplifying arrangement consists of stacking the plies in a symmetrical manner with respect to the laminate midplane. Such a symmetrical laminate gives $[B_{ij}]$ identically zero and thus

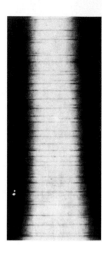

Fig. 14.3. Delaminations, as observed by contrast-enhanced X-ray radiography in a $[0°/\pm 45°/90°]_s$ carbon/epoxy laminate upon fatigue testing. [From Bergmann (1985), used with permission.]

eliminates stretching-bending and bending-torsion coupling. The phenomenon of edge effects, described in Section 11.7, should also be taken into account while arriving at a ply stacking sequence. Because of the edge effects, individual layers deforming differently under tension give rise to out-of-plane shear and bending in the neighborhood of the free edges. An arrangement that gives rise to compressive stresses in the thickness direction in the vicinity of the edges is to be preferred over one that gives rise to tensile stresses. The latter would tend to cause undesirable delamination in the composite. Figure 14.3 shows delaminations as observed by contrast-enhanced X-ray radiography, in a $[0°/\pm 45°/90°]_s$ carbon/epoxy laminate upon fatigue testing (Bergmann, 1985). The delaminations proceed from both sides toward the specimen center. The $[90°/\pm 45°/0°]_s$ laminate showed relatively less damage than the $[0°/\pm 45°/90°]_s$ laminate. Figure 14.4 compares the crack patterns at the free edges of $[0°/\pm 45°/90°]_s$ and $[90°/\pm 45°/0°]_s$ laminates shortly before failure. Note the greater degree of delaminations in $[0°/\pm 45°/90°]_s$ than in $[90°/\pm 45°/0°]_s$. The reason for this different behavior of these laminates, identical in all respects except the stacking sequence, is due to the difference in the state of stress near the free edges. Finite element calculations showed (Bergmann, 1985) that in the $[0°/\pm 45°/90°]_s$ sequence there were large interlaminar tensile stresses in the thickness direction while in the $[90°/\pm 45°/0°]_s$ sequence there were mostly compressive interlaminar stresses.

A little reflection will convince the reader that computers can be used extensively in the design and analyses of laminated fiber composites. In view of the rather tedious matrix calculations involved, this should not be surprising at all. Computer codes can be very effective in the analysis and design of fiber composite structures in a cost-effective manner. Most codes specify, among other things, the interply layers, laminate failure stresses, free edge stresses, and probable delamination locations around a hole. Murthy and

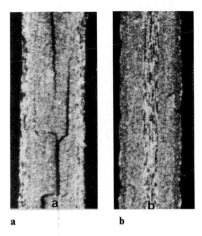

a   b

Fig. 14.4. Comparison of crack patterns at the free edges of a carbon/epoxy laminate: **a** $[0°/\pm 45°/90°]_s$, **b** $[90°/\pm 45°/0°]_s$. Note the greater severity of edge delaminations in **a** than in **b**. [From Bergmann (1985), used with permission.]

Chamis (1985, 1986) developed a computer code to analyze and design fiber composite structures. Many such codes are available commercially. An example of commercially available computer code is the FiberSIM suite of software tools that converts a CAD (computer-aided design) system into a composite design and manufacturing environment. Among its features are:

- Composite engineering environment: This allows engineers to work in their CAD environment with objects such as plies, cores, and tool surfaces. Forms and menus are used to enter and manage the geometry and other data required to define a composite part.
- Flat pattern/producibility: This module alerts the engineer to manufacturing problems such as wrinkling, bridging, and material width limitations. It also generates net flat patterns for woven and unidirectional complex curvature plies.
- Laminate properties: Exact fiber orientations are obtained for use in generating properties of a laminated composite.

14.5 Hybrid Composite Systems

Yet another degree of flexibility in fiber composites is obtained by making what are called *hybrid composites*, wherein one uses more than one type of fiber; see Figure 14.5. Cost-performance effectiveness can be increased by judiciously using different reinforcement types and selectively placing them to get the highest strength in highly stressed locations and directions. For example, in a hybrid composite laminate, the cost can be minimized by reducing the carbon fiber content, while the performance is maximized by

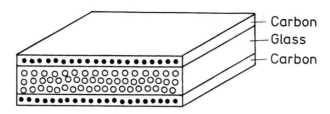

Fig. 14.5. Schematic of a hybrid laminate composite containing carbon and glass fibers.

"optimal placement and orientation of the fiber." Figure 14.6 shows the increases in specific flexural modulus (curve *A*) and specific tensile modulus (curve *B*) with weight % of carbon fiber in a hybrid composite (Riggs, 1985). Figure 14.7 shows the changes in flexural fatigue behavior as we go from 100% unidirectional carbon fibers in polyester (curve *A*) to unidirectional carbon faces over chopped glass core (curve *B*) to 100% chopped glass in polyester resin (curve *C*) (Riggs, 1985).

An interesting hybrid composite system called ARALL, an acronym for aramid aluminum laminates, has been developed (Vogelsang and Gunnik, 1983; Mueller et al., 1985). Initially developed at the Delft University of Technology in the Netherlands, the commercialization of ARALL was started by Alcoa and 3M Co. in the mid-1980s. ARALL consists of alternating layers of aluminum sheets bonded by adhesive containing aramid fibers. It is claimed that this unusual combination gives ARALL high strength, excellent fatigue and fracture resistance, combined with the advantages of metal construction such as formability, machinability, toughness, and impact resistance. It can be formed, punched, riveted, or bolted

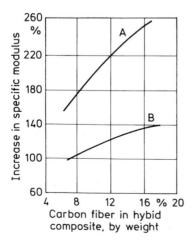

Fig. 14.6. Change in specific flexural modulus and specific tensile modulus with weight % carbon fibers. [From Riggs (1985), reprinted by permission.] *A* = specific flexural modulus; *B* = specific tensile modulus.

Fig. 14.7. Flexural fatigue behavior of 100% carbon fiber composite (curve *A*), carbon fiber facings with chopped glass fiber core (curve *B*), and 100% chopped glass fiber composite (curve *C*). [From Riggs (1985), reprinted by permission.]

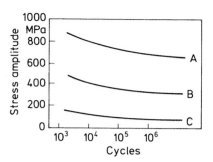

like a metal. Potential applications for ARALL include "tension-dominated fatigue and fracture critical structures" such as aircraft fuselage, lower wing, and tail skins. In these structures, the use of ARALL will result in 15 to 30% weight savings over conventional construction.

References

H. Bergmann (1985). In *Carbon Fibres and Their Composites*, Springer-Verlag, Berlin, p. 184.

W.T. Fujimoto and B.R. Noton (1973). In *Proceedings of the 6th St. Louis Symposium on Composite Material Engineering & Design*, American Society for Metals, Metals Park, OH, p. 335.

H.S. Kliger (Dec. 6, 1979). *Machine Des.*, **51**, 150.

L.N. Mueller, J.L. Prohaska, and J.W. Davis (1985). "ARALL: Introduction of a new composite material," Paper presented at the AIAA Aerospace Eng. Conf & Show, Los Angeles, CA.

P.L.N. Murthy and C.C. Chamis (1986). *J. Composite Tech. Res.*, **8**, 8.

P.L.N. Murthy and C.C. Chamis (1985). *ICAN: Integrated Composite Analyzer Users and Programmers' Manual*, NASA TP-2515.

J.P. Riggs (1985). In *Encyclopedia of Polymer Science & Engineering*, second ed., vol. 2, John Wiley & Sons, New York, p. 640.

J.F. Schier and R.J. Juergens (1983). *Astronautics and Aeronautics*, 44.

D.P. Seraphim, D.E. Barr, W.T. Chen, and G.P. Schmitt (1986). In *Advanced Thermoset Composites*, Van Nostrand Reinhold, New York, p. 110.

J.B. Sturgeon, (1978). In *Creep of Engineering Materials*, a Journal of Strain Analysis Monograph, p. 175.

L.B. Vogelsang and J.W. Gunnik (1983). *Delft University of Technology Report LR-400*, Delft, The Netherlands.

Suggested Reading

C.C. Chamis (Ed.) (1974). *Structural Design and Analysis, Parts I and II*, Academic Press, New York.

I.M. Daniel and O. Ishai (1994). *Engineering Mechanics of Composite Materials*, Oxford University Press, New York.

G. Lubin (Ed.) (1982). *Handbook of Composites*, Van Nostrand Reinhold, New York.

S.R. Swanson (1997). *Introduction to Design and Analysis with Advanced Composite Materials*, Prentice Hall, Upper Saddle River, NJ.

APPENDICES

Appendix A Matrices

Definition

A *matrix* is an array of numbers subject to certain operations. For example,

$$\begin{bmatrix} 3 & 2 & 9 & 7 \\ 0 & -63 & 49 & 5 \\ 1 & 5 & 4 & 8 \end{bmatrix} \qquad \text{3 rows, 4 columns}$$

$$\begin{bmatrix} a_{11} & a_{12} & \cdots & a_{1n} \\ a_{21} & a_{22} & \cdots & a_{2n} \\ \vdots & \vdots & \cdots & \vdots \\ a_{m1} & a_{m2} & \cdots & a_{mn} \end{bmatrix} \qquad m \text{ rows, } n \text{ columns}$$

The a_{ij} are called the *elements* of the matrix.

Notation

$$A = [A] = (a_{ij}) = [a_{ij}]_{m,n}$$

Null Matrix

In a *null matrix*, all the elements are zero, that is,

$$[N] = [0]$$

Any matrix multiplied by a null matrix results in a null matrix. Thus,

$$[A][N] = [N]$$

Square Matrix

The number of rows, called the *order* of the matrix, is equal to the number of columns. For example,

$$[A] = \begin{bmatrix} a_{11} & a_{12} \\ a_{21} & a_{22} \end{bmatrix}$$

is a square matrix of order 2.

Determinant

Associated with every square matrix $[A]$, there is a numerical value called its *determinant*, $|A|$.

$$|A| = a_{ij}(\text{cof } a_{ij})$$

where $\text{cof } a_{ij} = (-1)^{i+j}|M_{ij}|$. $\text{cof } a_{ij}$ is the *cofactor* of a_{ij} and M_{ij} is the *minor* of a_{ij}, obtained by eliminating the row and the column of a_{ij}.

Consider the matrix

$$[A] = \begin{bmatrix} a_{11} & a_{12} \\ a_{21} & a_{22} \end{bmatrix}$$

Determinant $|A| = \begin{vmatrix} a_{11} & a_{12} \\ a_{21} & a_{22} \end{vmatrix}$

$$\text{cof } a_{11} = (-1)^2|M_{11}| = M_{11} = a_{22}$$

$$\text{cof } a_{12} = (-1)^3|M_{12}| = -M_{12} = -a_{21}$$

$$\text{cof } a_{21} = (-1)^3|M_{21}| = -M_{21} = -a_{12}$$

$$\text{cof } a_{22} = (-1)^4|M_{22}| = M_{22} = a_{11}$$

$$|A| = \sum_{i=1}^{n} a_{ij}(\text{cof } a_{ij}) = a_{11}a_{22} + a_{12}(-a_{21})$$

$$= a_{11}a_{22} - a_{12}a_{21}$$

Another example is

$$[A] = \begin{bmatrix} 1 & 2 & -1 \\ 3 & 4 & 0 \\ 0 & 3 & 2 \end{bmatrix}$$

$$\text{cof } a_{11} = M_{11} = \begin{vmatrix} 4 & 0 \\ 3 & 2 \end{vmatrix} = 8 - 0 = 8$$

$$\text{cof } a_{12} = -M_{12} = \begin{vmatrix} 3 & 0 \\ 0 & 2 \end{vmatrix} = -6 + 0 = -6$$

$$\text{cof } a_{13} = M_{13} = \begin{vmatrix} 3 & 4 \\ 0 & 3 \end{vmatrix} = 9 - 0 = 9$$

$$|A| = 1 \times 8 + 2 \times (-6) - 1 \times 9 = 8 - 12 - 9 = -13$$

Diagonal Matrix

A square matrix whose nondiagonal elements are zero is called a *diagonal matrix*.

$$a_{ij} = 0 \quad \text{for } i \neq j$$

For example,

$$\begin{bmatrix} 3 & 0 & 0 \\ 0 & 2 & 0 \\ 0 & 0 & 1 \end{bmatrix}$$

Identity Matrix

A diagonal matrix whose elements are unity is called an *identity matrix*.

$$[I] = \begin{bmatrix} 1 & 0 & 0 \\ 0 & 1 & 0 \\ 0 & 0 & 1 \end{bmatrix}$$

$$[I][A] = [A]$$

Column or Row Matrix

All the elements in a column or row matrix are arranged in the form of a column or row. For example,

$$\text{column matrix} \quad \begin{bmatrix} a_1 \\ a_2 \\ \vdots \\ a_n \end{bmatrix}$$

$$\text{row matrix} \quad [b_1 \quad b_2 \quad \cdots \quad b_n]$$

Transpose Matrix

The substitution of columns by rows of the original matrix produces a *transpose matrix*. The transpose of a row matrix is a column matrix and vice versa. For example,

$$[A] = \begin{bmatrix} 0 & -5 & 1 \\ 8 & 4 & 2 \end{bmatrix} \qquad [A]^T = \begin{bmatrix} 0 & 8 \\ -5 & 4 \\ 1 & 2 \end{bmatrix}$$

$$[a] = \begin{bmatrix} 2 & 3 & -7 \end{bmatrix} \qquad [a]^T = \begin{bmatrix} 2 \\ 3 \\ -7 \end{bmatrix}$$

$$[AB]^T = [B]^T[A]^T$$

Symmetric Matrix

A square matrix is *symmetric* if it is equal to its transpose, that is,

$$A^T = A$$

$$a_{ij} = a_{ij} \quad \text{for all } i \text{ and } j$$

Antisymmetric or Skew-Symmetric Matrix

$$A^T = -A$$

$$a_{ij} = -a_{ji} \quad \text{for all } i \text{ and } j$$

For $i = j$, $a_{ii} = -a_{ii}$, that is, the diagonal elements of a skew-symmetric matrix are zero.

Singular Matrix

A square matrix whose determinant is zero is called a *singular matrix*.

Inverse Matrix

$[A]^{-1}$ is the inverse matrix of $[A]$ such that

$$[A][A]^{-1} = [I] = [A]^{-1}[A]$$

$$[A]^{-1} = \frac{1}{|A|}[\text{cof } A]^T$$

We now give some examples of inverse matrix determination.

Example 1

$$A = \begin{bmatrix} a_{11} & a_{12} \\ a_{21} & a_{22} \end{bmatrix}$$

$$[A]^{-1} = \begin{bmatrix} a_{22} & -a_{12} \\ -a_{21} & a_{11} \end{bmatrix} \frac{1}{a_{11}a_{22} - a_{12}a_{21}}$$

Example 2

$$A = \begin{bmatrix} a_{11} & 0 & \cdots & 0 \\ 0 & a_{22} & \cdots & 0 \\ \vdots & \vdots & \cdots & \vdots \\ 0 & 0 & \cdots & a_{nn} \end{bmatrix}$$

$$A^{-1} = \begin{bmatrix} 1/a_{11} & 0 & \cdots & 0 \\ 0 & 1/a_{22} & \cdots & 0 \\ \vdots & \vdots & \cdots & \\ 0 & 0 & \cdots & 1/a_{nn} \end{bmatrix}$$

Example 3

$$A = \begin{bmatrix} \cos\theta & -\sin\theta \\ \sin\theta & \cos\theta \end{bmatrix}$$

$$|A| = \cos^2\theta + \sin^2\theta = 1$$

$$A^{-1} = \begin{bmatrix} \cos\theta & \sin\theta \\ -\sin\theta & \cos\theta \end{bmatrix}$$

Example 4

$$T = \begin{bmatrix} m^2 & n^2 & 2mn \\ n^2 & m^2 & -2mn \\ -mn & mn & m^2 - n^2 \end{bmatrix} \qquad \begin{aligned} m &= \cos\theta \\ n &= \sin\theta \end{aligned}$$

There is a very important matrix called the *transformation matrix*:

$$|A| = 1$$

$$T^{-1} = \begin{bmatrix} m^2 & n^2 & -2mn \\ n^2 & m^2 & 2mn \\ mn & -mn & m^2 - n^2 \end{bmatrix}$$

Note that T^{-1} is obtained by replacing θ by $-\theta$ in T.

Some Basic Matrix Operations

The equality $[A] = [B]$ is valid if and only if $a_{ij} = b_{ij}$.

Addition and Subtraction

To perform the operations of addition or subtraction, the matrices must be *compatible*, that is, they must have the same dimension:

$$[A] + [B] + [C] = [D]$$

$$a_{ij} + b_{ij} + c_{ij} = d_{ij}$$

Similarly,

$$[E] = [A] - [B]$$

$$e_{ij} = a_{ij} - b_{ij}$$

Multiplication

To perform the operation of multiplication, the number of columns of one matrix must be equal to the number of rows of the other matrix:

$$[A]_{m \times n}[B]_{n \times r} = [C]_{m \times r}$$

$$c_{ij} = \sum_{k=1}^{n} a_{ik}b_{kj}$$

$$\begin{bmatrix} 3 & 5 \\ 9 & 8 \end{bmatrix} \begin{bmatrix} 1 & 0 \\ 4 & 3 \end{bmatrix} = \begin{bmatrix} 3 \times 1 + 5 \times 4 & 3 \times 0 + 5 \times 3 \\ 9 \times 1 + 8 \times 4 & 9 \times 0 + 8 \times 3 \end{bmatrix} = \begin{bmatrix} 23 & 15 \\ 41 & 24 \end{bmatrix}$$

Matrix multiplication is associative and distributive but not commutative. Thus,

$$[AB][C] = [A][BC]$$

$$[A][B + C] = [A][B] + [A][C]$$

$$[A][B] \neq [B][A]$$

The order of the matrix is very important in matrix multiplication.

Appendix B   Fiber Packing in Unidirectional Composites

Geometrical Considerations

Consider a composite containing uniaxially aligned fibers. Assume for the sake of simplicity that the fibers have the same cross-sectional form and area. Then for any uniaxial arrangement of fibers, we can relate the fiber volume fraction V_f, the fiber radius r, and the center-to-center spacing of fibers R as

$$V_f = \alpha \left(\frac{r}{R}\right)^2$$

where α is a constant that depends on the geometry of the arrangement of fibers.

Let S be the distance of closest approach between the fiber surfaces. Then

$$S = 2(R - r)$$

or

$$S = 2\left[\left(\frac{\alpha}{V_f}\right)^{1/2} r - r\right] = 2r\left[\left(\frac{\alpha}{V_f}\right)^{1/2} - 1\right]$$

Thus, the separation between two fibers is less than a fiber diameter when

$$\left[\left(\frac{\alpha}{V_f}\right)^{1/2} - 1\right] < 1$$

or

$$\left(\frac{\alpha}{V_f}\right)^{1/2} < 2$$

or

$$\frac{\alpha}{V_f} < 4$$

Table B.1 Geometrical fiber packing parameters

Arrangement	α	$\alpha/4$	$V_{f_{\max}}$
Hexagonal	$\dfrac{\pi}{2\sqrt{3}} = 0.912$	0.228	0.912
Square	$\dfrac{\pi}{4} = 0.785$	0.196	0.785

or

$$V_f > \frac{\alpha}{4}$$

The densest fiber packing $V_{f_{\max}}$ corresponds to touching fibers, that is,

$$\left[\left(\frac{\alpha}{V_{f_{\max}}} \right)^{1/2} - 1 \right] = 0$$

or

$$V_{f_{\max}} = \alpha$$

Table B.1 shows the values of α, $\alpha/4$, and $V_{f_{\max}}$ for hexagonal and square arrangements of fibers.

Appendix C Some Important Units and Conversion Factors

Stress (or Pressure)

1 dyn $= 10^5$ newton (N)
1 N m^{-2} $= 10$ dyn cm^{-2} $= 1$ pascal (Pa)
1 bar $= 10^5$ N m^{-2} $= 10^5$ Pa
1 hectobar $= 100$ bars $= 10^8$ cm^{-2}
1 kilobar $= 10^8$ N m^{-2} $= 10^9$ dyn cm^{-2}
1 mm Hg $= 1$ torr $= 133.322$ Pa $= 133.322$ N m^{-2}
1 kgf mm^{-2} $= 9806.65$ kN m^{-2} $= 9806.65$ kPa $= 100$ atmospheres
1 kgf cm^{-2} $= 98.0665$ kPa $= 1$ atmosphere
1 lb in^{-2} $= 6.89476$ kN m^{-2} $= 6.89476$ kPa
1 kgf cm^{-2} $= 14.2233$ lb in^{-2}
10^6 psi $= 10^6$ lb in^{-2} $= 6.89476$ GN m^{-2} $= 6.89476$ GPa
1 GPa $= 145\,000$ psi

Density

1 g cm^{-3} $= 62.4280$ lb ft^{-3} $= 0.0361$ lb in^{-3}
1 lb in^{-3} $= 27.68$ g cm^{-3}
1 g cm^{-3} $= 10^3$ kg m^{-3}

Viscosity

1 poise $= 0.1$ Pa s $= 0.1$ Nm^{-2} s
1 GN m^{-2} s $= 10^{10}$ poise

Energy per Unit Area

1 erg cm^{-2} $= 1$ mJ m^{-2} $= 10^{-3}$ J m^{-2}
10^8 erg cm^{-2} $= 47.68$ ft lb in^{-2} $= 572.16$ psi in

Fracture Toughness

1 psi in$^{1/2}$ = 1 lbf in$^{-3/2}$ = 1.11 kN m$^{-3/2}$ = 1.11 kPa m$^{-1/2}$
1 ksi in$^{1/2}$ = 1.11 MPa m$^{-1/2}$
1 MPa m$^{-1/2}$ = 0.90 ksi in$^{1/2}$
1 kgf mm^{-2} mm$^{1/2}$ = 3.16 × 10^4 N m^{-2} m$^{1/2}$

Problems

Chapter 1

1.1. Describe the structure and properties of some fiber composites that occur in nature.

1.2. Many ceramic based composite materials are used in electronics industry. Describe some of these electroceramic composites.

1.3. Describe the use of composite materials in the Voyager airplane that circled the globe for the first time without refueling in flight.

Chapter 2

2.1. Nonwoven fibrous mats can be formed through entanglement and/or fibers bonded in the form of webs or yarns by chemical or mechanical means. What are the advantages and disadvantages of such nonwovens over similar woven mats?

2.2. Glass fibers are complex mixtures of silicates and borosilicates containing mixed sodium, potassium, calcium, magnesium, and other oxides. Such a glass fiber can be regarded as an inorganic polymeric fiber. Do you think you can provide the chemical structure of such a polymer chain?

2.3. A special kind of glass fiber is used as a medium for the transmission of light signals. Discuss the specific requirements for such an optical fiber.

2.4. The compressive strength of Kevlar aramid fiber is about one-eighth of its tensile stress. Estimate the smallest diameter of a rod on which the Kevlar aramid fiber can be wound without causing kinks etc. on its compression side.

2.5. Several types of Kevlar aramid fibers are available commercially. Draw the stress strain curves of Kevlar 49 and Kevlar 29. Describe how much of the strain is elastic (linear or nonlinear). What microstructural processes occur during their deformation?

2.6. Kevlar aramid fiber, when fractured in tension, shows characteristically longitudinal splitting (micro-fibrillation) is observed. Explain why.

2.7. Describe the structural differences between Kevlar and Nomex (both aramids) that explain their different mechanical characteristics.

2.8. What is asbestos fiber and why is it considered to be a health hazard?

2.9. Describe the problems involved in mechanical testing of short fibers such as whiskers.

Chapter 3

3.1. Ductility, the ability to deform plastically in response to stresses, is more of a characteristic of metals than it is of ceramics or polymers. Why?

3.2. Ceramic materials generally have some residual porosity. How does the presence of porosity affect the elastic constants of ceramic materials? How does it affect the fracture energy of ceramics?

3.3. Explain why is it difficult to compare the stress-strain behavior of polymers (particularly thermoplastics) with that of metals.

3.4. The mechanical behavior of a polymer can be represented by an elastic spring and a dashpot in parallel (Voigt model). For such a model we can write for stress

$$\sigma = \sigma_{el} + \sigma_{visc} = E\varepsilon + \eta \, d\varepsilon/dt$$

where E is the Young's modulus, ε is the strain, η is the viscosity, and t is the time. Show that

$$\varepsilon = \sigma/E \left[1 - \exp\left(-E/\eta\right)t\right].$$

3.5. What is the effect of the degree of crystallinity on fatigue resistance of polymers?

3.6. Discuss the importance of thermal effects (internal heating) on fatigue of polymers.

3.7. Glass-ceramics combine the generally superior electrical and mechanical properties of crystalline ceramics with the processing ease of glasses. Give a typical thermal cycle involving the various stages for producing a glass-ceramic.

Chapter 4

4.1. Describe some techniques for measuring interfacial energies in different composite systems.

4.2. In order to study the interfacial reactions between the fiber and matrix, oftentimes one uses very high temperatures in order to reduce the time necessary for the experiment. What are the objections to such accelerated tests?

4.3. What are the objections to the use of short beam shear test to measure the interlaminar shear strength (ILSS)?

Chapter 5

5.1. Why are prepregs so important in polymer matrix composites? What are their advantages? Describe the different types of prepregs.

5.2. Randomly distributed short fibers should result in more or less isotropic properties in an injection molded composite. But this is generally not true. Why? What are the other limitations of injection molding process?

5.3. In a thermally cured PMC, the fiber surface treatments have been well established for certain systems. For example, silanes on glass fiber in an epoxy matrix or oxidizing treatment for carbon fiber in an epoxy matrix. What would be the effect of electron beam curing on the interface development in a PMC?

5.4. Describe the major differences in the processing of composites having a thermoset matrix and those having a thermoplastic matrix.

5.5. What are the important factors in regard to fire resistance of PMCs?

Chapter 6

6.1. Discuss the advantages and disadvantages of casting vis a vis other methods of fabricating metal matrix composites.

6.2. Silicon carbide (0.1 µm thick) coated boron fiber was used to reinforce a metallic matrix. The SiC coating serves as a diffusion barrier coating. Estimate the time for dissolution of this coating at 700 K if the diffusion coefficient at 700 K is 10^{-16} m^2 s^{-1}.

6.3. The metallic matrix will generally undergo constrained plastic flow in the presence of a moderately high volume fraction of high modulus fibers. Draw schematically the stress-strain curves of a constrained metal matrix (i.e. insitu behavior) and an unconstrained metal (i.e. 100% matrix metal). Explain the difference.

6.4. Discuss the problem of thermal stability of unidirectionally solidified eutectic (insitu) metallic composites.

Chapter 7

7.1. What are the sources of fiber degradation during processing of ceramic matrix composites?

7.2. Describe the advantages of using sol-gel and polymer pyrolysis techniques to process the ceramic matrix in CMCs.

7.3. Explain how a carbon fiber reinforced glass-ceramic composite can be obtained with an almost zero in-plane coefficient of thermal expansion.

Chapter 8

8.1. In view of the highly inert and non-leachable nature of carbon fibers, what recovery, recycling, or waste disposal practices can you recommend for carbon fibers and their composites?

8.2. List the various matrix resins that are commonly used with carbon fibers.

8.3. Distinguish between interphase and interface.

8.4. Nickel coated carbon fibers dispersed in a polymeric resin matrix are used for shielding against electromagnetic interference. Considering a Ni-coated carbon fiber to be a composite, make plots of density and modulus of the coated fiber as a function of volume fraction of nickel. Take $\rho_{Ni} = 8.9$ g/cm^3, $\rho_c = 1.7$ g/cm^3, $E_{Ni} = 200$ GPa, $E_c = 260$ GPa, where ρ is the density, E is the Young's modulus. *Hint: Consult chapter 10 for some expressions that you will need to do this problem.*

Chapter 9

9.1. There are many known superconducting A15 compounds. Of these Nb$_3$Al, Nb$_3$Ga, and Nb$_3$Ge have higher values of Tc and Hc_2 than do Nb$_3$Sn and

V_3Ga. How then does one explain the fact that only Nb_3Sn and to a lesser extent V_3Ga are available commercially?

9.2. It is believed that grain boundaries are the imperfections responsible for the flux-pinning in high-J_c materials like Nb_3Sn and V_3Ga. How does J_c vary with grain size?

9.3. What is the effect of any excess unreacted bronze leftover in the manufacture of Nb_3Sn superconductor composite via the bronze route?

9.4. Examine the Nb-Sn phase diagram. At what temperature does the A15 compound (Nb_3Sn) become unstable? Nb_3Sn is formed by solid state diffusion in Nb/Cu-Sn composites at $700\,°C$ or below. Is this in accord with information from the phase diagram? Explain.

9.5. Do you think it is important to study the effect of irradiation on superconducting materials? Why?

9.6. In the high magnetic field coils of large dimensions, rather tensile and compressive loads can be encountered during energizing and deenergizing. Discuss the effects of cyclic stress on the superconducting coil materials.

9.7. Superconducting composites in large magnets can be subjected to high mechanical loads. Describe the sources of these.

Chapter 10

10.1. Describe some experimental methods of measuring void content in composites. Give the limitations of each method.

10.2. Consider a 40% V_f SiC whisker reinforced aluminum composite. $E_f = 400$ GPa, $E_m = 70$ GPa, and $(l/d) = 20$. Compute the longitudinal elastic modulus of this composite if all the whiskers are aligned in the longitudinal direction. Use Halpin-Tsai-Kardos equations. Take $\xi = 2(l/d)$.

10.3. A composite has 40% V_f of a 150 μm diameter fiber. The fiber strength is 2. GPa, the matrix strength is 75 MPa, while the fiber/matrix interfacial strength is 50 MPa. Assuming a linear build up of stress from the two ends of a fiber, estimate the composite strength for (a) 200 mm long fibers and (b) 3 mm long fibers.

10.4. Derive the load transfer expression (Eq. 10.62) using the boundary conditions. Show that average tensile stress in the fiber is given by Eq. (10.63).

10.5. Consider a fiber composite system in which the fiber has an aspect ratio of 1000. Estimate the minimum interfacial shear strength τ_i, as a percentage of the tensile stress in fiber, σ_f, which is necessary to avoid interface failure in the composite.

10.6. Show that as $\xi \to 0$, the Halpin-Tsai equations reduce to

$$1/p = V_m/P_m + V_f/P_f \quad \text{while as } \xi \to \infty, \text{they reduce to}$$

$$p = V_m + V_f P_f$$

10.7. Consider an alumina fiber reinforced magnesium composite. Calculate the composite stress at the matrix yield strain. The matrix yield stress in 180 MPa, $E_m = 70$ GPa, and $v = 0.3$. Take $V_f = 50\%$

10.8. Estimate the aspect ratio and the critical aspect ratio for aligned SiC whiskers (5 μm diameter and 2 mm long) in an aluminum alloy matrix. Assume that the matrix alloy does not show may work hardening.

10.9. Alumina whiskers (density $= 3.8$ g cm^{-3}) are incorporated in a resin matrix (density $= 1.3$ g cm^{-3}). What is the density of the composite? Take $V_f = 0.35$. What is the relative mass of the whiskers?

10.10. Consider a composite made of aligned, continuous boron fibers in an aluminum matrix. Compute the elastic moduli, parallel and transverse to the fibers. Take $V_f = 0.50$.

10.11. Fractographic observations on a fiber composite showed that the average fiber pullout length was 0.5 mm. If $V_{fu} = 1$ GPa and the fiber diameter is 100 μm, calculate the strength of the interface in shear.

10.12. Consider a tungsten/copper composite with following characteristics: fiber fracture strength $= 3$ GPa, fiber diameter $= 200$ μm, and the matrix shear yield strength $= 80$ MPa. Estimate the critical fiber length which will make it possible that the maximum load bearing capacity of the fiber is utilized.

10.13. Carbon fibers ($V_f = 50\%$) and polyimide matrix have the following parameters:

$$E_f = 280\,\text{GPa} \qquad E_m = 276\,\text{MPa}$$

$$v_f = 0.2 \qquad v_m = 0.3$$

(a) Compute the elastic modulus in the fiber direction, E_{11}, and transverse to the fiber direction, E_{22}.
(b) Compute the Poisson ratios, v_{12} and v_{21}.

10.14. Copper or aluminum wires with steel cores are used for electrical power transmission. Consider a Cu/steel composite wire having the following data:
inner diameter $= 1$ mm
outer diameter $= 2$ mm

$$E_{Cu} = 150\,\text{GPa} \qquad \alpha_{Cu} = 16 \times 10^{-6}\,\text{K}^{-1}$$

$$E_{steel} = 210\,\text{GPa} \qquad \alpha_{steel} = 11 \times 10^{-6}\,\text{K}^{-1}$$

$$\sigma_{yCu} = 100\,\text{MPa} \qquad v_{Cu} = v_{steel} = 0.3$$

$$\sigma_{ysteel} = 200\,\text{MPa}$$

(a) The composite wire is loaded in tension. Which of the two components will yield plastically first? Why?
(b) Compute the tensile load that the wire will support before any plastic strain occurs.
(c) Compute the Young's modulus and coefficient of thermal expansion of the composite wire.

10.15. A composite is made of unidirectionally aligned carbon fibers in a glass-ceramic matrix. The following data are available:

$$E_{f1} = 280\,\text{GPa}, \quad E_{f2} = 40\,\text{GPa}, \quad E_m = 70\,\text{GPa}$$
$$v_{f1} = 0.2 \quad v_m = 0.3$$
$$G_{f12} = 18\,\text{GPa}$$

(a) Compute the elastic modulus in the longitudinal and transverse directions.
(b) Compute the two Poisson's ratios.
(c) Compute the principal shear modulus, G_{12}.

Chapter 11

11.1. An isotopic material is subjected to a uniaxial stress. Is the stain state also uniaxial? Write the stress and strain in matrix form.

11.2. For a symmetric laminated composite, we have, $\bar{Q}_{ij}(+z) = \bar{Q}_{ij}(-z)$, i.e., the moduli are even functions of thickness z. Starting from the definition split the integral and show that B_{ij} is identically zero for a symmetric laminate.

11.3. An orthotropic lamina has the following characteristics: $E_{11} = 210$ PGa, $E_{22} = 8$ GPa, $G_{12} = 5$ GPa, and $v_{12} = 0.3$. Consider a three-ply laminate made of such laminae arranged at $\theta = \pm 60°$. Compute the submatrices $[A]$, $[B]$, and $[D]$. Take the ply thickness to be 1 mm.

11.4. Enumerate the various phenomena which can cause microcracking in a fiber composite.

11.5. A thin lamina of a composite with fibers aligned at 45 degrees to the lamina major axis is subjected to the following stress system along its geometric axes:

$$[\sigma_i] = \begin{bmatrix} \sigma_x \\ \sigma_y \\ \sigma_s \end{bmatrix} = \begin{bmatrix} 10 \\ 2 \\ 3 \end{bmatrix} \text{MPa}.$$

Compute the stress components along the material axes (i.e., σ_1, σ_2, and σ_6).

11.6. A two-ply laminate composite has the top and bottom ply orientations of 45° and 0° and thicknesses of 2 and 4 mm, respectively. The stiffness matrix for the

$$[Q_{ij}] = \begin{bmatrix} 20 & 1 & 0 \\ 1 & 3 & 0 \\ 0 & 0 & 1 \end{bmatrix} \text{GPa}.$$

Find the $[\bar{Q}_{ij}]_{45}$ and then compute the matrices $[A]$, $[B]$, and $[D]$ for this laminate.

11.7. A two-ply laminate composite is made of polycrystalline, isotropic aluminum and steel sheets, each 1 mm thick. The constitutive equations for the two sheets are:

$$\text{Al}: \begin{bmatrix} \sigma_1 \\ \sigma_2 \\ \sigma_6 \end{bmatrix} = \begin{bmatrix} 70 & 26 & 0 \\ 26 & 70 & 0 \\ 0 & 0 & 26 \end{bmatrix} \begin{bmatrix} \varepsilon_1 \\ \varepsilon_2 \\ \varepsilon_6 \end{bmatrix} \text{MPa}$$

$$\text{Steel} : \begin{bmatrix} \sigma_1 \\ \sigma_2 \\ \sigma_6 \end{bmatrix} = \begin{bmatrix} 210 & 60 & 0 \\ 60 & 210 & 0 \\ 0 & 0 & 78 \end{bmatrix} \begin{bmatrix} \varepsilon_1 \\ \varepsilon_2 \\ \varepsilon_6 \end{bmatrix} \text{MPa}.$$

Compute the matrices $[A]$, $[B]$ and $[D]$ for this laminate composite. Point out any salient features of this laminate.

Chapter 12

12.1. For a ceramic fiber with $\mu = 12\%$, show that $\beta \simeq 10$. Show also that if the fiber length is changed by an order of magnitude, the corresponding drop in the average strength is about 20 percent.

12.2. The average stress in a fiber ($E_f = 400\,\text{GPa}$) is given by

$$\bar{\sigma}_f = \sigma_f[1 - (1 - \beta)l_c/l]$$

where β is the coefficient of load transfer from matrix to fiber, l_c, is the critical fiber length, l is the fiber length. The fiber behaves elastically and breaks at a strain of 0.1%. The matrix has a $\sigma_{mu} = 1$ MPa. Compute the V_{crit} for this system. Take $\beta = 0.5$, $\alpha = l/l_c = 10$, and the matrix stress corresponding to a strain of 0.1% equal to 0.5 MPa.

12.3. In a series of tests on boron fibers, it was found that $\mu = 10\%$. Compute the ratio $\bar{\sigma}_B/\bar{\sigma}$, where $\bar{\sigma}_B$ is the average strength of the fiber bundle and $\bar{\sigma}$ is the average strength of fibers tested individually.

12.4. Estimate the work of fiber pullout in a 40% carbon fiber/epoxy composite. Given $\sigma_{fu} = 0.2$ GPa, $d = 8\,\mu\text{m}$, and $\tau_i = 2$ MPa.

Chapter 13

13.1. List some of the possible fatigue crack initiating sites in particle, short fiber, and continuous fiber reinforced composites.

13.2. What factors do you think will be important in the environmental effects on the fatigue behavior of fiber reinforced composites?

13.3. Acoustic emission can be used to monitor damage in carbon fiber/epoxy during fatigue. Under steady loading conditions the damage is controlled by fiber failure and one can describe the acoustic emission by

$$\frac{dN}{dt} = \frac{A}{(t + T)^n}$$

where N is the total number of emissions, t is the time, T is a time constant, and A is constant under steady loading conditions. Taking $n = 1$, show that $\log t$ is a linear function of the accumulated counts. (Hint: see M. Fuwa, B. Harris, and A.R. Bunsell, J. App. Phys., 8 (1975) 1460).

13.4. Discuss the effects of frequency of cycling in PMCs and CMCs.

13.5. Discuss the fatigue behavior an aramid fiber reinforced PMC is subjected to fatigue at negative and positive stress ratio (R).

13.6. Which one will have a better creep resistance in air: an oxide/oxide composite or a nonoxide/nonoxide system? Explain your answer.

13.7. Diffusional creep involving mass transport becomes important at low stresses and high temperatures. Discuss the importance of reinforcement/matrix interface in creep of a composite under these conditions.

13.8. Assume that the creep of fiber and matrix can be described by a power-law and that a well bonded interface exists. Assume also that the strain rate of the composite is given be the volume weighted average of the strain rates of the fiber and matrix. Derive an expression for the strength of such a composite.

13.9. In some composites, residual thermal stress distribution obtained at room temperature on cooling from the high processing temperature results in compressive radial gripping at the interface. Discuss the effect of high temperatures on creep in such a composite.

Chapter 14

14.1. Following are the data for a 60% V_f, unidirectionally reinforced, carbon fiber/epoxy composite:

$$\text{Longitudinal tensile strength} = 1200\,\text{MPa}$$
$$\text{Longitudinal compressive strength} = 1000\,\text{MPa}$$
$$\text{Transverse tensile strength} = 50\,\text{MPa}$$
$$\text{Transverse compressive strength} = 250\,\text{MPa}$$

Calculate F_{11}, F_{22}, F_{66}, F_1, F_2, and F_{12}^*. Take $F_{12}^* = -0.5$. Compute the focal points plot the failure envelope for this composite.

14.2. For a carbon/epoxy composite, the strength parameters are:

$$F_{11}(\text{GPa})^{-2} = 0.45$$
$$F_{22}(\text{GPa})^{-2} = 101$$
$$F_{12}(\text{GPa})^{-2} = -3.4$$
$$F_{66}(\text{GPa})^{-2} = 215$$
$$F_1(\text{GPa})^{-1} = 0$$
$$F_2(\text{GPa})^{-1} = 21.$$

Compute the off axis uniaxial strengths of this composite for different θ and obtain a plot of σ_x vs. θ.

Author Index

Subject Index